Agri-Environment: Environmental Impacts of Agricultural Practices

Agri-Environment: Environmental Impacts of Agricultural Practices

Edited by **Farrell Waltz**

SYRAWOOD
PUBLISHING HOUSE
New York

Published by Syrawood Publishing House,
750 Third Avenue, 9th Floor,
New York, NY 10017, USA
www.syrawoodpublishinghouse.com

Agri-Environment: Environmental Impacts of Agricultural Practices
Edited by Farrell Waltz

International Standard Book Number: 978-1-68286-215-5 (Hardback)

Printed in China.

Contents

Preface

Agriculture is a staple economic activity. In the recent years agricultural practices have had certain adverse ecological impacts. This innovative and comprehensive book presents the well-developed theory in this area of study by highlighting the key environmental factors that affect agricultural properties. The topics included in this book on ecological impact of agricultural practices are based on various researches conducted across the world. Soil degradation, recycling of organic waste, fertilizers, contaminated water, effects of climate change, etc., are some topics that have been covered and are of utmost significance. This book is an invaluable resource for academicians and professionals alike, as it consists of various topics which can be of interest to a wide audience.

Various studies have approached the subject by analyzing it with a single perspective, but the present book provides diverse methodologies and techniques to address this field. This book contains theories and applications needed for understanding the subject from different perspectives. The aim is to keep the readers informed about the progress in the field; therefore, the contributions were carefully examined to compile novel researches by specialists from across the globe.

Indeed, the job of the editor is the most crucial and challenging in compiling all chapters into a single book. In the end, I would extend my sincere thanks to the chapter authors for their profound work. I am also thankful for the support provided by my family and colleagues during the compilation of this book.

Editor

Identification of QTL for *Striga hermonthica* Resistance Using Backcross Population Derived from a Cross between *Oryza sativa* (cv. Nipponbare) and *O. rufipogon*

Evans Atuti Atera[1], John Collins Onyango[2], Pham Thien Thanh[3], Takashige Ishii[3] & Kazuyuki Itoh[3]

[1] Department of Technical Services, Lake Basin Development Authority, Kenya

[2] Department of Botany, Maseno University, Maseno, Kenya

[3] Graduate School of Agricultural Science, Kobe University, Japan

Correspondence: Evans Atuti Atera, Department of Technical Services, Lake Basin Development Authority, P. O. Box 1516, Kisumu 40100, Kenya. E-mail: eatera@yahoo.com

Abstract

The obligate root hemiparasite, *Striga hermonthica* (Del.) Benth., native to sub-Saharan Africa causes serious economic constraint to cereal production. Studies on *Striga* spp. interactions with rice are desirable as it is a model monocot with high density molecular linkage maps. In this study, quantitative trait locus (QTL) analysis for *S. hermonthica* resistance was carried out using 141 backcross recombinant inbreed lines (BRILs) derived from a cross between *Oryza sativa* (cv. Nipponbare) and *O. rufipogon* W630. The population was grown in the field at Lake Basin Development Authority, Alupe farm in 2013 and infected with *S. hermonthica* from Alupe, Kenya. Putative QTL for *S. hermonthica* resistance was assumed using single-point analysis (qGene program) at $p < 0.01$ significance level. As a result, a single QTL explaining 6.6% of total phenotypic variance was detected near RM242 marker locus on chromosome 9, and the Nipponbare allele was found to have *S. hermonthica* resistance. The QTL chromosomal region can also be further studied to promote better understanding on the nature of resistance.

Keywords: quantitative trait loci (QTLs), rice, *Striga hermonthica*, resistance

1. Introduction

Rice is the most economically important food crop in sub-Saharan Africa (SSA). It is consumed widely in all countries and sub-regions of the continent and is mostly cultivated by resource-poor farmers under diverse ecosystems (Balasubramanian et al., 2007). Both cultivated rice species, *Oryza sativa* (L.) and *O. glaberrima* (Steud.), are grown in Africa. For the last 30 years, the harvested area has risen by 105% while production is by 170% (Rodenburg & Demont, 2009). However, the average rice production per unit is still low which is as a result of several production constraints, of which weed competition is regarded as the most severe. Weeds such as *Striga* spp. are the most problematic in upland conditions of SSA (Jamil et al., 2011). There are four *Striga* spp. considered to be serious pests to rice: *Striga hermonthica* (Del.) Benth., *S. asiatica* (L.) Kuntze, *S. aspera* (Willd.) Benth. and *S. forbesii* Benth. (Rodenburg et al., 2010; Atera et al., 2011). Among these species *S. hermonthica* is the most destructive, leading to severe yield losses of over 50% thereby affecting livelihoods of millions of farmers (Franke et al., 2006).

Cereal cultivars that offer resistance can post a major impact on limiting *Striga* infections and would dramatically improve yield (Rodenburg & Bastiaans, 2011). This would boost the morale of farmers in SSA as they view resistance as a desirable characteristic in cultivars. Several reports of resistance to *S. hermonthica* have been documented in rice (Harahap et al., 1993; Gurney et al., 2006; Jamil et al., 2011), sorghum [*Sorghum bicolor* (L.) Moench] (Vogler et al., 1996; Ezeaku & Gupta, 2004, Noubissie et al., 2012), pearl millet [*Pennisetum glaucum* (L.) R. Br.] (Kountche et al., 2013) and maize [*Zea mays* L.] (Amusan et al., 2008; Karaya et al., 2012). However, this information has not found its way to the hands of farmers as it is imbedded in books and journals which make it difficult for farmers to access.

In many studies, African rice species appears to offer better sources of resistance to *Striga* parasitism than Asian rice species (Johnson et al., 2000; Kaewchumnong & Price, 2008). Harahap et al. (1993) and Johnson et al. (2000) reported resistance to *S. hermonthica* observed in African rice cultivars in CG14, IG10, Makassa and ACC102196, and in Asian rice cultivars IR49255-B-B-5-2 and IR47255-B-B-5-4. The inter-specific New Rice for Africa (NERICA) cultivars offers a potentially interesting gene pool of resistant rice cultivars. NERICA 1 and NERICA 10 are said to be resistant to *S. hermonthica* infections (Cissoko et al., 2011; Atera et al., 2012).

A resistant phenotype is characterized by lack of ability of the parasite to penetrate through the endodermis and therefore the parasite cannot make xylem-xylem connections to establish a vascular for continuity after attachment to the host. This has been demonstrated by Gurney et al. (2006) with Nipponbare, *O. sativa* Japonica lowland rice cultivar containing sources of resistance of post-attachment to *S. hermonthica.* Advanced backcross inbred lines (BILs) from a cross of Nipponbare and Kasalath (*O. sativa*, indica cultivar) were screened for resistance to *S. hermonthica* from Kibos, Kenya. Seven QTLs were detected on chromosomes 1, 4, 5, 6, 7, 8 and 12. Interestingly, Nipponbare conferred greater resistance allele in six out of the seven QTLs. The two largest QTLs (an indication where phenotypic variance in a population is greatest) were on chromosome 4 associated with Kasalath allele and chromosome 12 of Nipponbare.

In the study of Swarbrick et al. (2008) of Koshihikari-Kasalath BILs, three *Striga* resistance QTLs were detected, two of which were from Kasalath alleles with the largest located on chromosome 4. In addition, Kaewchumnong and Price (2008) identified two other QTLs on chromosomes 1 and 8 from the population derived from a cross between cultivars Bala and Azucena which coincided with the QTLs found by Gurney et al. (2006) for post-attachment resistance to *S. hermonthica* in Nipponbare and Kasalath population. The impact of *Striga* spp. infestation in farmer's cereal crop fields must be reduced at all cost. This can be done through rice as it has played a central role in human nutrition and culture for the past 10,000 years. In addition, rice has a small genome (c. 389Mb) which can easily be used for QTL mapping (International Rice Genome Sequencing Project, 2005). In this study, we detected *S. hermonthica* resistance QTL in rice using BRILs derived from a cross between *O. sativa* cv. Nipponbare and *O. rufipogon* W630.

2. Materials and Method

2.1 Site Description

The field study was conducted in the long rains of March to August 2013 at Alupe farm of Lake Basin Development Authority (LBDA), near Busia town (0°29′N, 34°07′E) in western Kenya, where *S. hermonthica* is a serious limitation to cereal crop production. The site is located at 1189m above sea level and normally receiving a mean annual rain and temperature of 1028 mm and 30 °C, respectively. Prior to the trial, the site was under cultivation of maize.

2.2 Plant Materials

In this study, *O. sativa* Japonica cultivar Nipponbare and a wild annual accession *O. rufipogon* W630 from Myanmar were used. *O. rufipogon* W630 is quite susceptible to *S. hermonthica* (although not as susceptible as other cultivars such as Dourado precoce) collected from Alupe, Kenya in a pre-test pot experiment conducted at Maseno University (Figure 1). Nipponbare had previously been classified as a resistant cultivar to *S. hermonthica* from Kibos, Kenya (Gurney et al., 2006). The segregating population for QTL analysis consisted of 141 backcross recombinant inbred lines (BRILs) between Nipponbare (a recurrent parent) and *O. rufipogon* W630 (a donor parent) at BC_2F_{10} generation. BRILs were obtained from Kobe University, Japan. The *S. hermonthica* seeds used in this study were collected from maize host in 2011 at Kenya Agricultural Research Institute (KARI) Alupe, Kenya.

2.3 Genome Composition and Phenotypic Recording of BRILs

Estimation of wild chromosomal segments in each of the 159 BRILs at BC2F8 generation ranged between 0.0 to 23.6% (Thanh et al., 2011). The estimation was done using 180 microsatellite loci covering 1,362 cM of the 12 rice chromosomes. The BRILs have about 11.3% of wild genome on average in the genetic background of Nipponbare and between three (3) to thirty nine (39) lines of the BRILs were identified to have wild homozygous alleles at these marker loci.

In this study, the 141 BRILs were each inoculated with 0.5 g of *S. hermonthica* seed per row as described by Berner et al. (1997). This weight of *Striga* seeds contained about 3000 germinable seeds per hill. Before inoculation, the *Striga* seeds were thoroughly mixed with fine soil which was sieved through a screen of pore diameter of 250 μm to serve as a carry since *Striga* seeds are very small. Each hill was 30 cm by 30 cm of which soil of about 10 cm diameter and depth of 5 cm was dug, and sprinkled with a scoop full of *Striga*-soil mixture. Three rice seeds of each BRIL were planted per hill in ten hills (already artificially inoculated with *Striga*) per row. After germination, seedlings were thinned to one per hill. Fertilizer was applied at rate of 60 kg N ha-1. The infected field was weeded once with a hoe, after which the weeds were pulled by hand other than *Striga* to avoid damaging young *Striga* seedlings. Field data on the number of *S. hermonthica*, survived BRILs and infected hills per row were recorded.

Figure 1. *S. hermonthica* infecting (A) *O. rufipogon* W630; (B) Dourado precoce, a highly susceptible cultivar in a pot experiment at Maseno University, Kenya

2.4 QTL Analysis

Based on the infected rates (no. infected plants / total no. plants examined) by *Striga* in the BRILs (giving more than five plants examined), QTL analysis was carried out. Single marker analysis (SMA) was used to estimate QTLs for *S. hermonthica* resistance by qGene software (Nelson, 1997). The significance threshold for SMA was set at $p < 0.01$ level. The proportion of phenotypic variation explained by significant marker was estimated as a coefficient of determination (R^2) for the single locus model.

3. Results and Discussion

3.1 Evaluation of Striga Resistance in Parents and BRILs

O. sativa Japonica cv. Nipponbare is resistance while wild accession *O. rufipogon* W630 was susceptible to *S. hermonthica* from Alupe, Kenya infections. From the pre-test pot experiment at Maseno University, Nipponbare had no *Striga* plants while *O. rufipogon* had at least two to four *Striga* per pot (Atera & Itoh, communication). In the field at Alupe, Nipponbare was attacked by very few *Striga* plants compared with the other cultivars. This confirmed the classification of Nipponbare as a resistant cultivar as described by several authors (Kaewchumnong & Price, 2008; Swarbrick et al., 2008). Unfortunately *O. rufipogon* being an aquatic species, it could not grow in upland conditions as germination was very poor. Field observations showed that 50.4% (71 out of 141) of the BRILs had no *Striga* infections based on the number of *Striga* attached to rice plants above the ground. The frequency distribution of infected plants by *S. hermonthica* is as shown in Figure 2. This result suggested that the resistance for *S. hermonthica* was under polygenic control.

3.2 Detection and Analysis of QTL

The threshold to declare a QTL was at the significance level of $p < 0.01$ as revealed by the genome scan (Figure 3). The QTL for *S. hermonthica* resistance was detected on chromosome 9 with a P value of 0.0022 and a PV value (percentage of phenotypic variance explained by the QTL) of 6.6% (Table 1). At $p < 0.05$ significance level, other loci having weak effects were detected on chromosomes 2, 3, 7 and 12 but they may not be considered biologically significant unless they are validated (Figure 3).

The result showed that *O. rufipogon* W630 allele at QTL on chromosome 9 increased the infection rate in the genetic background of Nipponbare. The Nipponbare-derived allele conferred resistance at this QTL as expected. In our QTL detection, the BRILs were subjected to *Striga* infections in its natural condition with several environmental interferences. Previous QTL detections for *S. hermonthica* resistance were post-attachment in cultivated rice and in a controlled environment (Gurney et al., 2006; Kaewchumnong & Price, 2008; Swarbrick et al., 2008). In field experiments, crop varieties interactions between *Striga* and its host determine the reproductive success of the parasite. For incidence, *Striga* seeds only germinate when exposed to moisture, a favorable temperature and germination stimulant is required which is the host plant root exudates. These interactions sometimes provide opportunities to the host to resist the parasite.

Table 1. Putative QTL for *S. hermonthica* resistance detected in the back cross recombinant inbred lines derived from a cross between *O. sativa* Nipponbare and *O. rufipogon* W630

Chromosome	Marker	Source	P Value	[a]PV(%)	[b]Additive effect (%)
9	RM242	*O. rufipogon* W630	0.0022	6.6	7.6

Note. [a]Percentage phenotypic variance in the mapping population explained by the QTL; [b]Additive effect on mean resistance (arcsine transformed) of an allelic substitution from *O. rufipogon* W630 to Nipponbare allele. Significance level at $p < 0.001$.

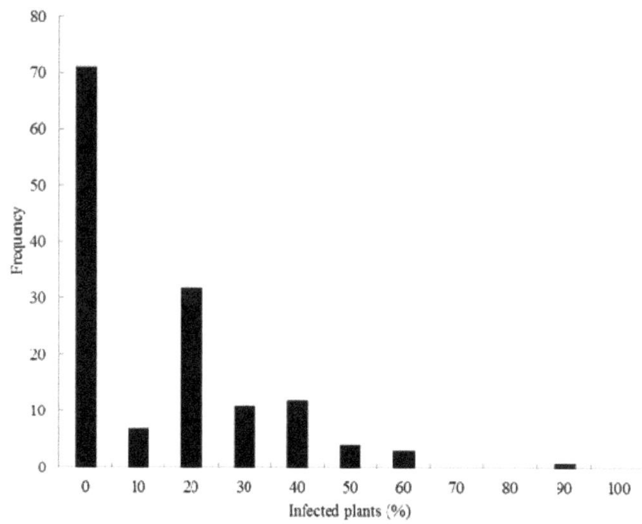

Figure 2. Frequency distribution of infected plants by *S. hermonthica* of the backcross recombinant inbred lines a cross between *O. sativa* Nipponbare and *O. rufipogon* W630. Frequency scores were calculated as the proportion of the lines in which *Striga* failed to attach to the BRILs

The detected major QTL can be further studied to see if there is a possibility of identifying candidates for resistance through fine mapping. It has been reported that genes that are expressed in infected plants tissues of two contrasting parents and able to map closely to a QTL are most likely to be appropriate candidates for host-resistance genes (Holloway et al., 2011). According to Swarbrick et al. (2008) resistance genes can be identified near the major QTL associated with resistance. If resistance genes can be identified in rice, orthologous genes may be estimated in other cereal crops such as maize and sorghum though the synteny of chromosomal gene order.

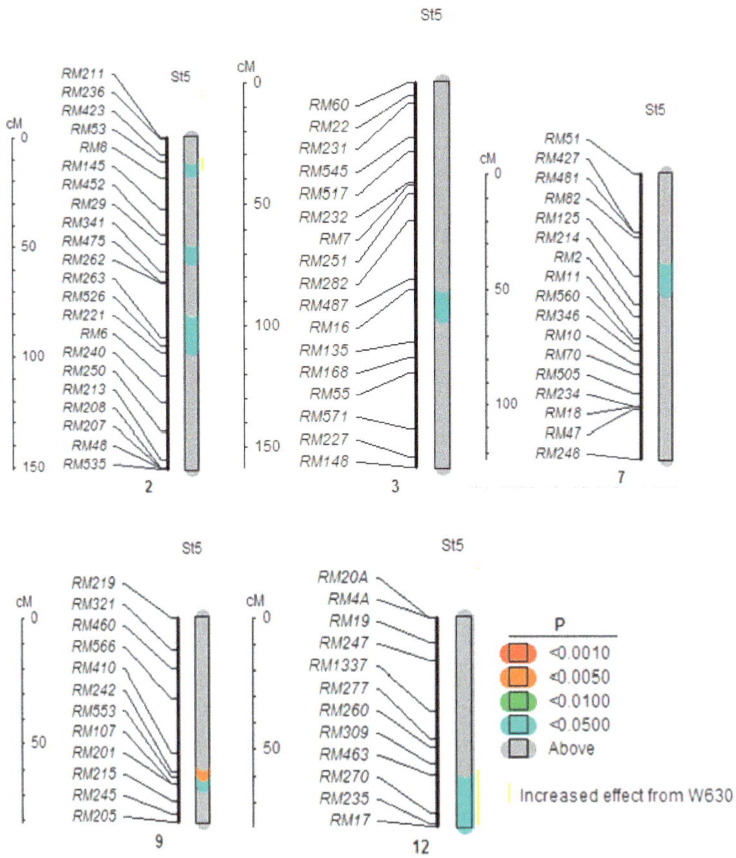

Figure 3. Genome scan of quantitative trait loci (QTL) for *Striga resistance* in *O. sativa* Nipponbare *and O. rufipogon* W630 population. The marker orders and the genomic regions associated with QTL scored at significance level at P < 0.001 and P < 0.05

Acknowledgements

This research was supported by a Grant-in-Aid for Science Research (No. 2340517) from the Ministry of Education, Culture, Sports, Science and Technology, Tokyo, Japan, of which we highly appreciate. The authors are grateful to B. Egesa and F. Rotich of Lake Basin Development Authority, Alupe integrated farm for their assistance in data collection.

References

Amusan, I. O., Rich, P. J., Menkir, A., Housley, T., & Ejeta, G. (2008). Resistance to *Striga hermonthica* in a maize inbred line derived from Zea diploperennis. *New Phytologist, 178*, 157-166. http://dx.doi.org/10.1111/j.1469-8137.2007.02355.x

Atera, E. A., Itoh, K., & Onyango, J. C. (2011). Evaluation of ecologies and severity of *Striga* weed on rice in sub-Saharan Africa. *Agriculture and Biology Journal of North America, 2*, 752-760. http://dx.doi.org/10.5251 /abjna. 2011.2.5.752.760

Atera, E. A., Itoh, K., Azuma, T., & Ishii, T. (2012). Response of NERICA rice to *Striga hermonthica* infections in western Kenya. *International Journal of Agriculture and Biology, 14*, 271-275.

Balasubramanian, V., Sie, M., Hijmans, R. J., & Otsuka, K. (2007). Increasing rice production in sub-Saharan Africa: challenges and opportunities. *Advances in Agronomy, 94*, 55-133. http://dx.doi.org/10.1016/S0065-2113(06)94002-4

Berner, D. K., Winslow, M. D., Awad, A. E., Cardwell, K. F., Mohan Raj, D. R., & Kim, S. K. (1997). Striga Research Methods - A Manual (pp. 1-9). International Institute of Tropical Agriculture, Ibadan, Nigeria.

Cissoko, M., Boisnard, A., Rodenburg, J., Press, M. C., & Scholes, J. D. (2011). New Rice for Africa (NERICA) cultivars exhibit different levels of post-attachment resistance against the parasitic weeds *Striga hermonthica* and *Striga asiatica*. *New Phytologist, 192*, 952-963.

http://dx.doi.org/10.1111/j.1469-8137 .2011.03846.x

Ezeaku, I. E., & Gupta, S. C. (2004). Development of sorghum populations for resistance to *Striga hermonthica* in the Nigerian Sudan Savanna. *African Journal of Biotechnology, 3*, 324-329.

Franke, A. C., Ellis-Jones, J., Tarawali, G., Schulz, S., Hussaini, M. A., Kureh, I., ... Olanrewaju, A. S. (2006). Evaluating and scaling-up integrated *Striga hermonthica* control technologies among farmers in northern Nigeria. *Crop Protection, 25*, 868-878. http://dx.doi.org/10.1016/j.cropro.2005.11.014

Gurney, A. L., Slate, J., Press, M. C., & Scholes, J. D. (2006). A novel form of resistance in rice to the angiosperm parasite *Striga hermonthica. New Phytologist, 169*, 199-208. http://dx.doi.org/10.1111/j.1469-8137.2005.01560.x

Harahap, Z., Ampong Nyarko, K., & Olela, J. C. (1993). *Striga hermonthica* resistance in upland rice. *Crop Protection, 12*, 229-231. http://dx.doi.org/10.1016/0261-2194(93)90114-X

Holloway, B., Luck, S., Beatty, M., Rafalski, J. A., & Li, B. (2011). Genome-wide expression quantitative trait loci (eQTL) analysis in maize. *BMC Genomics, 12*, 336. http://dx.doi.org/10.1186/1471-2164-12-336

International Rice Genome Sequencing Project. (2005). The map-based sequence of the rice genome. *Nature, 436*, 793-800. http://dx.doi.org/10.1038/nature03895

Jamil, M., Rodenburg, J., Charnikhova, T., & Bouwmeester, H. J. (2011). Pre-attachment *Striga hermonthica* resistance of New Rice for Africa (NERICA) cultivars based on low strigolactone production. *New Phytologist, 192*, 964-975. http://dx.doi.org/10.1111/j.1469-8137.2011.03850.x

Johnson, D. E., Riches, C. R., Jones, M. P., & Kent, R. (2000). The potential for host resistance to *Striga* on rice in West Africa. *Breeding for Striga resistance in cereals*. Proceedings of a workshop held at IITA, Weikersheim, Germany, Margraf Verlag.

Kaewchumnong, K., & Price, A. H. (2008). A study on the susceptibility of rice cultivars to *Striga hermonthica* and mapping of *Striga* tolerance quantitative trait loci in rice. *New Phytologist, 180*, 206-216. http://dx.doi.org/10.1111/j.1469-8137.2008.02568.x

Karaya, H., Kiarie, N., Mugo, S. N., Kanampiu, F. K., Ariga, E., & Nderitu, J. (2012). Identification of new maize inbred lines with resistance to *Striga hermonthica* (Del.) Benth. *J. Crop Protection, 1*, 131-142.

Kountche, B. A., Hash, C. T., Dodo, H., Laoualy, O., Sanogo, M. D., Timbeli, A., ... Haussmann, B. I. G. (2013). Development of a pearl millet *Striga*-resistant genepool: Response to five cycles of recurrent selection under *Striga*-infested field conditions in West Africa. *Field Crops Research, 154*, 82-90. http://dx.doi.org/10.1016/j.fcr.2013.07.008

Nelson, J. C. (1997). QGene: software for marker-based genomic analysis and breeding. *Molecular Breeding, 3*, 239-245. http://dx.doi.org/10.1023/A:1009604312050

Noubissie, J. B. T., Yadji, H. T., & Baldena, I. (2012). Screening Sorghum Populations for Resistance to *Striga hermonthica* (Del.) Benth. in Northern Cameroon. *Annals of Biology Research, 3*, 2357-2364.

Rodenburg, J., & Bastiaans, L. (2011). Host-plant defence against *Striga* spp.: reconsidering the role of tolerance. *Weed Research, 51*, 438-441. http://dx.doi.org/10.1111/j.1365-3180.2011.00871.x

Rodenburg, J., & Demont, M. (2009). Potential of herbicide-resistant rice technologies for sub-Saharan Africa. *AgBioForum, 12*, 313-325.

Rodenburg, J., Riches, C. R., & Kayeke, J. M. (2010). Addressing current and future problems of parasitic weeds in rice. *Crop Protection, 29*, 210-221. http://dx.doi.org/10.1016/j.cropro.2009.10.015

Swarbrick, P. J., Huang, K., Liu, G., Slate, J., Press, M. C., & Scholes, J. D. (2008). Global patterns of gene expression in rice cultivars undergoing a susceptible or resistant interaction with the parasitic plant *Striga hermonthica. New Phytologist, 179*, 515-529. http://dx.doi.org/10.1111/j.1469-8137.2008.02484.x

Thanh, P. T., Phan, P. D. T., Mori, N., Ishikawa, R., & Ishii, T. (2011). Development of backcross recombinant inbred lines between *Oryza sativa* Nipponbare and *O. rufipogon* and QTL detection on drought tolerance. *Breeding Science, 61*, 76-79. http://dx.doi.org/10.1270/jsbbs.61.76

Vogler, R. K., Ejeta, G., & Butler, L. G. (1996). Inheritance of low production of *Striga* germination stimulant in sorghum. *Crop Science, 36*, 1185-1191. http://dx.doi.org/10.2135/cropsci1996.0011183X003600050020x

2

The Economical Feasibility of Large Animal Composting

Joe L. Parcell[1], Joe Horner[1], Ryan Milhollin[1] & Jason Franken[2]

[1] Department of Agricultural Economics, University of Missouri, Columbia, United States

[2] Agricultural and Industrial Sciences, Sam Houston State University, Huntsville, United States

Correspondence: Joe L. Parcell, Department of Agricultural and Applied Economics, University of Missouri, Columbia, MO, 65211, United States. E-mail: parcellj@missouri.edu

Abstract

This study examines the economic feasibility of 4 alternative large animal mortality composting systems that a producer-owned entity could potentially operate. The 4 systems evaluated are a vertical mechanical composter (Dutch Composter), a horizontal mechanical composter (BIOvator™), an open static pile yard and a roofed or covered static pile yard. This study includes a 5-year pro forma financial analysis and suggests recommendations regarding the most economically viable and environmentally appropriate alternatives. Based on the financial analysis and the current regulatory environment, the recommended compost system for a regional animal mortality facility is static piles under roof. However, alternative technologies may be feasible in other scenarios.

Keywords: animal mortality, composting, dead animal disposal, economic feasibility, large animal

1. Introduction

In Missouri, all animal mortalities must be properly disposed within 24 hours. Five disposal options that have been acceptable, in order of preference, are rendering, composting, land filling, incineration, and burial (Fulhage, 1994). Based on the U.S. Department of Agriculture's 2007 *Census of Agriculture* livestock estimates, Missouri State Highway Patrol 2007 data on deer-related accidents, and assumptions regarding death losses and average animal sizes, Missouri annual beef, dairy, hog, horse, and deer mortalities total 264.9 million kg and require proper disposal (Figure 1). Burial is challenging due to equipment needs and geographic restrictions. Few Missouri renderers, commercial incinerators, and landfills allow dead animal disposal. Several renderers ceased operations due to lower hide values linked to the recession and higher disposal costs associated with U.S. Food and Drug Administration (FDA) regulations issued April 27, 2009 (Note 1). These events caused the industry to lack traditional animal disposal means.

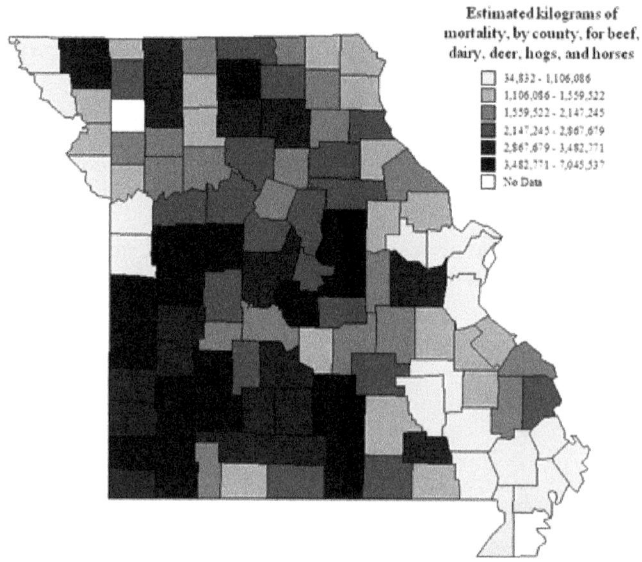

Figure 1. Kilograms of large animal mortalities in Missouri by county

Kilograms of large animal mortalities in Missouri by county. In Missouri, annual estimated beef, dairy, hog, horse, and deer mortalities total 264.9 million kg.

Legend: Estimated kilograms of mortality, by county, for beef, dairy, deer, hogs, and horses; 34,832 to 1,106,086; 1,106,086 to 1,559,522; 1,559,522 to 2,147,245; 2,147,245 to 2,867,679; 2,867,679 to 3,482,771; 3,482,771 to 7,045,537.

One study evaluated dead animal disposal costs as a future threat for Arkansas stocker cattle operators. On a 1 to 5 scale – 1 meaning not at all important and 5 meaning extremely important – stocker cattle operators on average scored dead livestock disposal costs a 3.2 (Troxel et al., 2007). Animal disposal acceptance is a potential challenge. A Pennsylvania study asked participating beef, veal, dairy, poultry, and swine producers whether they'd fielded comments or complaints about their operations within the past 2 years. On average, 25.9 percent had received comments or complaints, and 0.5 percent specifically noted receiving dead livestock disposal comments or complaints (Nordstrom et al., 1999).

Composting can be an environmentally compatible, low-capital option for dead animal disposal. Understanding composting processes, implementation, and costs is necessary before making a composting facility investment. This study evaluates the economic feasibility of 4 large animal composting systems.

2. Materials and Methods

To assess the economic viability of a dead animal composting facility, this study used interview and economic modeling methods. It considers the revenue and costs associated with operating 4 large animal composting systems: a vertical mechanical composter (Dutch Composter), a horizontal mechanical composter (BIOvator^{TM}), an open static pile yard, and a roofed or covered static pile yard. Figure 2 diagrams a possible composting business supply chain that a farmer cooperative could operate. Not only could farmers themselves use a composting facility, but livestock auction operators and small-scale meat processors are also potential customers. A well-designed composting facility that uses sawdust, woodchips, and other carbon sources could significantly increase efficiency and enable good throughput for dead animal mortalities.

In each of these systems, the costs and composting speeds differ, but each requires a carbon source (wood chips, sawdust, hay, etc.) to ensure a proper carbon-to-nitrogen ratio and enable carcasses to decompose into soil amendments. Mechanical systems are popular in Canada, which has enacted strict rendering and disposal regulations. They've also been adopted by some large livestock complexes where arrival of dead animal pickup vehicles entails bio-security concerns that outweigh the high capital costs of mechanical systems. Piles are typically less expensive, but they lengthen the decomposition timeline, particularly when cooler weather slows the biological processes that lead to carcasses decomposing. Open static piles are the simplest, cheapest system. Covered static piles are similar, except composting takes place under a roof and on an impervious layer of packed clay, asphalt, or concrete. A roofed system is more environmentally conscious because it eliminates runoff caused when rainfall reaches a pile. Under roof, compost moisture levels are easier to control, which creates more ideal composting conditions.

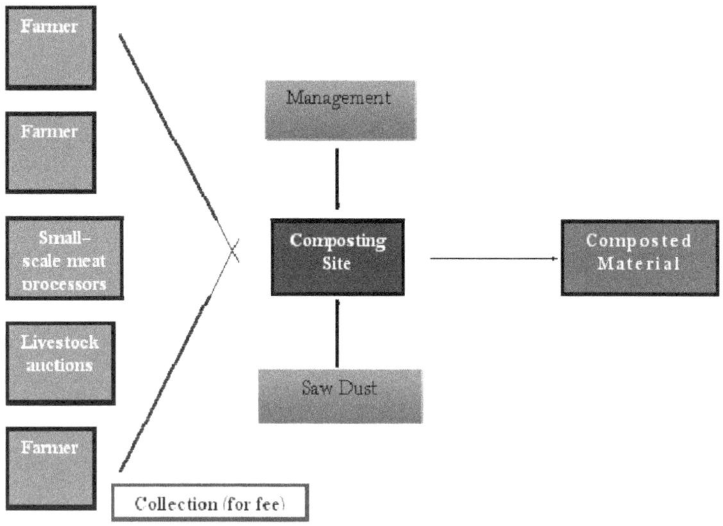

Figure 2. Composting business supply chain

Composting business supply chain. This schematic presents one possible composting business supply chain that a farmer cooperative could operate.

The composting process yields two main products: hides and the compost itself. The value of these products contributes to a large animal composting enterprise's economic viability. Skinning carcasses accelerates the composting process and provides additional value to the overall mortality composting enterprise. Hides usually sell for $40 to $55 per hide, but price may vary somewhat by region (Figure 3). The economic downturn in 2008 caused hide prices to plummet; however, they've since started to recover. The hide sales revenue stream would be influenced by economic conditions and could be uncertain.

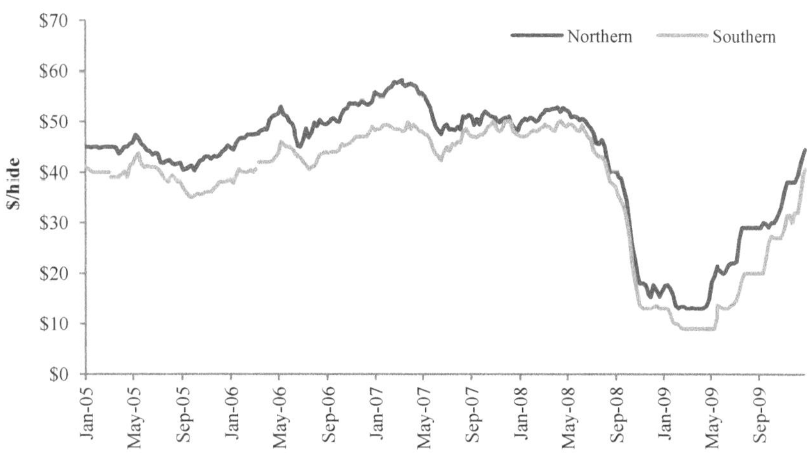

Figure 3. Heavy native cow U.S. hide prices ($/hide)

Heavy native cow U.S. hide prices ($/hide). Hide values would contribute to a large animal composting enterprise's economic viability.

Legend: Northern price per hide and southern price per hide.

Source: The Jacobsen Publishing Company, 2010.

High-value markets for compost, such as horticultural uses, typically require mixtures that deliver nutrients that promote plant growth. Table 1 presents compositional analyses for the composting mixture, which includes a carbon feedstock, before animal composting has occurred and after the process is complete.

Table 1. Composting mixture composition before and after composting

Nutrient	Before composting				After composting			
	Sample #1	Sample #2	Sample #3	Average	Sample #1	Sample #2	Sample #3	Average
Carbon (%)	42.8	47.8	4.3	44.8	17.9	18.1	28.9	21.6
Nitrogen (%)	4.2	3.4	.	3.8	2.3	1.3	2.3	2
C:N	10:01	14:01	.	12:01	8:01	14:01	13:01	11:01
Phosphorus (%)	0.01	0.01	.	0.01	0.01	0.01	0.04	0.02
Dry matter (%)	19.4	25.4	21.7	22.2	40.4	47.5	51.4	46.4
pH	8.5	8.4	8	8.3	8.6	8.7	8.2	8.5

Source: Looper, Fitzgerald, and Rogers (2002). Composting of bovine mortalities on New Mexico dairies. Paper presented at the annual meetings of the Western Section of the American Society of Animal Science, Fort Collins, CO.

Notably, the process does not yield high levels of nitrogen and phosphorus, which are key nutrient components important for plant growth. Instead, it mostly contributes dry matter. Table 2 shares nitrogen levels and pricing for organic ruminant compost compared with nitrogen content and pricing of other fertilizers.

Table 2. Economic comparison of organic ruminant compost with other fertilizers

Fertilizer	% nitrogen	kg nitrogen/100kg	price/100 kg of product	price/kg nitrogen	kg nitrogen/$1.00
Alfalfa meal	2.4%	2.4	$83.60	$34.83	0.03
Blood meal	12%	12	$198.00	$16.5	0.06
Fertrell 3-2-3	3%	3	$83.60	$27.87	0.04
Fish meal	8%	8	$114.40	$14.30	0.07
Soybean meal	7%	7	$123.20	$17.60	0.06
Chemical lawn fertilizer	10%	10	$55.00	$5.50	0.18
Ruminant compost	2%	2	$3.21[1]	$1.61[a]	0.62

Note. [1] Imputed from lowest value of commercial compost available in bulk (conversion is $31.90/tonne of product).

Relative to the other fertilizers analyzed, ruminant compost is among the least nitrogen-rich, and as such, it has the lowest assumed fertilizer value per unit. Although ruminant compost seems price competitive in terms of cost per part nitrogen, significant nitrogen quantities would be needed given the ruminant compost's high proportion of dry matter. Due to animal-based compost's low fertilizer value and potential zoonotic disease liability, its use in high-end horticultural markets is not advisable unless it is further processed. Such heat processing would add substantial costs, however. Still, the ruminant compost's nutrient composition is sufficient for nearby farmers to use such compost as a soil amendment if they can pick it up from the composting site for free. Thus, the compost would not be a revenue source, but it also would not pose a cost as farmers would provide free removal.

Table 3. Characteristic comparison of available carbon sources

	C:N	Moisture content (%)	Bulk density (kg/m^3)
Sawdust	276:1	41.0	237.31
Fescue hay	39:1	12.1	403.43
Straw	33:1	14.5	134.67
Corn stalks	67:1	12.0	18.98
Cardboard	563:1	8.0	153.66
Newsprint	625:1	6.0	129.33

Source: Zulovich (2009). Personal Communication. Division of Food Systems and Bioengineering, University of Missouri.

Composting is an aerobic process. Selecting the best biomass feedstock to use for composting depends on price, availability, and ease of use. Composting requires a carbon-to-nitrogen (C:N) ratio range between 20:1 and 40:1. The C:N ratios of all ingredients used in the composting recipe must be known and mixed in roughly correct proportions for the process to proceed efficiently. Moisture content in the 50 percent to 60 percent range is desirable for piles to efficiently compost material. Cattle's body composition is approximately 55 percent water. Efficient composting operations make special efforts to utilize recipes that achieve a moisture content and C:N ratio in the optimum ranges. Bulk density of composting ingredients is another useful property. The moisture content, C:N ratio, and bulk density of six carbon sources are shown in Table 3.

Halfway Packing Company, interviewed early in 2010, indicated that the cost to operate a covered truck pickup route that accumulates dead animals is about \$0.9321 per km. Respondents reported that established routes into livestock concentration areas usually yielded about 4,535.92 kg of dead animals. A typical route required about 563.27 km of travel to fill the truck. Mortality pickup costs, based on these surveys, would be approximately \$0.11 per kg. At this rate, a 544.31-kg carcass would incur an average \$60 pickup charge. Lower carcass pickup costs could be realized by shortening haul distances or loading multiple mortalities from a single site. Hauling costs may be a major economic factor that influences centralized dead animal disposal options, especially if the carcass rendering value is low and does not provide adequate income to offset the transportation costs.

In the following analysis, all four systems were assumed to be built in a facility capable of handling one dead cow per day. This was the minimum starting point thought to be practical. Specifying a facility that handles one cow per day allows for comparing each system on an equivalent basis. All systems are scalable simply by adding more mechanical compost units or adding more compost pile space. A \$75 processing fee per 680.39 kg of animal mortality composted is assumed. This fee is not adjusted for potential inflation in the financial analysis. The financial analysis does not explicitly include a route truck or account for transporting dead animals from their original location to the compost site. It does, however, provide an investment option for a covered trailer that could transport dead animals, and it discusses the effect of potential government subsidies.

Table 4 lists capital costs for each composting system considered. As noted earlier, the systems can manage one 680.39-kg dead cow per day, and they are scalable by adding more mechanical compost units or more compost pile space. For each system, a \$50,000 working capital investment is assumed as a beginning cash balance, and each system has the same basic land and improvements requirements. These requirements include two acres that are graded, landscaped, and graveled to control water run-off issues. The site must be fenced and gated to control access, and each system requires a connection to electricity and water. For the Dutch Composter system, a concrete pad, on which the Dutch Composter would be erected, is an additional cost. Each system also uses a shipping container for equipment storage and office space, and each requires a payloader to move carcasses and carbon sources. Similar costs are also assumed across systems for miscellaneous tools and equipment. Major differences in each system's capital costs are due to varying building and machinery costs.

As the most simplistic system, open static piles incur no building costs other than the aforementioned shipping container for equipment storage and office space. Both the Dutch Composter and BIOvator^{TM} systems include a 92.90 m^2 building with a concrete floor that would store finished compost up to 60 days. The covered composting system is operated within a 1,393.5 m^2 roofed facility, which is the largest cost of that system.

Machinery is the major cost involved in a mechanical composting system. The Dutch Composter completely composts a cow within 4 days, and hence, 4 units are required to handle an inflow of 1 cow per day. In comparison, a single BIOvator™ unit has the capacity to compost 680.39 kg of mortality per day, and thus, 3 BIOvator™ units are needed to handle an inflow of 1 cow per day. For static pile-based systems, another pile may be initiated daily to increase capacity. The Dutch Composter base unit costs $48,300, excluding shipping warranty, optional features, and installation costs. This analysis assumes the estimated installed cost to be $60,000 per unit. A single BIOvator™ unit costs $50,000. Both mechanical systems require an auger elevator to safely move finished compost from the machines to storage. The Dutch Composter, however, does not incur a screener expense required by the other systems because it fully grinds bones into thumbnail-sized or smaller pieces. As summarized in Table 4, the Dutch Composter has the highest startup costs, the BIOvator™ and roofed piles have similar costs, and open piles have the lowest startup costs.

Table 4. Comparison of capital costs across composting systems

Capital investments	Dutch composter	Biovator™	Open piles	Roofed piles
Land and improvements				
-Land (0.8 hectare × $4,940 per hectare)	$4,000	$4,000	$4,000	$4,000
-Fence and gate	$1,500	$1,500	$1,500	$1,500
-Landscaping	$1,000	$1,000	$1,000	$1,000
-Gravel area	$4,000	$4,000	$4,000	$4,000
-Land grading	$2,000	$2,000	$2,000	$2,000
-Electrical connection	$1,500	$1,500	$1,500	$1,500
-Water source (PWS connection, pond or well)	$7,000	$7,000	$7,000	$7,000
-Concrete pad for Dutch Composter (2.44 m by 2.44 m by 0.15 m)	$1,200			
Buildings				
- Shipping container (storage, office, etc.)	$1,500	$1,500	$1,500	$1,500
- Composting facility (1,393.55 m^2 × $10)				$13,935.50
- Covered (finished compost) storage	$10,000	$10,000		
Machinery				
- Biovater™ (3 units)		$150,000		
- Dutch Composter (4 units)	$240,000			
- Screener		$1,100	$1,100	$1,100
- Auger elevator	$1,000	$1,000		
- Payloader	$20,000	$20,000	$20,000	$20,000
- Misc. tools and equipment	$1,000	$1,000	$1,000	$1,000
Total capital investment	$295,700	$205,600	$44,600	$194,600
Working capital investments	$50,000	$50,000	$50,000	$50,000
Total startup investment	$345,700	$255,600	$94,600	$244,600
Per 680.39 kg of mortalities	$81.01	$56.33	$12.22	$53.32

Table 5 displays 2011 operational expenses for each composting system. Each system is configured to accept one 680.39-kg cow per day or 365 mature cows per year. A $10 per hour labor rate is assumed for an employee working 1 hour per day to operate the facility. Other operational inputs necessary for composting the animal mortalities are electricity and a carbon source. Electricity expenses are computed based on usage information supplied by the mechanical composter manufacturers and local electricity rates (Energy Information Administration). Electricity expenses for the Dutch Composter are notably higher than the BIOvator™. No electricity expenses are assumed for static pile-based systems. Carbon source expenses reflect identical compost system requirements of 226 kg per animal composted and an average price of $35 per tonne for local carbon sources. An annual $500 miscellaneous fuel expense covers regional travel to secure carbon source materials and other administrative travel expenses. This fuel expense does not include the expense of operating a route truck, as noted earlier. Other operational expenses shown in Table 5 are estimates. An annual inflation rate of 3 percent is applied to each expense in the second year to fifth year shown in the financial analysis. Operational costs for the Dutch Composter are about 3 times the operational costs of the other composting systems.

Table 5. Comparison of operating costs across composting systems

Cash operating costs	Dutch Composter	BIOvator™	Open piles	Roofed piles
Employee (1 hour per day×$10 per hour)	$3,650	$3,650	$3,650	$3,650
Electricity[a]	$21,900	$600		
Carbon ($38.50/tonne×0.23 tonnes of carbon/day×365 days)	$3,194	$3,194	$3,194	$3,194
Fuel (excluding route truck)	$500	$500	$500	$500
Maintenance/repairs	$1,500	$1,000	$1,000	$1,000
Insurance	$500	$500	$500	$500
Taxes	$500	$500	$500	$500
Total (annual)	$31,744	$9,944	$9,344	$9,344
Per 680.39 kg of mortalities	$86.97	$27.24	$25.60	$25.60
Depreciation and operating expenses (per 680.39 kg)	$167.98	$83.57	$37.82	$78.91
Composting service charge ($50 per 92.9 kg)	$75.00	$75.00	$75.00	$75.00
Net income (per 680.39 kg of moralities)	-$92.98	-$8.57	$37.18	-$3.91

Note. [a] Electricity costs are $60 per animal to operate the Dutch Composter and $1.65 per animal for the BIOvator™.

3. Results and Discussion

A 5-year financial statement summary, based on the previously defined assumptions and projections, is given in Table 6. The initial investment is assumed to occur in 2011, and a 3 percent annual inflation rate was applied to all expenses and revenue from 2012 to 2015.

Table 6. Five year financial analysis based on assumptions presented in Tables 4 and 5

Financial parameter	Dutch Composter	BIOvator™	Static compost piles (unroofed)	Static compost piles (under roof)
Initial capital investments	$345,700	$255,600	$94,600	$244,600
Operating expenses (year 5)	$30,696	$21,686	$5,586	$20,585
Net income (year 5)	($37,923)	($4,376)	$12,399	($2,601)
Net cash flow (year 5)	($8,353)	$16,183	$16,858	$16,858

Operating expenses entail general expenses (i.e., insurance and taxes) and depreciation, which is calculated using the straight-line method, a 10-year useful life, and $0 salvage value. Income is assumed to be equivalent for each system. It represents the revenues from the $75 processing fee charged on each 680.39-kg carcass received. Thus, at one carcass per day, annual revenue is $27,375 (= $75 × 365 days). Due to low nutrient (nitrogen and phosphorus) value and potential zoonotic disease liability associated with animal-based compost, the compost is not assumed to have commercial resale value. However, the compost's nutrient profile is sufficient for nearby farmers to utilize the compost as a soil amendment if they can pick up it for free from the composting site. Hence, the analysis assumes no cost for removing finished compost.

Net income is computed as the difference between revenue and operating and direct (i.e., electricity and carbon) costs. Similarly, net cash flow is the difference between actual cash receipts and expenditures. The Dutch Composter system is financially infeasible given these assumptions, as indicated by negative net income and cash flow reported in Table 6. Its infeasibility is mostly due to high electricity and depreciation costs. Positive cash flows make operating the BIOvator^{TM} more feasible, even considering the net income loss due to depreciation expenses. Both net income and net cash flow indicate that open static piles are financially feasible. Cash flow in year 5 for roofed or covered static piles is the same as open piles, but net cash flow is negative due to costs of a roofed facility. However, the covered facility may be preferable due to environmental reasons, particularly if some initial investment could be offset by government subsidies.

Currently, a few federal and state programs offer subsidies for this type of operation. An animal composting operation may qualify for a government subsidy. A partially subsidized animal composting operation would reduce the startup investment requirement because the owner(s) investment would be reduced by the amount of subsidies received. Subsidies would not affect annual net income, cash flows, or net worth, but they would supplement the owner investment for capital expenditures required at startup.

Although not shown in the financial analysis, a route truck could be used to bring carcasses to the facility. Whether the added route truck expense would provide a positive economic advantage for the mortality composting facility is questionable. Purchasing a 1.83-meter by 3.66-meter enclosed utility trailer with a winch capable of handling 907.18 kg may be better option. A trailer and winch system could cost less than $5,000. The composting operation could rent the trailer to customers who need to transport animal mortalities. A $5 to $10 daily rental fee would cover depreciation and mileage expenses associated with using the trailer, but it would not provide a significant revenue stream that would affect the operation economically.

The full feasibility study, with engineering recommendations, can be obtained from the authors upon request. Based on the financial analysis and possible future environmental regulatory issues, the recommended compost system for a regional animal mortality facility is static piles under roof. This system can operate without regard to rain, snow, or other bad weather, and it minimizes environmental risks and handling of dead animals. Furthermore, roofed barns should not present public acceptance problems. This system is the most robust in its ability to be implemented correctly with minimal management, and it offers the most flexibility of scale and dead animal throughput volume. In addition, its capital investment and operating costs are less significant than those of the 2 mechanical options. Finally, roofed buildings may be constructed with public assistance and potentially operated for decades with minimal additional repairs or investments. The results of this financial analysis depend on regulations, competitiveness, climate, and access to inputs including carbon sources. Alternative technologies may be feasible in other scenarios.

4. Implications

This study's results may inform industry participants of the most viable, economical dead animal disposal options available to livestock producers and handlers. Based on the financial analysis and possible future environmental regulations, the recommended compost system for a regional animal mortality facility is static piles under roof. This system can operate without regard to weather, it minimizes environmental and dead animal handling risks, and roofed facilities should not create public acceptance problems. Given other circumstances, however, an alternative composting technology may be more feasible.

References

Energy Information Administration. (2010). *Average retail price of electricity to ultimate customers by end-use sector, by state*. Retrieved January 20, 2010, from http://www.eia.doe.gov/cneaf/electricity/epm/table5_6_b.html

Fulhage, C. D. (1994). *Dead animal disposal laws in Missouri*. University of Missouri-Columbia, Cooperative Extension Service. Retrieved October 20, 2009, from http://extension.missouri.edu/publications/DisplayPub.aspx?P=WQ216

Looper, M. L., Fitzgerald, A. C., & Rogers, C. A. (2002). *Composting of bovine mortalities on New Mexico dairies*. Paper presented at the annual meetings of the Western Section of the American Society of Animal Science, Fort Collins, CO.

Missouri State Highway Patrol. (2007). Analysis of deer involvement in Missouri traffic crashes. Retrieved February 20, 2010, from http://www.mshp.dps.missouri.gov/MSHPWeb/SAC/pdf/2007DeerReport.pdf

Nordstrom, P. A., Wilson, L. L., Richards, M. J., Kelsey, T. W., & Pitts, C. W. (1999). Self-Perceptions and Self-Assessments of Livestock Producers in Pennsylvania. *Prof. Anim. Sci., 15*, 156-163.

The Jacobsen Publishing Company. (2010). *Heavy hide prices*. Retrieved January 26, 2010, from http://www.thejacobsen.com

Troxel, T. R., Lusby, K. S., Gadberry, M. S., Barham, B. L., Poling, R., Riley, T., … Justice, T. (2007). The Arkansas Beef Industry – A Self-Assessment. *Prof. Anim. Sci., 23*, 104-115.

United States Department of Agriculture. (2007). *Census of Agriculture* (Vol. 1, p. 51). Retrieved November 21, 2009, from http://www.agcensus.usda.gov/Publications/2007/Full_Report

United States Food and Drug Administration. (2009). *FDA issues final guidance for renderers on substances prohibited from use in animal food or feed.* Retrieved April 30, 2009, from http://www.fda.gov/AnimalVeterinary/NewsEvents/CVMUpdates/ucm144008.htm

Zulovich, J. (2009). *Personal Communication.* Division of Food Systems and Bioengineering, University of Missouri.

JEL Code: Q13

Notes

Note 1. U.S. regulatory actions to safeguard against bovine spongiform encephalopathy (BSE) have significantly changed rendering companies' business practices, value of products, and costs incurred to pick up animals. Regulations require rendering operations to remove brains and spinal cords of all cattle older than 30 months if the carcass is processed for animal feed.

3

Chemical and Biological Properties of Paddy Soil Treated with Herbicides and Pyroligneous Acid

H. R. L. Benzon[1,2], M. R. U. Rubenecia[1], Jr. V. U. Ultra[3] & Sang Chul Lee[1]

[1] School of Applied Biosciences, Kyungpook National University, Daegu, Korea

[2] National Institute of Molecular Biology and Biotechnology, UPLB, College, Laguna, Philippines

[3] College of Natural Sciences, Catholic University of Daegu, Gyeongsan, Korea

Correspondence: Sang Chul Lee, School of Applied Biosciences, Kyungpook National University, Daegu 702-701, Korea. E-mail: leesc@knu.ac.kr

Abstract

In this study, the effect of different herbicide x pyroligenous acid mixtures on soil chemical and biological properties was evaluated in order to elucidate its potential impacts on nutrient availability and soil quality. The experiment was conducted under greenhouse conditions consisted of 100-fold diluted wood vinegar (100 WV), 50% BCB (Bentazone cyhalofop-butyl), 100% BCB, 50% BCB+100 WV, 50% BCB+250 WV and 50% BCB+500 WV applied against *Echinochloa crusgalli.* Ten days after herbicide application, the chemical properties were altered and the changes were influenced by the amount of WV dilutions. The soil pH was near neutral to slightly alkaline coupled with decreased electrical conductivity (EC). The total carbon, available phosphorous, exchangeable magnesium, exchangeable sodium, exchangeable calcium, exchangeable potassium and cation exchange capacity (CEC) were relatively lower compared to the control. Biolog™ assay showed that different treatments resulted to variable microbial activity, Shannon-Weaver index and richness of carbon utilization potential. Principal component analysis showed segregation of samples based on extracted principal components indicating variable carbon utilization potential due to different treatments combination. Correlation analysis between soil and microbial properties revealed that Principal Component 2 has negative correlation with pH and available P. On the other hand, Principal Component 3 was positively correlated with total C, exch. Ca, Mg, Na, and CEC (0.475, 0.490, 0.555, 0.489, and 0.517, respectively). Application of pyroligneous acid combined with herbicides resulted to changes in soil chemical and biological characteristics which may have unique implications on nutrient availability and overall soil quality.

Keywords: bentazone cyhalofop-butyl, Biolog™, pyroligneous acid, soil chemical and biological properties

1. Introduction

The continuous use of environmentally persistent herbicides and other synthetic agricultural chemicals posed great risks to soil and water contamination. A viable alternative pesticide that is equally effective and less harmful to the environment is in demand. The role of pyroligneous acids (PA), also as plant and soil treatments gained the attention of research enthusiasts. Pyroligneous acid is a dark brown solution obtained as a by-product of wood carbonization. Large number of substances has been found in the pyrolysis liquids from different resources. The detected substances from the acids belong to different classes of organic compounds, namely, aldehydes, ketones, alcohols, organic acids, esters, derivatives of furan and pyran, phenolics, hydrocarbons and nitrogen compounds, in which the major ones are organic acids and phenolics (Souza et al., 2012). These compounds are known to increase herbicide efficacy (Kim et al., 2001).

The acidic nature of PA may have an effect on displacement of ions from soil exchange complex. An advantage of this would be improved nutrient availability. However, increased displacement would possibly result to leaching of substances through the soil profile (Jones, 1998; Andrade, 2003). For that reason, studies on the impact of the use of this kind of product are of great agricultural and environmental importance, and absolutely necessary to support the correct and safe application of chemicals (synthetic or alternative agricultural supplies) on the soil.

Soil microorganisms play an important role in soil-plant-herbicide-fauna-man interactions. They participate in herbicide degradation and activity, they may serve as bioindicators of changes in soil biological activity and, some microbial species may be used as bioherbicides. The microbial biomass can also represent a large pool of nutrients especially on less fertile sites (Giardina et al., 2000). Microorganisms can act as sinks, during immobilization and sources, during mineralization of labile nutrients (Garcia-Oliva et al., 1998; Stenstrom et al., 1998). The soil biota activity can be stimulated as an indirect result of the chemical changes in the soil.

There is no or limited scientific information with regard to the influence of pyroligneous acids coupled with synthetic herbicides on soil chemical and biological properties. Hence, the objective of this study was to evaluate the combined effect of different herbicide and pyroligenous acid mixtures on soil chemical and biological properties in order to elucidate its potential impacts on nutrient availability and soil quality.

2. Methods

2.1 Experimental Set Up

The experiment was conducted under greenhouse conditions at the Agricultural Experiment Station and Research Facility, College of Agriculture and Life Sciences, Kyungpook National University, Daegu, South Korea. The experimental set up was in a Completely Randomized Design (CRD) with three replications.

The treatments were: T1-Control, T2-100 WV (100-fold diluted wood vinegar), T3-50% BCB (Bentazone cyhalofop-butyl), T4-100% BCB, T5-50% BCB+100 WV, T6-50% BCB+250 WV (250-fold diluted wood vinegar) and T7-50% BCB+500 WV (500-fold diluted wood vinegar). The herbicide BCB and the WV was applied singly or consecutively by spraying. The wood vinegar (Table 1) was obtained from a local distributor. The planting media was composed of 1:1 paddy and sandy soil (ph = 6.69, EC = 72.4 uS/cm). Water depth in pots was maintained at 3cm on the soil surface during the experiment period. Treatment application was done at 2 to 3 leaf growth stage, approximately 6 to 8 days after transplanting of barnyard grass. The experiment was terminated 10 days after treatment application.

Table 1. Physical and chemical components of wood vinegar

Item	Wood vinegar
pH	2.97
Electrical conductivity	2313 uS/cm
Transparency	Transparent
Color	Golden brown
Methanol	0.31%
Phenolic compounds	2.44%
Propanoic acid	0.89%
Acetic acid	9.95%

Typically, wood vinegar is high in low-molecular weight acids (formic and acetic), alcohols (methanol) and aldehydes, which can serve as a carbon and energy resource for prototrophic bacteria occurring in the soil (Focht, 1999 as cited by Hagner, 2013). However, the high acidity, methanol and phenol content have strong bactericidal effect at a high concentration. Therefore, wood vinegar if applied in correct concentration, can enhance the soil biological properties.

2.2 Soil Sample Preparation and Analysis

Immediately after sampling, soil samples for microbial characterization were stored at 4 °C until analysis while samples for chemical analyses were air-dried, ground, and passed through 2 mm sieve. Soil pH and EC were determined in 1:5 soil-water ratio (Thermo Scientific Orion Star A215 Benchtop pH/EC meter). Total organic C was determined using an elemental analyzer (ThermoFisher Flash 2000). Available P was determined following the method described by Pierzynski (2000). Air-dried soil samples (2.0 g) were added in 20 mL Bray and Kurtz P-1 extracting solution (0.025 M HCl in 0.03 M NH_4F) and were placed in a mechanical shaker (200 or more epm) for 5 minutes at room temperature. The extracts were filtered through Whatman No. 42 filter paper and were analyzed for phosphorus by ICP-MS (PerkinElmer NexION 300X). Soil exchangeable bases were extracted through ammonium acetate method according to Brix (2008). Air-dried soil samples (2.0 g) were added in 20 mL 1 M NH_4OAc and were placed in a mechanical shaker for 2 hours. The extracts were filtered through Whatman

No. 42 filter paper and were also analyzed by ICP-MS (PerkinElmer NexION 300X). Soil microbial activity was evaluated based on substrate utilization profiles that were established using BIOLOG EcoMicroPlate™ (Biolog, Hayward, CA, USA). Soil sample (5 g) was suspended in 0.1 M NaH_2PO_4 solution (pH 6) at a ratio of 1:9 (w/v). The soil suspension was diluted 1,000-fold with 0.15 M NaCl and 100 μL of the diluted suspension was inoculated into each well. The microplates were incubated at 28 °C and their absorbance was measured at an optical density of 590 nm using a Multiskan™ Go Microplate Spectrometer (ThermoFisher, USA) every 24 h for 6 days. The absorbance of the 31 substrates was used to calculate the average well color development (AWCD) and the values were plotted against the incubation period of the plate (Ultra et al., 2012). The Shannon-Weaver index and richness of bacterial communities were calculated based on the absorbance readings of the microplate wells after 96 h of incubation.

2.3 Statistical Analysis

The data on soil properties were analyzed statistically following the analysis of variance (ANOVA) and the mean differences among treatments were compared by Fisher's Least Significant Difference (LSD) using the statistical computer package program, R-Software. On the other hand, AWCD, richness, and Shannon-Weaver index were statistically analyzed using the IBM SPSS Statistics ver. 21 for Windows. Mean values per treatment were compared using two-way analysis of variance. Tukey's honestly significant difference was performed to determine significant differences ($P < 0.05$). The optical density data from the Biolog EcoMicroPlate™ were subjected to principal component analysis using SPSS.

3. Results

3.1 Soil Chemical Characteristics

The pH observed for all treatments was neutral to slightly alkaline and relatively lower compared to the control. Noticeable decrease in pH compared to the control was seen in soils treated with 50% BCB and 50% BCB + 250 WV. The EC was highest in soils treated with 100% BCB (Table 2). However, the result was comparable to other treatments except that of 50% BCB.

A general trend was seen in total C, available P, exch. Mg, exch. Na, exch. Ca, exch. K and CEC (Table 2) with the highest values obtained in control treatment. In terms of total C content, there were no significant differences observed among treatments. The data on available P, exch. Mg, exch. Na, exch. Ca, exch. K and CEC revealed similar trends. Relatively lower values were obtained compared to the control (Table 2).

Table 2. Effect of herbicide and pyroligneous acid application on the chemical properties of soils under greenhouse conditions

Treatments	ph	EC (uS/cm)	Total C (%)	avail. P	exch. Ca	exch. Mg	exch. K	exch. Na	CEC ($cmol_c kg^{-1}$soil)
				-----------	-----------	ppm	-----------	-----------	
Control	7.70 a	162.83 ab	0.39 a	10.32 a	4.57 a	1.41 a	0.12 a	0.40 a	6.49 a
100 WV	7.30 ab	141.10 ab	0.32 a	7.87 b	3.69 b	1.02 b	0.10 ab	0.33 b	5.14 b
50% BCB	7.28 b	119.37 b	0.29 a	6.89 b	3.48 b	0.98 b	0.10 ab	0.31 b	4.87 b
100% BCB	7.50 ab	163.10 a	0.29 a	7.80 b	3.90 ab	1.05 b	0.08 b	0.33 b	5.37 ab
50% BCB + 100 WV	7.46 ab	131.50 b	0.29 a	7.72 b	3.68 b	1.02 b	0.10 ab	0.31 b	5.11 b
50% BCB + 250 WV	7.17 b	122.73 ab	0.31 a	6.14 b	3.54 b	0.96 b	0.11 ab	0.31 b	4.93 b
50% BCB + 500 WV	7.30 ab	128.57 ab	0.31 a	7.66 b	3.26 b	0.92 b	0.09 ab	0.30 b	4.57 b

3.2 Soil Biological Characteristics

The highest AWCD on the Biolog EcoMicroplate™ was observed under 50% BCB + 500 WV treatment after 4 days of incubation which indicates high microbial activity (Figure 1). In addition, an increased activity of soil microorganisms was also observed in the control. On the other hand, lower AWCD was observed in the 100% BCB treatment. A noticeable decrease in microbial activity was observed in 50% BCB+250 WV. This suggests that treatment combination synergistically affected the soil microorganisms. In terms of Shannon-Weaver index, there were no noticeable differences among treatments (Figure 2). However, a significant difference was observed in richness which was attributed to the 50% BCB treatment (Figure 3).

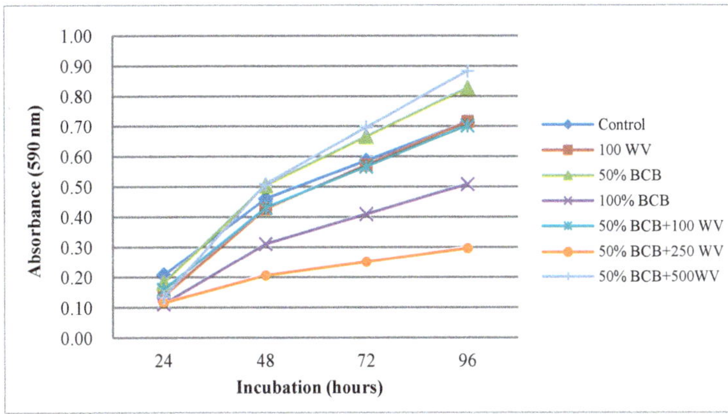

Figure 1. Effect of herbicide x pyroligneous acid application on the Average Well Color Development (AWCD) based on Biolog Ecomicroplate^{TM} absorbance data of soils under greenhouse conditions

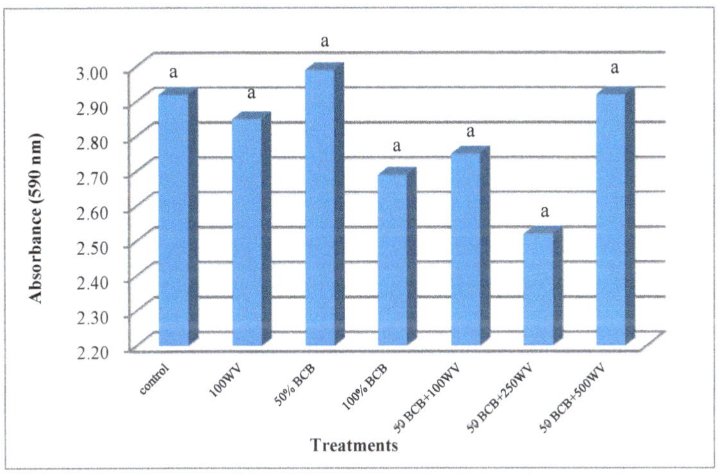

Figure 2. Influence of herbicide x pyroligneous acid application on the Shannon-Weaver index of soils based on Biolog EcoMicroplate^{TM} absorbance data at 96 h of incubation. Same letters indicate non significant differences according to Tukey's test ($P \leq 0.05$)

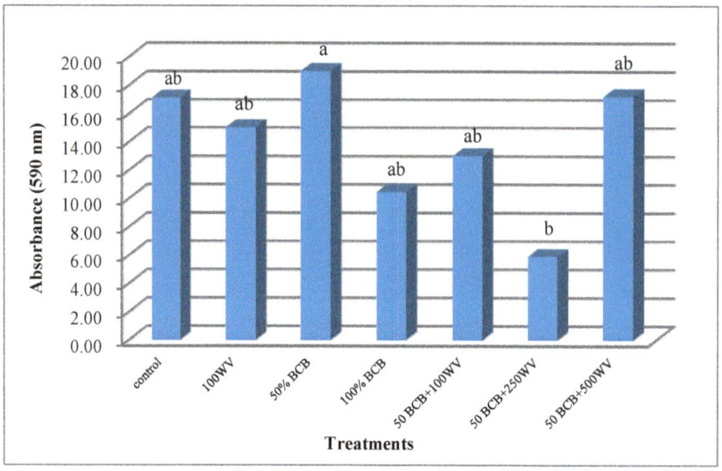

Figure 3. Richness of metabolized carbon of soils as affected by herbicide x pyroligneous acid application based on Biolog EcoMicroplate™ absorbance data at 96 h of incubation. Same letters indicate non significant differences according to Tukey's test ($P \leq 0.05$)

The Biolog EcoMicroplate™ readings at 96 hours were subjected to principal component analysis (PCA) to determine the extent of differentiation of microbial functional structure based on the carbon source utilization among soils from different treatments. This resulted in the extraction of 3 principal components (PC) contributing about 25%, 17%, and 10% to the data variance, corresponding to PC1, PC2, and PC3, respectively (Figure 4). The plots between PC1 and PC2 scores were able to differentiate the soil treated with 50% BCB+250 WV from other treatments (Figure 4a). On the other hand, the plots of PC1 and PC3 scores were able to differentiate the soils from the control and 50% BCB from other treatments (Figure 4b).

The carbon substrates were most heavily loaded based on the principal component analysis (Table 3). Twelve carbon substrates contributed to the separation of treatments along PC1. The highest utilization of these substrates was observed in 50% BCB+500 WV treatment. The carbohydrate D-mannitol was highly utilized by the microorganisms in all treatments except for 50% BCB+250 WV treatment. The carbon substrate distinct to PC2 was D-cellobiose which was highly utilized under the treatment combination 50% BCB+250 WV. The substrates unique to PC3 were 4-hydroxy benzoic acid, glycyl-L-glutamic acid, α-cyclodextrin, and glycogen, which in general, were less utilized as seen in all treatments.

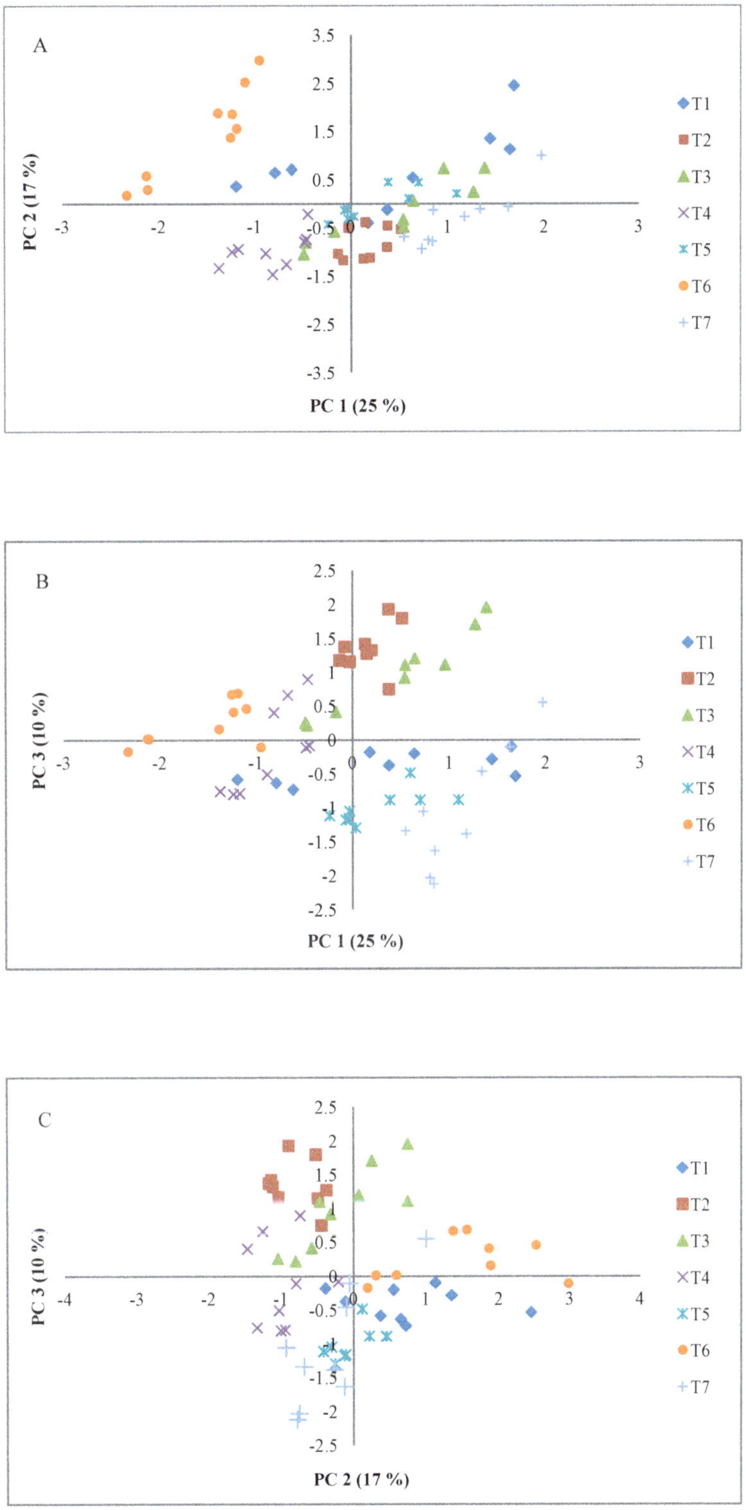

Figure 4. Principal component analysis based on the carbon source utilization among soils from different treatments (T1 – 100 WV, T2 – Control, T3 – 50% BCB, T4 –100% BCB, T5 – 50% BCB+100 WV, T6 – 50% BCB+250 WV and T7 – 50% BCB+500 WV)

Table 3. Carbon substrates most heavily loaded based on the principal component analysis

Carbon substrates	PC 1*	PC 2*	PC 3*
Carboxylic acids			
γ-hydroxy butyric acid	.580		
D-galactonic acid γ-lactone		.616	
2-hydroxy benzoic acid		.710	
α-keto butyric acid		.727	
4-hydroxy benzoic acid			.752
Carbohydrates			
β-methyl-D-glucoside	.738		
i-erythritol	.527		
D-mannitol	.707		
N-acethl-D-glucosamine	.846		
α-D-lactose	.597	.527	
D-xylose		.798	
D-cellobiose		.505	
Amino acids			
L-arginine	.516		
L-asparagine	.776		
L-serine	.777		
L-phenylalanine		.659	
Glycyl-L-glutamic acid			.720
Polymers			
Tween 40	.687		
α-cyclodextrin			.512
Glycogen			.761
Amines			
Phenylethyl-amine	.625		
Putrescine		-.578	
Phosphorylated chemical	.707		

Note. *PC 1, PC 2 and PC 3 are the principal components extracted from the principal component analysis (PCA) of the microbial community substrate utilization potential of herbicide-wood vinegar treated soils.

Table 4. Correlation analysis between the chemical and biological properties of soils applied with herbicide and wood vinegar

Properties	PC1	PC2	PC3	Richness	Shannon-Weaver Index
EC	-.003	-.372	.226	.056	.086
pH	.091	-.547*	.246	.242	.255
Total C	.352	.157	.475*	.320	.404
P	.136	-.519*	.231	.276	.175
Ca	.090	-.073	.490*	.125	.236
Mg	.147	-.178	.555**	.234	.309
Na	.122	-.077	.489*	.171	.282
K	.085	.388	.243	.048	.119
CEC	.107	-.089	.517*	.155	.260

Note. *$P < 0.05$; **$P < 0.01$.

4. Discussion

4.1 Soil Chemical Characteristics

The experimental set up mimicked a paddy soil condition wherein the soil was under waterlogged conditions. The soil system was generally quite buffered to the extent that waterlogging was not effective in changing the pH, though it has been known that in waterlogged soil, anaerobic conditions take place and as a result fermentation of carbohydrates occur releasing acids which shift the pH to the acidic side (Taha et al., 1967). In addition, the decrease in pH can be attributed to the acidic nature of wood vinegar and herbicides.

The lower values obtained in total C, available P, exch. Mg, exch. Na, exch. Ca, exch. K and C can be attributed to wood vinegar. Wood vinegar has the ability to reduce the cluster value of water. This means that the water is activated and can be easily absorbed by the plants because water with a low cluster value is in a very small mass. Each of these masses will hold one or few mineral elements which can be easily taken up by plants (http://www.agrowingculture.org).

Lower P contents observed on the treatments could be related to fixation in the soil, since it was alkaline in reaction (Taha et al., 1967). Phosphorous can be fixed in the form of calcium, magnesium, iron or aluminum phosphates and clay-adsorbed phosphorous (Mahdi et al., 2012). When alkaline soils were waterlogged, calcium phosphates and aluminum phosphates increased while iron phosphates decreased (IRRI, 1978). The best availability of phosphorous is in the range of pH 6.0 to 7.0. According to Foth (1978), calcium phosphates begin to precipitate at about pH 6.0. On the contrary, above pH 7.0, there is a reduction in phosphorus solubility or availability. Increases in pH above 7.0 create sufficient OH^- to react with $H_2PO_4^-$ to form HPO_4^{2-} and water to cause the latter form of phosphorus to become the most abundant. Since HPO_4^{2-} is less readily taken up by plants than $H_2PO_4^-$, one can conclude that part of the reduced availability of phosphorus in alkaline soils is due to the presence of hydroxyl ions and formation of HPO_4^{2-}. The chemical components of the wood vinegar may have contributed to the production of OH^- reacting with $H_2PO_4^-$ resulting to unavailable forms of phosphorus (Foth, 1978). In addition, soluble $H_2PO_4^-$ quickly reacts with calcium to form a sequence of products of decreasing solubility (Mahdi et al., 2012).

4.2 Soil Biological Characteristics

The high microbial activity observed under 50% BCB+500 WV treatment could be attributed to low levels of herbicide and wood vinegar. The low concentrations of alcohols, acids, and aldehydes in the wood vinegar served as carbon and energy substrates for soil microorganisms (Steiner et al., 2008). The increased microbial activity under the control treatment was probably due to root exudates released by barnyard grass which provide nourishment to the soil microflora. In contrast, application of 100% BCB decreased the activity of soil microorganisms. This reduction in microbial activity, coinciding with withering plant biomass, could have resulted from drastically reduced root exudates that serve as a resource of the microorganisms in the soil. A shortage in root exudates can lead the rhizosphere microbes to enter a dormant, inactive stage (Hagner, 2013). Therefore, a decline in microbial activity can be observed. This also indicates that 100% application of herbicides is detrimental to soil microorganisms. The Shannon-Weaver index and richness were obtained after 96 hours of incubation. The Shannon-Weaver index is an indicator of the microbial functional diversity in soil. On the other hand, richness refers to the carbon utilization by microorganisms. Based on the results, all treatments did not significantly affect the microbial functional diversity in soil. However, a significant difference was observed in richness which was attributed to 50% BCB treatment. This can be a reflection of the availability of nutrients in the soil that were utilized by microorganisms. In general, herbicides affect soil microbes indirectly. They can be a source of nutrition for microbes or when applied in high doses, may kill microorganisms. The lower rate of BCB application and also its combination with 500 WV favored microbial activity in the soil. Related studies showed that microbes propagate well when higher dilution of wood vinegar was applied. An experiment conducted by Steiner et al. (2008) revealed significant increase in microbial biomass and population growth rate when PA was applied in highly weathered soils. This is mainly due to the effect on the metabolism by acetic acid. Acetyl co-enzyme is produced by plants and microbes from acetic acid. Through the TCA cycle, acetyl co-enzyme is converted into citridic acid, malic acid, fumaric acid, succiric acid and other elements that are necessary for the plants and microbes (Nelson & Cox, 2012). This is the main reason behind the propagation of microbes. The herbicide and wood vinegar may have worked synergistically for microbial growth and multiplication (http://www.agrowingculture.org).

The variability in PC1 can be explained by the high utilization of five carbohydrates and three amino acids. The carbohydrates include β-methyl-D-glucoside, i-erythritol, D-mannitol, N-acetyl-D-glucosamine and α-D-lactose. On the other hand, L-arginine, L-asparagine and L-serine are the amino acids responsible for the variation in

PC1. Variability in the second PC can be attributed by responses to carboxylic acids, carbohydrates and amine (Table 3).

Correlation analysis between soil and microbial properties were also evaluated (Table 4). Results revealed that PC2 had negative correlation with pH and available P. This implies that the set of particular microbial communities that utilized the carbon substrates in the Biolog Microecoplate^{TM} that contributed to PC 2 are pH-dependent and can be influenced by the availability of P. As cited by Bunemann et al. (2011) microorganisms require or preferentially use inorganic forms of P tying up the nutrient in microbial biomass, a process called immobilization. Microbial P immobilization can affect P availability by removing inorganic P from the soil solution, especially when soluble carbon is available for microbial growth. On the other hand, PC3 was positively correlated with total carbon, exch. Ca, Mg, Na, and CEC. The microbial communities that utilized the substrates that were distinct to PC 3 could be related to the availability of total carbon, exch. Ca, Mg, Na, and CEC. This implies the role of mineral nutrition on microbial growth. The alteration of substrate quality due to the addition of wood vinegar triggered the shift of soil microbial functional structure.

References

Andrade, F. V. (2003). Adição de ácidos orgânicos e húmicos em Latossolos e adsorção de fosfato. *Revista Brasileira de Ciência do Solo, 27*(6), 1003-1011. http://dx.doi.org/10.1590/S0100-06832003000600004

Brix, H. (2008). *Soil Exchangeable Bases (ammonium acetate method)*. Retrieved May 2, 2013, from http://mit.biology.au.dk/~biohbn/Protocol/Soil_Exchangeable_Bases_CEC_20081127.pdf

Bunemann, E. K., Oberson, A., & Frossard, E. (2011). Phosphorus in Action. *Soil Biology, 26*, 37-57. http://dx.doi.org/10.1007/978-3-642-15271-9_2

Focht, U. (1999). The effect of smoke from charcoal kilns on soil respiration. *Environmental Monitoring and Assessment, 59*, 73-80. http://dx.doi.org/10.1023/A:1006010629482

Foth, H. D. (1978). *Fundamentals of Soil Science*. John Wiley & Sons, Inc. USA.

Garcia-Oliva, F., Sanford, R. L., & Kelly, E. (1998). Effect of burning of tropical deciduous forest soil in Mexico on the microbial degradation of organic matter. *Plant Soil, 206*, 29-36. http://dx.doi.org/10.1023/A:1004390202057

Giardina, C. P., Sanford, R. L., Dockersmith, I. C., & Jaramillo, V. J. (2000). The effects of slash burning on ecosystem nutrients during the land preparation phase of shifting cultivation. *Plant Soil, 220*, 247-260. http://dx.doi.org/10.1023/A:1004741125636

Hagner, M. (2013). *Potential of the slow pyrolysis products birch tar oil, wood vinegar and biochar in sustainable plant protection-pesticidal effects, soil improvement and environmental risks*. Department of Environmental Sciences, University of Helsinki, Lahti, Finland.

International Rice Research Institute. (1978). *Soils and rice*. Los Baños, Philippines.

Jones, D. L. (1998). Organic acids in the rhizosphere – A critical review. *Plant and Soil, 205*(1), 25-44. http://dx.doi.org/10.1023/A:1004356007312

Kim, Y. H., Kim, S. K., Kim, K. S., & Lee, Y. H. (2001). Composition of constituents of commercial wood vinegar. *Korea J. Agric. Chem. and Biotech., 44*(4), 262-268.

Mahdi, S. S., Talat, M. A., Hussain Dar, M., Hamid, A., & Ahmad, L. (2012). Soil phosphorus fixation chemistry and role of phosphate solubilizing bacteria in enhancing its efficiency for sustainable cropping – A review. *Journal of Pure and Applied Microbiology, 6*(4), 1-7.

Nelson, D. L., & Cox, M. M. (2012). *Lehninger Principles of Biochemistry* (6th ed.). W. H. Freeman & Company. New York.

Pierzynski, G. M. (2000). Methods of Phophorus Analysis for Soils, Sediments, Residuals, and Waters. *Southern Cooperative Series Bulletin, 396*, 13-14.

Souza, J. B. G., Re-Poppi, N., & Raposo Jr., J. L. (2012). Characterization of pyroligneous acid used in Agriculture by Gas Chromatography-Mass Spectrometry. *J. Brazil. Chem. Soc., 23*(4), 610-617.

Steiner, C., Das, K. C., Garcia, M., Forster, B., & Zech, W. (2008). Charcoal and smoke extract stimulate the soil microbial community in a highly weathered xanthic Ferralsol. *Pedobiologia, 51*, 359-366. http://dx.doi.org/10.1016/j.pedobi.2007.08.002

Stenstrom, J., Stenberg, B., & Johanson, M. (1998). Kinetics of substrate-induced respiration (SIR): Theory.

Ambio, 27, 35-39.

Taha, S. M., Mahmoud, S. A. Z., & Ibrahim, A. N. (1967). Microbiological and Chemical Properties of Paddy Soil. *Plant and Soil, 26*, 1. http://dx.doi.org/10.1007/BF01978674

Ultra, V. U., Han, S. H., & Kim, D. H. (2012). Soil properties and microbial functional structure in the rhizosphere of *Pinus densiflora* (S. and Z.) exposed to elevated atmospheric temperature and carbon dioxide. *Journal of Forest Research, 18*(2), 149-158. http://dx.doi.org/10.1007/s10310-012-0333-6

4

Impact of *Gnetum africanum* and *Amaranthus rectroflexus* Plant Food on Urea, Creatinine and Plasma Electrolytes

Guy Stéphane Padzys[1], Joseph Privat Ondo[2] & Linda Priscilla Omouendze[1]

[1] Département de Biologie, Faculté des Sciences, Université des Sciences et Technique de Masuku, Franceville, Gabon

[2] Département de Chimie, Faculté des Sciences, Université des Sciences et Technique de Masuku, Franceville, Gabon

Correspondence: Guy Stéphane Padzys, Département de Biologie, Faculté des Sciences, Université des Sciences et Technique de Masuku, Franceville 943, Gabon. E-mail: padzys@gmail.com

Abstract

Amaranthus rectroflexus (*AmE*) and *Gnetum africanum* (*GnA*) is the most popular green leafy vegetable in Gabon and is gaining equal popularity as a delicious food leaf in other African countries such as Cameroon, Nigeria, Congo and Angola. The purpose of this study was to assess the impact of these plants in modulating urea, creatinine and plasma electrolytes. 50 mg/kg of aqueous extract were given to rats in the experimental groups for two week, while rats in the control group received an appropriate volume of water. Urea creatinine, and Plasma electrolytes were measured by spectrophotometric method. Chronic oral administration for two week of 50 mg/kg of *AmE* and *GnA* aqueous extracts significantly decreased plasma sodium concentration benefit to potassium ($P < 0.05$). No significant harmful changes ($P > 0.05$) in weight, blood glucose, creatinine and urea plasma concentration.

Keywords: *Amaranthus rectroflexus*, *Gnetum africanum*, electrolytes, urea, creatinine

1. Introduction

Plants, particularly green leafy vegetables, serve as a major dietary reservoir of the essential PUFAs, dietary fiber, antioxidants and other bioactive compounds that exert cardioprotective biological mechanisms (Veer et al., 2000; Kris-Etherton et al., 2000; Hu, 2003; Slavin, 2003) . Several cross-sectional and prospective studies have found that consumption of legumes and legume-based diets are inversely associated with the prevalence and incidence of obesity, cardiovascular disease, type 2 diabetes (Trinidad et al., 2010; Villegas et al., 2008; Venn et al., 2004; Kolonel et al., 2000; Flight et al., 2006).

Amaranthus extrafolus (*AmE*) and *Gnetum africanum* (*GnA*) are the most popularity plant food in Gabon and is gaining equal popularity in other African countries such as Cameroon, Nigeria, Congo and Angola (Eyo & Abel, 1983). *AmE*, is an erect, annual herb reaching a maximum height near 3 m. The leaves are nearly 15 cm long on large individuals, the ones higher on the stem having a lance shape and those lower on the plant diamond or oval in shape (Figure 1A). The seed of *GnA* is oval in shape and small in size, about 0.5 cm in diameter. They are greenish in color when and randomly from a thick bush around the rain unripe and reddish when ripe (Figure 1B).

(A) (B)

Figure 1. Photo of *Gnetum africanum* (A) *and Amaranthus rectroflexus* (B)

Little or no information is available on the impact of *AmE* and *GnA* chronic consumption on physiological parameters.Topics will show firstly the impact of regular consumption of *GnA* and *AmE* on body weight, blood glucose, urea and creatinine. Finally study the impact of *GnA* and *AmE* consumption on plasma concentration of Cloride (Cl), Potassium (K) and Sodium (Na).

2. Materials and Methods

2.1 Animals

Twenty eight males Wistar rats weighing 255-279 g (IFFA-CREDO, France) were used for this experiment. The animals were born in the laboratory from twenty litters, culled to 7 pups per litter to ensure normal body growth. The animals were housed in standard cages under controlled temperature conditions (22 ± 1 °C). Food and water were available ad libitum throughout the experiment. From birth, the rats were kept on a reversed 12:12 light–dark cycle (dark period 08:00–20:00).

2.2 Preparation of AmE and GnA Aqueous Extracts

Fresh plant samples of, *AmE* and *GnA* were obtained from market. The preparation of plant materials was as previously reported (Olatunji et al., 2006). The extract was then filtered through Whatman No.1 filter paper and the residue was discarded. The resultant filtrate was then evaporated to dryness and stored in capped bottles at 4 °C until use. The desired amount of extract was dissolved in water to make 50 mg/ml of stock solutions for lower and higher doses respectively.

2.3 Chronic Consumption of AmE and GnA Aqueous Extracts

All experiments conformed to the Guide for the Care and Use of Laboratory Animals published by the National Institutes of Health (no. 85-23, revised 1996), the recommendations of the European Community Council for the Ethical Treatment of Animals (no. 86/609/EEC) - and the regulations of the University of Masuku. All efforts were made to minimise animal suffering.

Rats in the control group received an appropriate volume of water per day as vehicle by gavage. The experimental rats received 50 mg/kg of the *AmE* and *GnA* aqueous extract by gavage. The control group or of *Ame* and *GnA* aqueous extract administration was stopped 24 h before the animals were killed. Weighed and an intracardiac blood sample (500–1000 ml) was taken between 11h and noon for electrolytes, urea and creatinine measurements. Blood was collected over 1–2 min into tube heparin or EDTA.

2.4 Biochemical Assays

Plasma concentrations of urea, creatinine, and electrolytes were performed by spectrophotometric method. These studies were made respectively using automata Bs 400 mindray chemistry analyzer (Biorad, France) for urea and creatinine. Electrolytes plasma were analysed by Spotlyte Na K Cl analyzer (Medica Corporation, USA).

Concentration of blood glucose was determined using method described by (Padzys et al., 2011). This approach using glucometer (Glucose-test, CarenSensN Sud Korea)

2.5 Preparation of Extracts for Phytochemical Screening

Air-dried powdered leaves (10 g) of *AmE* and *GnA* were separately extracted with 100 ml of water (Aq), by maceration for 24 h. Extracts were filtered and the filtrate was used for photochemical screening.

2.6 Phytochemical Screening

The extract, were screened for their classes of bioactive compounds using standard procedures (Culei, 1982; Harbone, 1984; Adegunloye, 1993; Sofowora, 1983; Trease & Evans, 2002) The extracts were tested qualitatively for the presence of chemical constituents such as tannins, terpenes, saponins, flavonoids, alkaloids, anthraquinones etc.

For gallic tannins, 2 ml of 1% ferric chloride solution was added to 2 ml of the filtrate (Stiasny's test). Dark-greenish coloration indicated their presence. Tannins catechic, 2 ml of a solution of hydrochloric n-butanol are added to 2 ml of filtrate, and then heating in a water bath for 5 to 10 minutes (Bate-Smith's test). Intense red coloration indicated the presence of the catechin tannins. For total flavonoids and anthocyanes, 1 ml of the sulfuric acid was added to 2 ml of the filtrate, then 1 ml NaOH. There shown a dark color after adding acid, indicating the presence of flavonoids, the color changes to purple after addition of NaOH, indicating the presence of anthocyanins. 2 ml of the filtrate were added magnesium strips followed by hydrochloric alcoholic (cyanidine test). A rose-orange effervescence showed the presence of flavones, rose-purplish indicated of flavanones and red denoted of flavonols. We had applied the Folin's test to determine polyphenols contents. 1 ml of the Folin reagent was added to 2 ml of the filtrate, then 1 ml NaOH. Dark green coloration indicated the presence of polyphenols.

For coumarins, 2 ml of filtrate combined with 2 ml of NH_4OH, then, lookin at UV lamp (366 nm). The fluorescence presence indicated the presence of coumarins. 2 ml NH_4OH solution was added to 2 ml of the filtrate (Borntrager's test). A rose pink colour in the ammonia layer indicated the presence of anthraquinones.

For alkaloids, some drops of sulfuric Dragendorff's reagent were added to 2 ml of the filtrate. Orange precipitate formed had showed the presence of alkaloids.

To determine terpenes, test such as Salkowski's and Lieberman's test were applied. 2 ml of concentrated H_2SO_4 were added to 2 ml of filtrate, a reddish-brown ring indicated the presence of steroid, an aglycone part of the cardiac glycoside (Salkowski's test). Another part of the filtrate (2 ml) was added with 2 ml of acetic anhydride and cooled well in ice and concentrated H_2SO_4 (2 ml) was carefully added. A colour change from blue to green indicated the presence of terpenes (Lieberman's test).

Saponins were determined through frothing test. The filtrate was vigorously shaken. Frothing which persisted on warming for about 15 min indicated the presence of saponins.

2.7 Statistical Analyses

The results were expressed as group means ± SEM. Student's t-test was used to establish the comparison between control and treatment animals since all data were normally distributed. Specific mean comparisons were then made using t-test with the Bonferroni correction. Differences were considered significant at $P < 0.05$.

3. Results and Discussion

3.1 Effects of GnA and AmE Aqueous Extract on Body Weight and Blood Glucose

Administration of *GnA and AmE* aqueous extract did not produce any significant effects on the body weight ($P > 0.05$).To determine if the administration of *GnA and AmE* aqueous extract was associated with change in glycemia, blood glucose levels were obtained (Table 1). No difference was observed in blood glucose level between control and experimental group rats: ($P > 0.05$). These results to be explain by the legumes are low-glycemic-index low-energy-dense foods that contain high amounts of dietary fiber, vegetable proteins, oligosaccharides, and phenolic and other bioactive compounds (Kromhout et al., 1996; Messina et al., 1999; Anchana, 2004).

Table 1. Effects of *AmE and GnA* aqueous extracts on body weight and blood glucose

Group	Body weight (g)	Blood glucose (mg/dl)
Control	255 ± 15,64	131.25 ± 12.8
AmE	279 ± 5,73	132 ± 11.5
GnA	272 ± 6,23	132.6 ± 10.1

Values are means ± SEM (n = 7/group).

3.2 Effects of GnA and AmE Aqueous Extract on Urea and Creatinine

We also assessed the possible effects of *GnA and AmE* on the kidney by urea and creatinine measurement. No significant changes were seen in these indices after *GnA and AmE* administration Table 2 shows that no significant change ($P > 0.05$).

Table 2. Effects of *AmE and GnA* aqueous extracts on plasma urea and creatinine

Group	Urea (mmol/l)	Creatinine (μmol/l)
Control	11.1 ± 2.1	102 ± 5
AmE	12 ± 1.1	104 ± 4.5
GnA	11 ± 1.2	106 ± 3.2

Values are means ± SEM (n = 7/group).

3.3 Effects of AmE and GnA Aqueous Extract on Plasma Electrolytes

We also measured Na and K to assess any electrolyte disturbance; analyses showed significant changes in Na and K after plant administration. Chronic administration of *GnA* and *AmE* aqueous extract produced a significant decrease ($P < 0.05$) of plasma sodium concentration (Table 3). These results show that a decrease plasma sodium concentration was associated with an increase plasma potassium concentration ($P < 0.05$). No significant change was observed in level of plasma chloride concentration after consumption of *GnA and AmE* extract aqueous ($P > 0.005$). These findings support the previous findings on the effect of legumes on plasma electrolytes in rats (Chen et al., 2003). Vegetable consumption was increased with a consequent increase in potassium intake from 37 mmol/day to 71 mmol/day, showed a large fall in blood pressure despite sodium intake being fixed at a low intake of 130 mmol/day (Appe et al., 1997). Published study by the same group clearly showed an additive effect of increasing potassium and reducing sodium intake (Sacks et al., 2001).

Table 3. Effects of *AmE and GnA* aqueous extracts on plasma electrolytes

Group	Sodium	Potassium	Chloride
Control	137.66 ± 0.982	9.48 ± 0.3	112 ± 2.1
AmE	133.1 ± 0.294*	12.02 ± 0.2*	113 ± 0.8
GnA	133.02 + 1.3*	13.44± 0.86*	113.5 ± 1.3

Values are means ± SEM (n = 7/group).*Significantly different from control group at $P < 0.05$.

The phytochemical screening of the extracts was first performed to detect the major chemical groups occurring in the extracts. The results of this screening were shown in Table 4. The qualitative analyze of extracts had shown the presence of triterpenoids, polyphenol and more coumarins in *AmE* and *GnA* in the phytochemical constituent of the aqueous extracts.

Table 4. Phytochemical screening of *AmE and GnA* water extracts

Chemical constituents		***AmE***	***GnA***
Saponins		+++	–
Tannins	*Gallic*	–	–
	Catechin	–	–
Alkaloids		+++	–
Triterpenoids		+	++
Polyphenols		+	+
Flavonoids	*Flavonols*	–	–
	Flavones	+	–
	Flavanones	–	–
Free anthraquinones		+	–
Coumarine		++	+++
Proanthocyane		–	–
Cardiac glycosides	*Digitoxine*	++	+++
	Digitoxigenine	–	–
	Gitoxine	–	–
	gitoxigenine	–	–
Reducing sugars		–	–

Legend: –: Not detected; +: Rare; ++: Abundant; +++: Very abundant.

4. Conclusions

The objective of this study was to demonstrate if the intake of *Amaranthus rectroflexus* and *Gnetum africanum,* is associated with modulation of urea, creatinine and plasma electrolytes in rat. We have demonstrated for the first time that, intake of *Amaranthus rectroflexus* and *Gnetum africanum* induces modulation of potassium and sodium plasma concentration in rat. This modulation is necessary for the prevention of cardiovascular disease (Appel et al., 1997; Sacks et al., 2001; Chang et al., 2006; Cook et al., 2007). The present diet may serve as a model of human alternative therapy of cardiovascular disease.

Acknowledgements

Funding provided by grants to Dr Isaac Mouaragadja Institut of Biotechnology, University of Masuku for assistance.

References

Anchana, C., Aphiwat, T., & Nuansri, R. (2005). Screening of antioxidant activity and antioxidant compounds of some edible plants of Thailand. *Food Chemistry, 92,* 491-497. http://dx.doi.org/10.1016/j.foodchem.2004.07.035

Appel, L. J., Moore, T. J., Obarzanek, E., Vollmer, W. M., Svetkey, L., & Sacks, F. M. (1997). A clinical trial of the effects of dietary patterns on blood pressure. *N. Engl. J. Med., 336,* 1117-24. http://dx.doi.org/10.1056/NEJM199704173361601

Chang, H. Y., Hu, Y. W., & Yue, C. S. J. (2006). Effect of potassium-enriched salt on cardiovascular mortality and medical expenses of elderly men. *Am. J. Clin. Nutr., 83,* 1289-1296.

Chen, C. C., Hsu, J. D., & Wang, S. F. (2003). Hibiscus Sabdariffa extract inhibits the development of atherosclerosis in cholesterol-fed rabbits. *Journal of Agricultural and Food Chemistry, 51,* 5472-5477. http://dx.doi.org/10.1021/jf030065w

Cook, N. R., Cutler, J. A., & Obarzanek, E. (2007).The long-term effects of dietary sodium reduction on cardiovascular disease outcomes: observational follow-up of the Trials of Hypertension Prevention. *Br Med J, 334,* 885-888. http://dx.doi.org/10.1136/bmj.39147.604896.55

Eyo, E., & Abel, U. (1983). Chemical composition of amino-acid content of *G. africanum* leaves. *Nig J Nutr Sci,*

4, 52-5.

Flight, I., & Clifton, P. (2006). Cereal grains and legume in the prevention of coronary heart disease and stroke: a review of the literature. *Eur J Clin Nutr, 60*, 1145-59. http://dx.doi.org/10.1038/sj.ejcn.1602435

Hu, F. (2003). Plant-based foods and prevention of cardiovascular disease: an overview. *Am J Clin Nutr, 78*, 544S-551S.

Kolonel, L. N., Hankin, J. H., Whittemore, A. S., Wu, A. H., Gallagher, R. P., Wilkens, L. R., ... West, D. W. (2000). Vegetables, fruits, legume and prostate cancer: a multiethnic casecontrol study. *Cancer Epidemiol Biomarkers Prev, 9*,795-804.

Kris-Etherton, P. M, Taylor, D. S., Yu-Poth, S., Huth, P, Moriarty, K., Fishell, V., ... Etherton, T. D. (2000). Polyunsaturated fatty acids in the food chain in the United States. *Am J Clin Nutr, 71*, 179S-188S.

Kromhout, D., Bloemberg, B. P., Feskens, E. J., Hertog, M. G., Menotti, A., & Blackburn, H. (1996). Alcohol, fish, fibre and antioxidant vitamins intake do not explain population differences in coronary heart disease mortality. *Int J Epidemiol, 25*, 753-759. http://dx.doi.org/10.1093/ije/25.4.753

Mirmiran, P., Noori, N., Zavareh, M., & Azizi, F. (2009). Fruit and vegetable consumption and risk factors for cardiovascular disease. *Metabolism, 58*, 460-468. http://dx.doi.org/10.1016/j.metabol.2008.11.002

Olatunji, L. A., Adebayo, J. O., Adesokan, A. A., Olatunji, V. A., & Soladoye, A. O. (2006). Chronic administration of aqueous extract of Hibiscus sabdariffa enhances Na+-K+-ATPase and Ca2+-Mg2+-ATPase activities of rat heart. *Pharm Biol, 44*, 213-216. http://dx.doi.org/10.1080/13880200600686129

Padzys, G. S., Thornton, S. N., Martrette, J. M., & Trabalon, M. (2011). Effects of short term forced oral breathing in rat pups on weight gain, hydration and stress. *Physiology & Behavior, 102*, 175-180. http://dx.doi.org/10.1016/j.physbeh.2010.10.018

Culei, I. (1982). Methodology for the analysis of vegetable drugs. *Practical manual on the industrial utilization of medicinal and aromatic plants* (pp. 67-81). Center Building, Romania.

Harbone, J. B. (1984). *Phytochemical methods* (2nd ed., pp. 84-196). London: Champion and Hall Publishers. http://dx.doi.org/10.1007/978-94-009-5570-7

Sofowora, A. (1993). *Medicinal plants and traditional medicine in Africa* (2nd ed., pp. 1-153). Spectrum Books limited, Ibadan, Nigeria.

Trease, G. E., & Evans, W. C. (2002). *Pharmacognosy* (15th ed., pp. 214-393). Saunders.

Adegunloye, B. J., Omoniyi, J. O., Owolabi, O. A., Ajagbonna, O. P., Sofola, O. A., & Coker, H. A. B. (1993). Petals extract of Hibiscus sabdariffa elicits relaxation of rat aorta by stabilizing the membrane and stimulating Na^{+}-K^{+}-ATPase. *Nig J Physiol Sci., 9*, 74-79.

Slavin, J. (2003). Why whole grains are protective: biological mechanisms. *Proc Nutr Soc, 62*, 129-134. http://dx.doi.org/10.1079/PNS2002221

Trinidad, T. P., Mallillin, A. C., Loyola, A. S., Sagum, R. S., & Encabo, R. R. (2010). The potential health benefits of legume as a good source of dietary fibre. *Br J Nutr, 103*, 569-74. http://dx.doi.org/10.1017/S0007114509992157

Veer, P., Jansen, M., Klerk, M., & Kok, F. (2000). Fruits and vegetables in the prevention of cancer and cardiovascular disease. *Public Health Nutr, 3*, 103-107.

Venn, B. J., & Mann, J. I. (2004). Cereal grains, legume and diabetes. *Eur J Clin Nutr, 58*, 1443-61. http://dx.doi.org/10.1038/sj.ejcn.1601995

Villegas, R., Gao, Y. T., Yang, G., Li, H. L., Elasy, T. A., Zheng, W., & Shu, X. O. (2008). Legume and soy food intake and the incidence of type 2 diabetes in the Shanghai Women's Health Study. *Am J Clin Nutr, 87*, 162-7.

5

An Experimental Study in Simulated Greenroof in Mediterranean Climate

Antonio Vestrella[1], Robert Savé[1] & Carmen Biel[1]

[1] IRTA, (Research and Technology, Food and Agriculture Institute) Torre Marimon, Caldes De Montbui, Barcelona, Spain

Correspondence: Antonio Vestrella, IRTA, (Research and Technology, Food and Agriculture Institute) Torre Marimon, 08140, Caldes De Montbui, Barcelona, Spain. E-mail: antoniovestrella@gmail.com

Abstract

The use of green roofs in the Mediterranean area is becoming more widespread. Use of the species which are most commonly used across the world can be limited by the specific conditions of the Mediterranean climate. An experimental green roof simulation tested the performance of 12 species divided into three groups according to growth form (ground cover, sub-shrub, rhizomatous-herbaceous). Different irrigation criteria were applied to each group: in two sections potential evapotranspiration (ET0) levels of 20% and 40% were applied, while the third section was rainfed. Mortality, green coverage, blooming and final biomass were measured.

The amount of water received and the low winter temperatures affected plant mortality. Significant differences in coverage were observed when comparing the rainfed section with the two irrigated areas. Ground cover plants in the irrigated sections achieved the best results in terms of coverage. Significant differences in biomass were observed only when comparing the irrigated plots with the rainfed plot. The influence of different irrigation protocols on flowering was observed in five species only.

The growth form groups responded in different ways to drought stress: rhizomatous-herbaceous developed slowly and consistently; ground cover developed quickly and relied more on the water supply; sub-shrubs showed lower resistance to drought stress under the present experimental conditions. Of the 12 plants tested for green-roof adaptation in a Mediterranean climate, those which performed best were: *Centranthus ruber* (L.) DC., *Santolina rosmarinifolia* L., *Helichrysum stoechas* (L.) Moench, *Iris lutescens* Lam..

Keywords: green roofs, drought stress, green coverage, mortality, flowering

1. Introduction

The remarkable growth of green roofing in Central and Northern Europe in recent years has not been followed by similar developments in nearby Mediterranean countries. Green roof experiments first took place in Germany in 1950, and other countries followed during the 1970s (Köhler, 2006). In 1982, the German Landscape Research, Development and Construction Society (FLL) published regulatory guidelines for green roofing (FLL, 2008). Many technical aspects of the issue have subsequently been investigated, with the aim of improving the efficiency of these systems with designs for lighter and safer materials and for reduced building costs. At the same time, urban ecology studies have looked at the impact of green roofs in terms of urban biodiversity (Lundholm, 2006; Bass, 2009; Lundholm & Peck, 2008; Oberndorfer et al., 2007), energy consumption (Andersson, 2006; Carter & Butler, 2008; Abram, 2011; Spala et al., 2008) and the psychological and social impact on individuals and small communities (Samangooei, 2006; So-Young et al., 2008). Over the past few years the use of green roofs has spread to Italy and Spain and other areas with continental and atlantic climates, but only very recently have green roofs begun to spread to Mediterranean coastal areas (Abram, 2006; Briz Escribano, 2004; Neila et al., 2008; Damas et al., 2010).

The genus *Sedum* represents the most widely used plant species in Mediterranean green roofs across the world. The species of this genus do perform well in terms of survival rate, but they can lack homogeneity of coverage, and have low growth rates and poor biodiversity. Studies have shown that a combination of different species results in improved resilience under stress conditions when compared to mono-species cultivations (Gedge & Kadas, 2005; Lundholm et al., 2010; Nagase & Dunnet, 2010; Bretzel et al., 2011).

Effectively combining different species in extensive Mediterranean green roofs is not an easy task. They should be selected for adaptability to the Mediterranean bioclimate as well as to urban environments, and to environmental conditions specific to green roofs (drought, poor nutrients, high temperature stress) (Caneva et al., 2013; Papafotiou et al., 2013). The species should also have low allergenicity, attract insects and birds, and display ornamental and aesthetic characteristics throughout the seasons (Benvenuti & Bacci, 2010; Provenzano et al., 2010; Brenneisen, 2006).

The urban planners of many cities integrate green roofs into the design of their ecological corridor projects, since they represent an effective method with which to extend green surface areas and increase biodiversity (Agencia de Ecología Urbana de Barcelona, 2010; Ajuntament de Barcelona, 2010; Carter & Butler, 2008; Damas et al., 2010; Lundholm & Peck, 2008). The aim of this study is to analyse the ecological behaviour of twelve plant species, which have been grouped into three different categories by taking into account their growth patterns and evaluating their performance under three different irrigation protocols (two protocols applying minimum watering and one with none) in a Mediterranean environment.

2. Materials and Methods

2.1 Study Site

The trials were carried out in Caldes de Montbui (205 m.a.s.l.) (41°63′N 2°16′E) - 30 km from Barcelona (Spain), where the Catalan mountains meet the Mediterranean coast. During the trial period (June 2009-May 2010) average monthly temperatures were between 6.1 °C and 25.1 °C. Average monthly rainfall was between 5.2 mm and 125.2 mm, potential evapotranspiration (ET0) ranged from 20.4 mm to 139.7 mm. High ET0 values were recorded during summer when rainfall was scarce. In June, ET0 reached 135.3 mm with precipitation up to 10.7 mm. In July maximum ET0 was 139.7 mm, with maximum precipitation at 10.7 mm. In August it only rained once, for a total of 27.4 mm, and ET0 reached 131 mm. Between June 2009 and May 2010 a total of 640 mm precipitation and 925 mm ET0 were recorded. Unusually, temperature values in December 2009 and March 2010 fell below 0 °C, reaching -8 °C and damaging some species while destroying others. Evapotranspiration values (for the calculation of irrigation patterns) and all other climate data was provided by the Caldes de Montbui weather station, located 150 m from the site of the experiment. Table 1 shows climate data related to the trial period along with a summary of the period between 1991 and 2010.

Table 1. Climatic monthly averages and absolute data* during the trial period (June 2009 - May 2010) and during the twenty year period from 1991 to 2010 (historic monthly climate data*)

Monthly data		Temp. (°C)			Absolute Temp. (°C)		Rainfall (mm)	ETO (mm)
		Mean	Max	Min	Max	Min		
Trial period	JUN 09	22.4	29.2	15.6	35	11.8	10.7	135.3
Historic (1991–2010)	JUN	21.0	27.3	14.7	33.8	9.1	44.6	124.9
Trial period	JUL 09	24.2	30.3	17.9	37.2	11.4	11.6	139.7
Historic (1991–2010)	JUL	23.5	29.7	17.2	35.4	11.5	24.7	130.5
Trial period	AUG 09	25.0	31.8	19.0	37.4	16.8	27.4	131.0
Historic (1991–2010)	AUG	22.4	28.2	16.8	33.6	11.0	50.7	120.7
Trial period	SEP 09	20.2	26.5	14.8	30	10.9	103.4	85.5
Historic (1991–2010)	SEP	18.9	24.6	13.8	31.2	7.7	89.1	77.3
Trial period	OCT 09	16.5	23.1	11.2	29.3	3.6	72.2	62.6
Historic (1991–2010)	OCT	15.7	21.5	10.9	27.7	3.7	81.9	53.1
Trial period	NOV 09	11.7	18.4	6.6	25	2	5.2	34.0
Historic (1991–2010)	NOV	10.1	16.4	5.0	22.3	-1.2	39.9	25.9
Trial period	DEC 09	7.4	13.0	2.8	19.2	-7.9	47.5	20.4
Historic (1991–2010)	DEC	7.2	13.4	2.4	19.1	-3.4	52.2	16.5
Trial period	JAN 10	6.2	10.8	2.5	16.5	-3.1	59	24.5
Historic (1991–2010)	JAN	6.7	13.5	1.4	19.6	- 4.0	40.4	22.3
Trial period	FEB 10	7.0	12.4	1.9	18	-5.9	89.5	36.6
Historic (1991–2010)	FEB	7.7	14.7	1.9	20.0	- 2.8	29.2	31.3
Trial period	MAR 10	8.8	14.7	3.4	20.9	-4.4	66	66.4
Historic (1991–2010)	MAR	10.5	17.2	4.4	24.4	-1.2	33.0	58.2
Trial period	APR 10	12.9	19.4	7.0	25.6	3.6	32.8	86.3
Historic (1991–2010)	APR	12.8	19.3	6.6	25.8	1.3	50.0	79.1
Trial period	MAY 10	15.3	21.2	9.6	30.1	4.1	125.2	106.9
Historic (1991–2010)	MAY	16.7	22.8	10.6	29.8	4.6	63.6	112.9

Note. *from the weather station of Caldes de Montbui (MeteoCat http://www.meteo.cat/servmet/radar/).

2.2 Simulation of a Green Roof System

The conditions found on extensive green roofs were replicated by constructing walled containers with prefabricated concrete panels. The other materials used were produced by ZinCo®: the container was covered with a synthetic water-retentive polypropylene SSM45 protection mat, having a water-retention capacity of approximately 5 m^{-2}. A drainage and water storage sheet made of recycled polyethylene Floradrain FD25-E was placed over the polypropylene sheet, having a volume of approximately 10 m^{-2}. The resulting structure was then covered with a polypropylene geo-textile, followed by a substrate of Floral Zincoterra. The substrate thickness was 11 ± 1 cm and the total system thickness was approximately 15 cm. The ZinCo Floral substrate was composed mainly of brick fragments with organic matter making up approximately 10% of its content (Zinco, 2010). Apparent density of the substrate was 90%, with a total porosity of 66%, pH values ranged between 7.95 and 8.08 and electrical conductivity values ranged from 158.4 to 194.2 μS/cm (microSiemens/cm). Porosity and density levels were calculated in the laboratory using samples from the substrate, while pH and electrical conductivity were measured every three months using a laboratory pH meter.

2.3 Plants

Twelve species were selected according to their ecological and structural characteristics (see Table 2) and their availability in commercial nurseries. The flora of different environments was taken into consideration, including rocky environments, dunes, and dry coastal areas. The flora of conglomerate soils were also examined, as well as surfaces without an edaphic layer such as walls and cliffs (Caneva et al., 2013). Species were classified according to three different growth-forms: ground cover, sub-shrub, and rhizomatous-herbaceous. Such a classification is mainly related to the morphological behaviour of plants growing in a Mediterranean environment, although some species can display different features in different environments due to phenotypic plasticity.

Table 2. Selected species: ecological and structural information

Species	Botanic family	Chorotype	Environment	Growth form
Armeria maritima (P. Mill.) Willd.	Plumbaginaceae	Sub-cosmopolita	Coastal areas, dry environment, sandy soils, marshland	Rhizomatous-herbaceous
Asteriscus maritimus (L.) Less.	Compositae	W. Medit.	Rocky coastal Sandy like soils	Sub-shrub
Centranthus ruber (L.) DC.	Valerianaceae	Steno Medit.	Calcareous and rocky soils	Sub-shrub
Dymondia margaretae Compton	Asteraceae	South Africa	Sandy coastal soils	Ground cover
Drosanthemum floribundum (Haw.) Schwantes	Aizoaceae	South Africa	Cultivated and sub-spontaneous in Mediterranean area	Sub-shrub
Frankenia laevis L.	Frankeniaceae	Steno Medit.-Central Asia South Africa	Sandy coastal soils and marshland	Ground cover
Helichrysum stoechas (L.) Moench	Compositae	W. Medit.	Sandy, rocky dry soils	Sub-shrub
Iris lutescens Lam.	Iridaceae	NW-Steno-Medit.	Dry and rocky fields	Rhizomatous-herbaceous
Limonium virgatum (Willd.) Fourr.	Plumbaginaceae	Euri-Medit.	Costal retro dunes and salty cliffs	Rhizomatous-herbaceous
Lotus creticus L.	Leguminosae	Steno Medit.	Coastal sandy soils and dune systems environments	Sub-shrub
Santolina rosmarinifolia L.	Compositae	W. Medit. North Africa	Dry rocky, sandy, conglomerate like soils	Sub-shrub
*Thymus serpyllum** L.	Labiatae	Euri-Medit.	Arid grasslands, rocky soils and mesophytic forest borders	Sub-shrub/ground cover

Note. *Differential behavior due to phenotypic plasticity.

2.4 Experimental Design and Irrigation

The experimental design consisted of a split-plot: the main influencing factor was irrigation, with the application of three different irrigation protocols; the second influencing factor was plant species; twelve different species were included in the experiment. Three plots were planted for each irrigation protocol for a total of nine plots. Each of the nine main plots had a surface area of 12 m^2 and was divided into twelve sub-plots of 1 m^2 each. Nine plants of a single species were planted in each sub-plot resulting in a density of nine plants per square meter. The placement of the twelve species was the same for every plot in each irrigation protocol. Three different irrigation protocols were used: rain fed (no use of artificial irrigation, 0% ET0); replenishment of 20% of water lost to potential evapotranspiration (20% ET0); and replenishment of 40% of the same (40% ET0). The quantity of water supplied was calculated on a weekly basis, taking into account the previous week's data and subtracting the total amount of precipitation from the total value of ET0 (evapotranspiration minus rainfall). The resulting amount of water was distributed evenly throughout the week. The water distribution system was composed of drip emitting tubes placed on the surface with a diameter of 16 mm, and a 30 cm space between tubes, with a flow rate of 22.8 litres/m^2/hour. Over the whole period of the trial, 216 litres/m^2 of water were supplied to the 40 % ET0 sub-plots and 108 litres/m^2 to the 20% ET0 sub-plot; the average weekly values were, respectively, 4.9 and 2.5 litres/m^2 for the two irrigation protocols. The water supplied during the hottest months (from June to September) was 184 litres for the 40% ET0 irrigation protocol and 92 litres for the 20% ET0 protocol, equivalent to 85% of the total water supplied throughout the year.

2.5 Green Roof Performance and Measures

The following parameters were measured: mortality, coverage, timing, duration and percentage of flowering, and hypogeal and epigeal biomass. The results for each species were analysed according to their functional group and to the irrigation protocol assigned to them. The survival rate of the plants was measured each month; frequency and duration of flowering were recorded every fifteen days. Plants lacking leaves and living branches were considered dead and left *in situ*. Flowering was recorded throughout the trial period. In order to evaluate the entire flowering period of a plant the duration and number of flowering branches were recorded. The number of flowering branches was also recorded as a percentage of the total number of branches on a single plant. With regard to coverage, the experimental system was photographed every two months, from a height of 2.5 m with a Nikon EOS 500 camera using Nikon EOS specific software. Photographs were taken around midday (11 am - 2 pm) in order to reduce the effect of shadows. The camera was placed on a tripod equipped with a horizontal extendable arm. All photographs were taken from the same distance and height, and the same image and camera settings were used. Images were then processed with the Greenpix software (developed by the Institut de Recerca i Tecnologia Agroalimentàries - IRTA), for the analysis of digital images (Casadesús et al., 2005, 2007). A Hue range of 0 to 180 was used, making it possible to identify the number of yellow, brown and green pixels, and to calculate the percentage of each color portion in relation to the total image pixels, and to transform the pixels into cm^2 for an evaluation of total cover (green, yellow and brown). With regard to the plant growth parameter, only the green surface corresponding to pixels within the 60-180 Hue range was taken into account. The evaluation of biomass was carried out by sorting the different species according to location and to irrigation protocol. All plants were then carefully separated from the earth In some cases portions of roots were sifted out from the substrate. Finally, the epigeal parts (leaves and stalk) were separated from the roots. The roots were rinsed thoroughly with water, and all plants were placed in a paper bag allowing permeation of air and humidity. All plant material was then dried in an oven for a period of seven days at 65 °C before being weighed.

2.6 Statistical Analysis

An experiment was designed to measure the following factors: irrigation, species, and sampling. In order to study the spread of the coverage area for each species, the ANOVA mixed models statistical analysis was applied using the sampling and irrigation factors as fixed data, and the block as a variable factor; three levels were set for the irrigation factor, eight for the sampling factor, and three for the block factor. To determine significant outcome differences between the irrigation protocols and the sampling schedule, a post hoc test was applied using Tukey adjustments. The same analysis was carried out by grouping the species according to their growth form. Biomass analysis for each species was carried out using ANOVA mixed models, setting a fixed variable (irrigation protocol) and a random variable (block). The overall analysis was conducted using version 9.2 of the SAS software. For each irrigation protocol plot, mean biomass and mean coverage, both according to growth form, were compared using the Tukey Kramer HSD test and version 10 of the JMP software.

3. Results

3.1 Plant Mortality

Average mortality results of each growth form are reported in Table 3: mortality rates of all species, categorised by irrigation protocol, are shown in Figure 1. It can be seen that the mortality rate of *Asteriscus maritimus*, *Drosanthemum floribundum* and *Lotus creticus* was above 81% in winter, regardless of the irrigation protocol applied. The remaining species to which water was supplied artificially resulted in a mortality rate ranging from 0% to 38%, whereas the species in the rain fed group showed mixed results: *Asteriscus maritimus*, *Drosanthemum floribundum*, *Lotus creticus* and *Dymondia margaretae* had an almost 100% mortality rate; *Frankenia laevis* and *Limonium virgatum* had a mortality rate of 70% and 74% respectively; *Thymus serpyllum* 48%; *Helichrysum stoechas* 37%; the mortality rates of *Armeria maritima*, *Centranthus ruber*, *Iris lutescens* and *Santolina rosmarinifolia* were between 14% and 18%.

3.2 Coverage

Coverage values of each species were categorised according to irrigation protocol and to growth form. The results of each group's coverage performance throughout the trial are shown in Table 4. Table 5 shows total averages and significant differences: all growth form groups displayed significant differences in terms of coverage when comparing the irrigated plots with the rain fed plot. Irrigated plants reached greater coverage in May and November; the lowest values were recorded in March. The results of each species were as follows (Figure 2): *Armeria maritima*, *Iris lutescens* and *Santolina rosmarinifolia* showed steady growth without seasonal variation. *Thymus serpyllum*, *Frankenia laevis*, *Limonium virgatum* and *Centranthus ruber* are the species that performed best in the irrigated protocols when the trial was completed, despite showing seasonal variations. The worst final results, irrespective of irrigation protocol, were recorded for *Asteriscus maritimus*, *Lotus creticus* and *Drosanthemum floribundum*. In the rain fed protocol *Centranthus ruber*, *Helichrysum stoechas* and *Limonium virgatum* showed the best results.

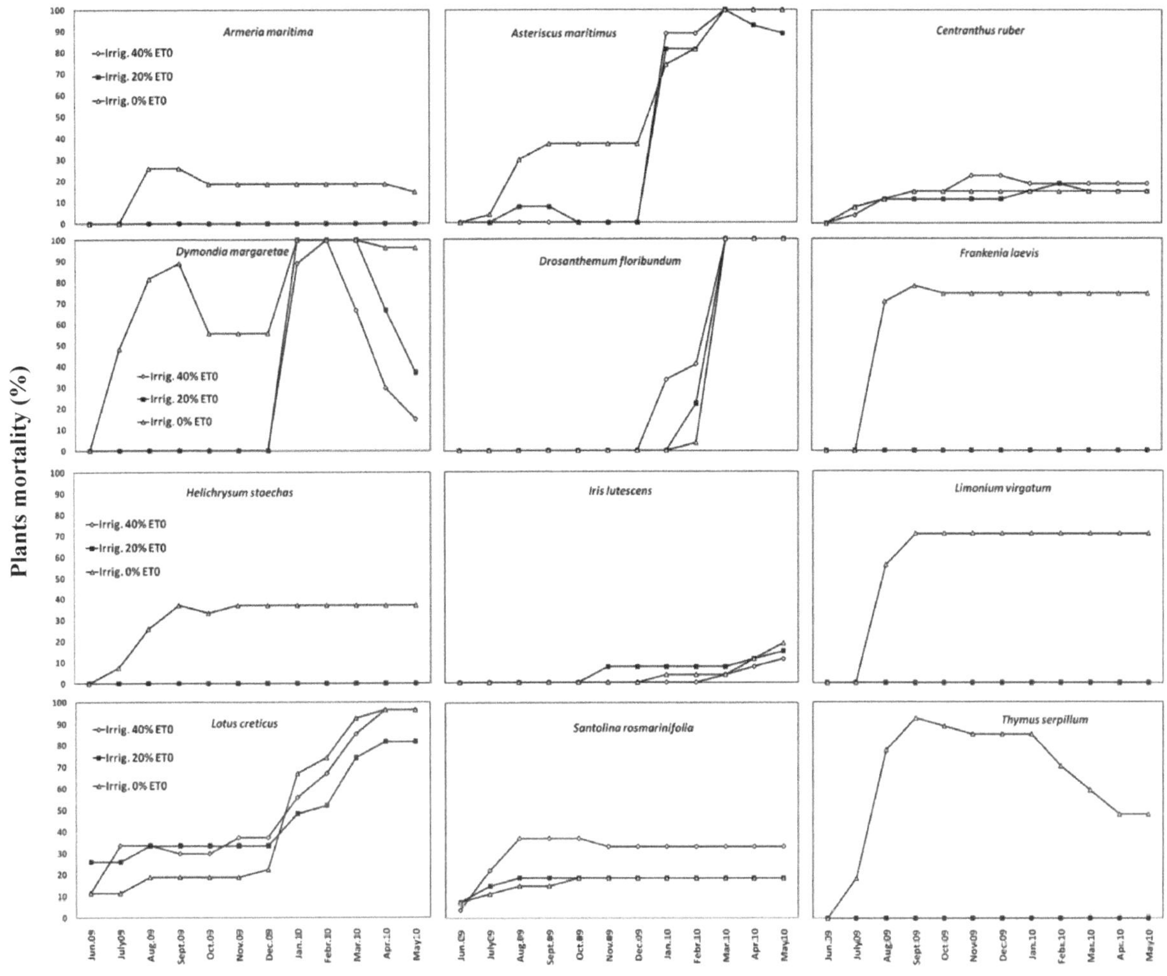

Figure 1. Mortality rate of species from June 2009 to May 2010 according to rain fed (0% ET0), 20% ET0 and 40% ET0 irrigation protocols

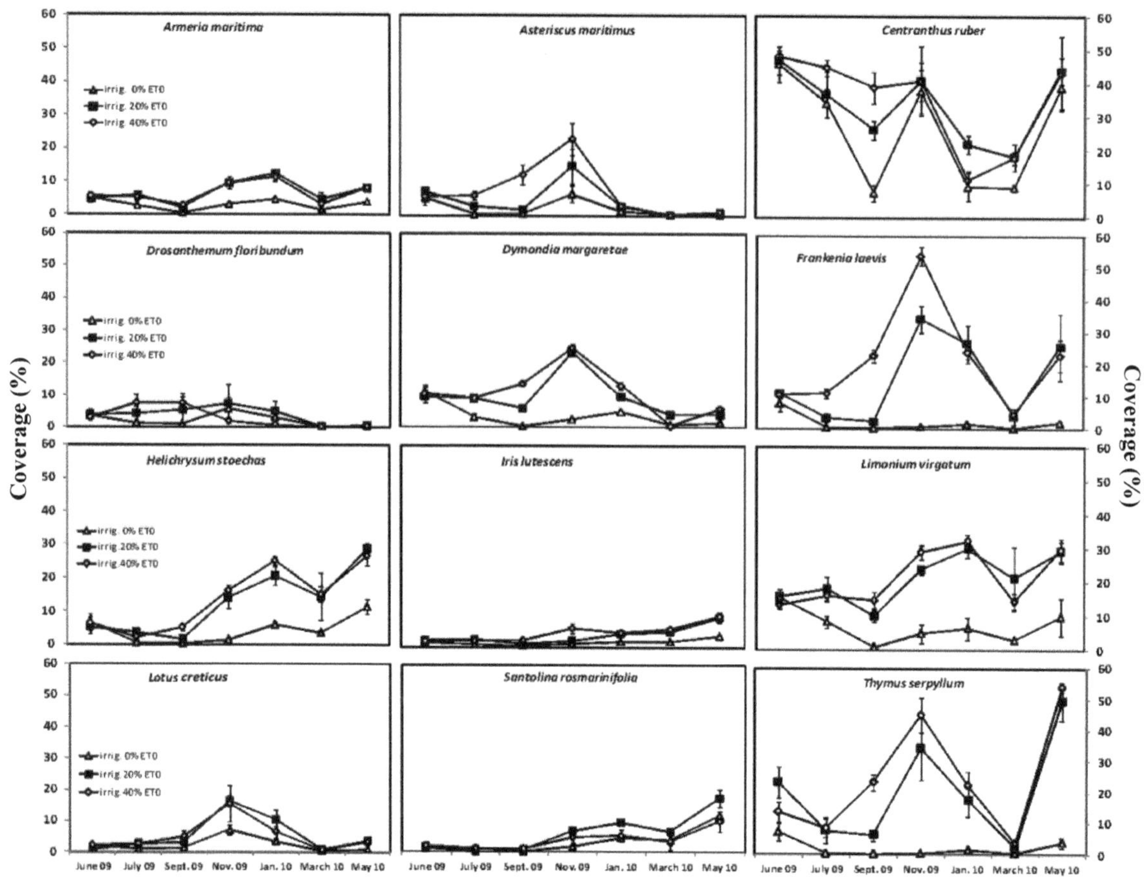

Figure 2. Green cover from June 2009 to May 2010 according to the different irrigation protocols 0%, 20%, and 40% of ET0. Final values are the result of 3 mean values ± standard error

Table 3. Summary of plant mortality by percentage (± standard error) for three irrigation treatments 0%, 20%, 40% ET0 (average figures for all values under the same irrigation protocol) and for three life-form groups (sub-shrubs, ground cover, rhizomatous-herbaceous) categorized by irrigation protocol 0%, 20%, 40% ET0

% Plants Mortality						
Date	***Irrigation 0% ET0***	***Irrigation 20% ET0***	***Irrigation 40% ET0***	***Sub-shrubs irrigation 0% ET0***	***Sub-shrubs irrigation 20% ET0***	***Sub-shrubs irrigation 40% ET0***
June 2009	1.54±1.06	2.78±2.19	1.23±0.95	2.65±1.75	4.76±3.68	2.12±1.59
July 2009	8.95±3.96	4.01±2.39	4.94±3.17	8.47±2.24	6.88±3.83	8.47±5.16
Aug. 2009	34.26±8.51	5.86±3.04	6.79±3.94	25.40±9.47	10.05±4.69	11.64±6.28
Sept. 2009	39.81±9.79	5.86±3.04	6.79±3.81	30.69±11.45	10.05±4.69	11.64±6.02
Oct. 2009	35.80±8.61	5.25±3.07	6.79±3.81	30.16±10.83	8.99±4.91	11.64±6.02
Nov. 2009	35.80±8.44	5.86±3.04	7.72±4.14	30.16±10.40	8.99±4.91	13.23±6.46
Dec. 2009	36.11±8.38	5.86±3.04	7.72±4.14	30.69±10.31	8.99±4.91	13.23±6.46
Jan. 2010	46.91±10.12	22.53±10.11	26.54±9.92	42.33±12.52	23.28±11.68	32.80±11.97
Feb. 2010	47.22±9.95	25.00±9.95	29.01±10.79	42.86±12.12	27.51±11.14	35.45±12.60
Mar. 2010	57.41±10.75	34.57±12.84	33.95±12.10	60.32±14.27	43.92±17.29	48.15±17.24
April 2010	57.10±10.51	32.10±11.67	32.10±12.09	59.26±14.59	43.92±17.10	49.74±17.87
May 2010	57.41±10.39	29.63±11.06	31.17±12.13	59.26±14.59	43.39±16.85	49.74±17.87
Date	***Ground cover irrigation 0% ET0***	***Ground cover irrigation 20% ET0***	***Ground cover irrigation 40% ET0***	***Rhizomatou-herbaceous irrigation 0% ET0***	***Rhizomatou-herbaceous irrigation 20% ET0***	***Rhizomatous-herbaceous irrigation 40% ET0***
June 2009	0.00±0.00	0.00±0.00	0.00±0.00	0.00±0.00	0.00±0.00	0.00±0.00
July 2009	22.22±14.02	0.00±0.00	0.00±0.00	0.00±0.00	0.00±0.00	0.00±0.00
Aug. 2009	76.54±3.27	0.00±0.00	0.00±0.00	27.16±16.05	0.00±0.00	0.00±0.00
Sept. 2009	86.42±4.45	0.00±0.00	0.00±0.00	32.10±20.55	0.00±0.00	0.00±0.00
Oct. 2009	72.84±9.64	0.00±0.00	0.00±0.00	29.63±21.06	0.00±0.00	0.00±0.00
Nov. 2009	71.60±8.64	0.00±0.00	0.00±0.00	29.63±21.06	2.47±2.47	0.00±0.00
Dec. 2009	71.60±8.64	0.00±0.00	0.00±0.00	29.63±21.06	2.47±2.47	0.00±0.00
Jan. 2010	86.42±7.51	33.33±33.33	29.63±29.63	30.86±20.21	2.47±2.47	0.00±0.00
Feb. 2010	81.48±9.32	33.33±33.33	33.33±33.33	30.86±20.21	2.47±2.47	0.00±0.00
Mar. 2010	77.78±11.91	33.33±33.33	22.22±22.22	30.86±20.21	2.47±2.47	1.23±1.23
April 2010	72.84±13.91	22.22±22.22	9.88±9.88	33.33±18.64	3.70±3.70	2.47±2.47
May 2010	72.84±13.91	12.35±12.35	4.94±4.94	34.57±17.93	4.94±4.94	3.70±3.70

Table 4. Summary of coverage by percentage (± standard error) for three irrigation protocols 0%, 20%, 40% ET0 (average figure for all values of the same treatment) and three life-form groups for each irrigation protocol 0%, 20%, 40% ET0 (sub-shrubs, ground cover, rhizomatous- herbaceous)

	Coverage %					
Date	**Irrigation 0% ET0**	**Irrigation 20% ET0**	**Irrigation 40% ET0**	**Sub-shrubs irrigation 0% ET0**	**Sub-shrubs irrigation 20% ET0**	**Sub-shrubs irrigation 40% ET0**
June 09	10.42±3.52	12.12±3.72	11.35±3.64	10.91±0.87	13.57±6.27	12.38±6.16
July 09	4.72±2.76	8.93±2.91	10.66±3.39	5.55±4.76	8.78±4.71	11.16±5.73
Sept. 09	1.25±0.57	6.07±2.03	14.03±3.36	1.72±0.96	6.73±3.31	14.60±5.11
Nov. 09	6.72±2.95	21.49±3.78	25.65±5.08	9.45±4.90	21.26±4.72	23.63±6.55
Jan. 10	4.73±0.77	15.58±2.38	14.66±2.86	4.80±1.10	13.70±2.70	11.43±3.42
Mar. 10	2.43±0.80	7.59±2.06	6.30±1.82	2.94±1.34	6.78±2.85	6.28±2.76
May 10	8.30±3.22	20.06±4.80	19.27±4.84	10.87±5.38	22.40±7.76	20.86±8.04
Date	**Ground cover irrigation 0% ET0**	**Ground cover irrigation 20% ET0**	**Ground cover irrigation 40% ET0**	**Rhizomatous-herbaceous irrigation 0% ET0**	**Rhizomatous-herbaceous irrigation 20% ET0**	**Rhizomatous-herbaceous irrigation 40% ET0**
June 09	11.10±3.51	16.58±3.57	13.83±1.92	4.55±1.10	16.25±7.04	16.81±7.82
July 09	2.02±1.55	8.72±3.32	11.48±1.81	4.05±2.17	9.09±4.51	8.01±4.02
Sept. 09	0.43±0.08	6.13±2.11	22.92±0.27	2.39±0.48	11.11±5.14	8.79±3.25
Nov. 09	1.76±1.14	35.69±1.48	46.58±3.82	6.41±1.81	17.27±6.23	17.63±6.32
Jan. 10	3.71±2.10	19.97±3.36	22.63±0.72	3.23±0.97	12.09±6.13	15.90±6.56
Mar. 10	0.64±0.28	4.17±1.21	2.99±1.17	7.54±4.03	7.94±4.01	7.24±3.18
May 10	2.72±0.60	27.15±2.39	28.75±13.17	0.69±0.17	4.56±2.78	6.96±3.92

Table 5. Summary of percentage of coverage for three irrigation protocols 0%, 20%, 40% ET0 (average figure for all values of the same treatment) and three life-form groups for each irrigation protocol 0%, 20%, 40% ET0 (sub-shrubs, ground cover, rhizomatous-herbaceous). Average figures not associated with the same letter are significantly different at $p < 0.05$ (TEST TUKEY KRAMER HSD)

Irrigation			Sub-shrubs			Ground cover			Rhizomatous-herbaceous		
0% ET0	4.64	B	*0% ET0*	6.60	B	*0% ET0*	3.19	B	*0% ET0*	4.12	B
20% ET0	13.80	A	*20% ET0*	13.31	A	*20% ET0*	16.91	A	*20% ET0*	11.18	A
40% ET0	15.75	A	*40% ET0*	14.33	A	*40% ET0*	21.30	A	*40% ET0*	11.62	A

3.3 Flowering

Most species flowered during the Summer of 2009 and Spring of 2010 (see Table 6). The only species that presented differences in the duration of its flowering period between one irrigated plot and another was *Thymus serpyllum,* which flowered until October in the 40% ET0 protocol plots, two months longer than the plants in the 20% ET0 protocol plots; and *Asteriscus maritimus* which flowered an additional month when being watered with the 40% ET0 system compared to plants in the 20% ET0 protocol plots. *Lotus creticus* and *Santolina rosmarinifolia* did not flower at all, whereas *Frankenia laevis,* (butonly the plants in the 20% ET0 protocol plots)presented limited flowering in May. *Dymondia margaretae* flowered only in August. *Asteriscus maritimus,Centranthus ruber, Helichrysum stoechas, Limonium virgatum* and *Thymus serpyllum* presented limited flowering in terms of time and quantity in the rain fed plots when compared to the plants in the irrigated plots.

Table 6. Flowering periods of the trial species for three different irrigation protocols: 0% ET0, 20% ET0 and 40% ET0

Plants species	Irrigation	Flowering periods											
		June	July	Aug	Sept	Oct	Nov	Dec	Jan	Feb	Mar	Apr	May
Armeria maritima	0% ET0		= 50%									> 50%	> 50%
	20% ET0		= 50%									> 50%	> 50%
	40% ET0		= 50%									> 50%	> 50%
Asteriscus maritimus	0% ET0		> 50%	> 50%									
	20% ET0		> 50%	> 50%		= 50%							
	40% ET0		= 50%	= 50%	< 50%	= 50%							
Centranthus ruber	0% ET0		= 50%	> 50%								= 50%	> 50%
	20% ET0		= 50%	> 50%	> 50%							= 50%	> 50%
	40% ET0		= 50%	> 50%	> 50%							= 50%	> 50%
Drosanthemum floribundum	0% ET0	= 50%											
	20% ET0	= 50%											
	40% ET0	= 50%											
Dymondia margaretae	0% ET0			= 50%									
	20% ET0			= 50%									
	40% ET0			= 50%									
Frankenia laevis	0% ET0												
	20% ET0												= 50%
	40% ET0												
Helichrysum stoechas	0% ET0		= 50%										
	20% ET0		= 50%	= 50%									
	40% ET0		= 50%	= 50%									
Iris lutescens	0% ET0										= 50% (2nd half)	> 50%	
	20% ET0										> 50% (2nd half)	> 50%	
	40% ET0										> 50% (2nd half)	> 50%	
Limonium virgatum	0% ET0		= 50%	> 50%									
	20% ET0		= 50%	> 50%	> 50%								
	40% ET0		= 50%	> 50%	> 50%								
Lotus creticus	0% ET0												
	20% ET0												
	40% ET0												
Santolina rosmarinifolia	0% ET0												
	20% ET0												
	40% ET0												
Thymus serpyllum	0% ET0			> 50%									
	20% ET0		= 50%	> 50%									
	40% ET0		= 50%	> 50%		= 50%							

Note. Quantity of bloomed branches: [light] < 50%; [medium grey] = 50%; [dark] > 50%.

3.4 Biomass

In all growth forms, the biomass of the upper portion was greater than that of the root system; all groups presented significant differences in biomass production when comparing the irrigated plots to the rain fed plots, with the exception of the roots of rhizomatous-herbaceous. No group displayed significant differences in biomass production between the two irrigation protocols (20% and 40% ET0) (Table 7). Table 8 shows the average biomass values of each species and its portions (shoot system, root system, and total biomass) and the ratio between root system and shoot system for each irrigation protocol. Only nine out of the twelve species were used for this analysis, because the remaining three (*Asteriscus maritimus*, *Drosanthemum floribundum*, *Lotus creticus*) had an insufficient biomass for sampling. Sampling of *Dymondia margaretae* was performed only for the plants in the irrigated plots, because all plants in the rain fed plots died. Of the nine species recorded, five (*Armeria marítima*, *Frankenia laevis*, *Helichrysum stoechas*, *Iris lutescens* and *Thymus serpyllum*) displayed significant differences between shoot system and total biomass. *Centranthus ruber*, *Limonium virgatum* and *Santolina rosmarinifolia* did not display any significant differences between irrigation protocols in terms of epigeal development. Four species developed a significantly greater root system biomass for the irrigated plants than for the rain fed plants: *Armeria maritima*, *Dymondia margaretae*, *Frankenia laevis*, *Helichrysum stoechas* and *Thymus serpyllum*. For *Limonium virgatum*, there was a significant difference between root system biomass values in the rain fed and 20% ET0 protocol plots and those in the 40% ET0 protocol plots.

Table 7. Average for g (± standard error) of the different fragments and total dry biomass (g) of the plants under three irrigation protocols 0%, 20%, 40% ET0 (average figure for all values of the same treatment), and of the three life-form groups for each irrigation protocol 0%, 20%, 40% ET0 (sub-shrubs, ground cover, rhizomatous-herbaceous). Average figures not associated with the same letter are significantly different ($p < 0.05$) (TEST TUKEY KRAMER HSD)

BIOMASS				
Irrigation				
	Nº	Shoot system	Root system	Total
0% ETO	138	24.11± 2.30 b	6.61± 0.62 b	30.72± 2.55 b
20% ETO	222	39.50± 1.82 a	8.94± 0.49 a	48.36± 2.01 a
40% ETO	219	40.79± 1.83 a	8.57± 0.49 a	49.37± 2.02 a
Sub-shrubs				
	Nº	Shoot system	Root system	Total
0% ETO	80	27.95± 2.80 b	8.12± 0.89 b	36.07± 3.51 b
20% ETO	99	45.20± 2.52 a	11.60± 0.80 a	56.68± 3.16 a
40% ETO	93	40.92± 2.60 a	10.42± 0.82 ab	51.34± 3.26 a
Ground cover				
	Nº	Shoot system	Root system	Total
0% ETO	26	4.79± 4.92 b	0.71± 0.88 b	5.50± 4.76 b
20% ETO	73	29.80± 2.94 a	6.88± 0.52 a	36.69± 2.84 a
40% ETO	77	34.21± 2.86 a	7.81± 0.51 a	42.03± 2.77 a
Rhizomatous-herbaceous				
	Nº	Shoot system	Root system	Total
0% ETO	51	19.64± 3.76 b	4.96± 0.92 a	24.60± 3.75 b
20% ETO	77	37.55± 3.06 a	7.07± 0.75 a	44.53± 3.05 a
40% ETO	76	44.56± 3.08 a	6.37± 0.75 a	50.93± 3.07 a

Table 8. Average dry weight (g) of the trial species under three irrigation protocols (40% ET0, 20% ET0, and 0% ET0). Average figures not associated with the same letter are significantly different ($p < 0.05$)

Plants species	Irrigation	Shoot system (g)	Root system (g)	Total (g)	Root/Shoot
Armeria maritima	0% ET0	13.64 b	1.31 b	14.96 b	0.1 a
	20% ET0	26.14 a	3.04 a	29.17 a	0.1 a
	40% ET0	27.97 a	2.88 a	30.85 a	0.1 a
Centranthusruber	0% ET0	46.61 a	17.97 a	64.58 a	0.4 a
	20% ET0	56.94 a	20.05 a	76.99 a	0.4 a
	40% ET0	49.86 a	20.10 a	69.96 a	0.4 a
Dymondia margaretae	0% ET0				
	20% ET0	8.11 b	8.47 b	16.57 a	1.0 a
	40% ET0	2.83 a	13.13 a	15.96 a	4.6 a
Frankenia laevis	0% ET0	12.92 c	1.35 b	14.27 b	0.1 b
	20% ET0	46.19 b	4.91 a	51.10 a	0.1 b
	40% ET0	62.10 a	4.50 a	66.60 a	0.1 a
Helichrysu stoechas	0% ET0	29.99 b	5.40 b	35.39 b	0.2 a
	20% ET0	52.32 a	9.53 a	61.84 a	0.2 a
	40% ET0	50.36 a	8.83 a	59.19 a	0.2 a
Iris lutescens	0% ET0	12.46 b	8.88 a	21.34 b	0.7 a
	20% ET0	18.66 ab	14.37 a	33.03 ab	0.8 a
	40% ET0	26.35 a	14.08 a	40.43 a	0.5 a
Limonium virgatum	0% ET0	50.14 a	6.04 a	56.18 a	0.1 c
	20% ET0	65.25 a	5.32 a	70.56 a	0.1 b
	40% ET0	76.01 a	3.59 b	79.60 a	0.0 a
Santolina rosmarinifolia	0% ET0	29.53 a	6.44 a	35.97 a	0.2 a
	20% ET0	44.98 a	9.93 a	54.91 a	0.2 a
	40% ET0	26.85 a	6.50 a	33.35 a	0.2 a
Thymus serpyllum	0% ET0	1.79 b	0.48 b	2.27 b	0.3 a
	20% ET0	28.65 a	7.76 a	36.41 a	0.3 a
	40% ET0	33.04 a	6.60 a	39.64 a	0.2 a

4. Discussion

4.1 Mortality

Two main factors affected plant mortality: the amount of water received and the low winter temperatures. The mortality rate of all rain fed species rose during the trial. Irrigated plants experienced a higher mortality rate in January than in other months due to low temperatures.

4.2 Coverage

As shown in the literature, the growth form and morphology of a plant affect the interception of rain by the substrate as well as its thermal and hydrological characteristics (Del Barrio, 1998; Dunnett et al., 2008; Nagase & Dunnett, 2012; Mentens et al., 2005; Buccola & Spolek, 2010; Scott MacIvor et al., 2011; Olivieri et al., 2013). In this trial the sub-shrub form was shown to be affected by low temperatures more than other forms; ground cover species suffered more than others from drought; and rhizomatous-herbaceous plants were the least affected by drought and cold weather.

Green cover had two growth peaks, in autumn and spring, corresponding to typical seasonal changes in the Mediterranean climate. In these seasons the increased availability of rainwater associated with low evapotranspiration can reduce the importance of irrigation. In spring and autumn the rain fed plants (grouped by species or by growth form) extended their percentage of coverage, achieving higher growth rates than those in the irrigated plots, although they did not exceed a maximum value of 11%; a better performance may be advantageous if they are to be used in green roofs. Some species, despite low final coverage, developed slowly and constantly, showing that the speed of coverage development and the colonisation of ecological niches varies for each species; this depends mostly on stress tolerance and ecological characteristics (Benvenuti, 2014). In order to obtain greater cover with those species which displayed low development, it is advisable to use higher plant density.

Notable differences in terms of coverage arise only when comparing rain fed plots with irrigated plots (regardless of the specific irrigation protocol applied). The results of this study did not achieve the official requirements established for temperate climate countries, which require a minimum coverage of 80% of the total surface (FLL, 2008) after the initial period. Given requirements are not always applicable in a Mediterranean environment, thermal and rainfall conditions are very different in the Mediterranean area compared to Northern Europe (Van Mechelen et al., 2014). In this study, irrigated ground cover species achieved higher coverage values compared to other growth forms, this could be due to morphological difference in these plants. The development of ground cover plants was mainly horizontal and they produced root systems which allowed them to spread out and multiply; sub-shrubs developed mainly in volume. Rhizomatous-herbaceous and sub-shrubs displayed minor seasonal variations and achieved lower coverage values compared to the ground cover plants. All rain fed plots resulted in a lower percentage of coverage. In order to reach a wider coverage as quickly as possible, and to achieve ornamental heterogeneity, all three growth forms should be used, with a prevalence of ground cover plants.

4.3 Flowering

Irrigation influenced the flowering of five species. The flowering period of rain fed plants, in some cases, was delayed or ended early, and they displayed a lower percentage of flowering branches. In Mediterranean environments plants often flower in spring, and occasionally a new outbreak is expected in autumn. The evaluation of the percentage of flowering branches for each plant and of their flowering period makes it possible to analyse responses to stress conditions, but these parameters do not include the size of the plant, making it difficult to carry out a complete aesthetic evaluation. This study provides information about species which are able to produce flowers in conditions of little to no artificial watering. The results also suggest that irrigation could be required in order to obtain even minimal flowering on Mediterranean green roofs, due to the severe summer drought.

4.4 Biomass

Final average biomass values back up most of the parameters analysed: no significant difference arises between results from the two irrigation protocols. The experimental settings may have been responsible for the fact that significant differences appeared only when comparing irrigated plots with rain fed plots: it is not possible to detect a difference in outcomes between the two minimal irrigation protocols, although it is possible to compare their results with the rain fed outcome. A study carried out in a Mediterranean-Australian climatic zone obtained similar results: plants with similar irrigation protocols did not present any difference in their final biomass (Williams et al., 2010). In terms of species, five out of nine displayed significant differences in total biomass values, and four species also displayed significant differences in their root systems between irrigated and rain fed plots. The ratio between root biomass and upper portions did not present significant differences between one irrigation protocol and another, apart from two species (*Frankenia laevis* e *Limonium virgatum*), suggesting that biomass was distributed homogeneously between roots and upper parts. On the other hand, a study carried out by Farrell et al. (2013) on the possibility of using species which grow on a thin granite substrate, observed that most species responded to water shortages by increasing the supply to the roots and reduced their leaf surface area.

4.5 Species

The analysis of single species shows that *Centranthus ruber* employs a strategy that causes a reduction of the shoot system during winter and a greater vegetative growth in spring, coinciding with the season where no water was supplied although the reduced water supply in summer affected coverage, which decreased in plants receiving 20% of ET0 and in rain fed plants. Flowering also suffered a negative effect in rain fed plants, whose flowering duration was shortened by one month. A study carried out in Tuscany, Italy (Benvenuti & Bacci, 2010) using *Centranthus ruber* on an igroperlite®substrate at different depths (15 cm and 20 cm), obtained a coverage of 62% and 100% respectively. This species is well adapted to Mediterranean conditions. The high mortality rate of rain fed *Frankenia laevis*, and variations in its coverage throughout the seasons, show the high dependence of this species on irrigation. *Armeria maritima* and *Iris lutescens* developed slowly, reaching minimal coverage values which remained consistent throughout the trial. Nevertheless, their mortality rate was low, making their use in green roofs advisable: increasing the density of initial planting could compensate for reduced growth. Provenzano et al. (2010), in a study carried out in Italy, confirmed these same results for *Armeria maritima,* for which they obtained a coverage of 40% along with a low growth rate on an 8 cm deep irrigated substrate, where the ground humidity was lower than 20%. *Santolina rosmarinifolia* and *Helichrysum stoechas* showed high adaptability to the Mediterranean climate, presenting little variation in coverage and low mortality rates. The growth rate of *Santolina* rosmarinifolia was not improved by the water supply, as opposed to *Helichrysum stoechas*. Papafotiou et al. (2013) have observed that two species of *Helichrysum*, tested with limited irrigation, were found to be appropriate for mediterranean green roofs. *Thymus serpyllum*, when irrigated, reached high coverage values of about 50%. This growth performance has been confirmed by two separate studies: Provenzano et al. (2010), in a Mediterranean environment, on an 8 cm deep substrate, reported coverage of 85% some months after winter transplanting; in Japan a trial was conducted on a 10 cm deep substrate, receiving 5 litres of water every two days, the coverage increased by more than 70% between May and September, reaching a final value of 90% (Takahiro et al., 2010). In this trial, rain fed plants reached 4% coverage. Irrigated plants benefitted in terms of flowering, mortality and coverage (mainly during the summer), this species therefore seems to rely on watering. Rain fed *Dymondia margaretae* did not tolerate drought, resulting in high mortality rates in summer, whereas irrigated plants of this species did not tolerate low temperatures (their leaves dried). However, it presented good vegetative growth over the following spring. This is the only species in which the root system biomass exceeded that of the shoot system; growth was limited and flowering scarce. Mortality for *Asteriscus maritimus*, *Lotus creticus* and *Drosanthemum floribundum* was caused by low winter temperatures, regardless of the irrigation protocol – meaning that the water supplied throughout the trial did not improve resistance. Irrigated plants of the *Limonium virgatum* species had a 100% survival rate, and little variation in flowering and coverage; growth was constant throughout the trial. Rain fed plants had a reduced flowering period (one month less compared to irrigated plants). The rosette plant structure does not allow rapid extension in coverage. However, the ability to produce viable seeds, their germination, and the survival of seedlings demonstrated efficient adaptation to the Mediterranean climate.

5. Conclusion

This study shows that some species can be used with minimal irrigation for green roofs in the Mediterranean area. The results provide indications on growth form functionality: different growth form groups responded in different ways to stresses of different seasons. Rhizomatous-herbaceous developed slowly and consistently; ground cover plants developed quickly and relied more on the water supply; sub-shrubs showed lower resistance to stresses under the present experimental conditions. *Centranthus ruber*, *Santolina rosmarinifolia*, *Helichrysum stoechas* and *Iris lutescens* performed well under the experimental conditions; the performance of *Limonium virgatum*, *Armeria maritima*, *Frankenia laevis* and *Thymus serpyllum* was good, although the results are dependent on minimal irrigation (growth in *Armeria maritima* and *Iris lutescens* was slower than in other species); *Asteriscus maritimus*, *Lotus creticus*, *Drosanthemum floribundum* and *Dymondia margaretae* failed to adapt to the experimental conditions of this trial. This study is a useful tool for the selection of plants for use in Mediterranean green roofing, providing information on survival and the development of plant types and species. An appropriate mixture of all three growth forms could be an interesting subject for future investigation, especially in relation to the amount of water required.

Acknowledgements

This paper shows partial results developed in the frame of CUMED project, which has been partially economically supported by MINECO of Spanish Government.

References

Abram, P. (2006). *Il verde pensile in Italia e in Europa*. Milano: Il Verde Editoriale.

Abram, P. (2011). *Verde pensile*. Pozzuoli: Sistemi Editoriali SE.

Agencia de Ecología Urbana de Barcelona. (2010). Cobertes i Murs Verds a Barcelona. *Estudi sobre les existentes, el potencial i les estratègies de implantació*. Retrieved from http://bcnecologia.net/sites/default/files/proyectos/doc_cobertes_i_murs_verds_01m2010_0.pdf

Ajuntament de Barcelona. (2010). Els corredors verds urbans. *Exemples i criteris de disseny. Área de Medi*

Ambient. Retrieved from http://www.sostenibilitatbcn.cat/attachments/article/123/CorredorsVerdsUrbans.pdf

Andersson, E. (2006). Urban landscapes and sustainable cities. *Ecology and Society, 11(*34). Retrieved from http://www.ecologyandsociety.org/vol11/iss1/art34/

Benvenuti, S., & Bacci, D. (2010). Initial agronomic performances of Mediterranean xerophytes in simulated dry green roofs. *Urban Ecosystems, 13*, 349-363. http://dx.doi.org/10.1007/s11252-010-0124-129

Benvenuti, S. (2014). Wildflower green roofs for urban landscaping, ecological sustainability and biodiversity. *Landscape and Urban Planning, 124*, 151-161. http://dx.doi.org/10.1016/j.landurbplan.2014.01.004

Bass, B. (2009). *Biodiversity Research Protocol*. Green Roofs for Healthy Cities – Draft Biodiversity Research Protocol. Retrieved from http://www.greenroofs.org/resources/Biodiversity_Research_on_Green_Roofs_Protocol_2009.pdf

Brennesein, S. (2006). Space for Urban Wildlife: Designing Green Roofs as Habitats in Switzerland. *Urban Habitats, 4*(1), 27-36.

Bretzel, F., Pezzarossa, B., Benvenuti, S., Bravi, A., & Malorgio, F. (2011). Soil influence on the performance of 26 native herbaceous plants suitable for sustainable Mediterranean landscaping. *Acta Oecologica, 35*, 657-663. http://dx.doi.org/ 10.1016/j.actao.2009.06.008

Britz-Escribano, J. (2004). *Naturación urbana y mejora medioambiental: Cubiertas ecológicas*. Madrid: Mundi-prensa Libros.

Buccola, N., & Spolek, G. (2011). A Pilot-Scale Evaluation of Greenroof Runoff Retention, Detention, and Quality. *Water Air Soil Pollution, 216*, 83-92. http://dx.doi.org/10.1007/s11270-010-0516-8

Caneva, G., Kumbaric, A., Savo, V., & Casalini, R. (2013). Ecological approach in selecting extensive green roof plants: A data set of Mediterranean plants. *Plant Byosystem*, 1-10. http://dx.doi.org/10.1080/11263504.2013.819819

Carter, T., & Butler, C. (2008). Ecological impacts of replacing traditional roofs with green roofs in two urban areas. *Cities and the Environment, 1*(2), 1-17.

Casadesús, J., Biel, C., & Savé, R. (2005). *Turf color measurement with conventional digital cameras*. Proceedings of EFITA/WCCA Joint Congress: On it in Agriculture 25-28 July 2005, Vila Real, Portugal.

Casadesus, J., Kaya, Y., Bort, J., Nachit, M. M., Araus, J. L., Amor, S., ... Villegas, D. (2007). Water Use Efficiency. Using vegetation indices derived from conventional digital cameras as selection criteria for wheat breeding in water-limited environments. *Ann. Appl. Biol., 150*, 227-236. http://dx.doi.org/10.1111/j.1744-7348.2007.00116.x

Damas, O., Donvez, J., Ferrando, D., Ferre, A., Marqueyssat, P., & Delhommeau, P. (2010). *Identification of plant ranges adapted to water limited conditions of green roofs: A case study from France*. Proceedings of World Green Roof Congress: Green roofs for a changing climate, 15-16 September 2010, London, UK.

Del Barrio, E. P. (1998). Analysis of the green roofs cooling potential in buildings. *Energy and Buildings, 27*, 179-193. http://dx.doi.org/10.1016/S0378-7788(97)00029-7

Dunnett, N., Nagase, A., Booth R., & Grime, P. (2008). Influence of vegetation composition on runoff in two simulated green roof experiments. *Urban Ecosystems, 11*(4), 385-398. http://dx.doi.org/10.1007/s11252-008-0064-9

Farrell, C., Szota, C., Williams, N. S., & Arndt, S. K. (2013). High water users can be drought tolerant: using physiological traits for green roof plant selection. *Plant Soil, 372*(1-2), 177-193. http://dx.doi.org/10.1007/s11104-013-1725-x

FLL, German Landscape Research, Development and Construction Society. (2008). *Guidelines for the Planning, Construction and Maintenance of Green Roofing*. Bonn, Germany.

Gedge, D., & Kadas, G. (2005). Green roofs and biodiversity. *Biologist, 52*(3), 161-169.

Köhler, M. (2006). Long-Term Vegetation Research on Two Extensive Green Roofs in Berlin. *Urban Habitats, 4*(1), 3-27. Retrieved from http://www.urbanhabitats.org

Lundholm, J. (2006). Green roofs and facades: A habitat template approach. *Urban Habitats, 4*(1), 87-101.

Lundholm, J., & Peck, S. W. (2008). Introduction: Frontiers of green roof ecology. *Urban Ecosystems, 11*,

335-337. http://dx.doi.org/10.1007/s11252-008-0070-y

Lundholm, J., MacIvor, J. S., MacDougall, Z., & Ranalli, M. (2010). Plant Species and Functional Group Combinations Affect Green Roof Ecosystem Functions. *PLoS ONE, 5*(3), http://dx.doi.org/10.1371/journal.pone.0009677

Mentens, J., Raes, D., & Hermy, M. (2005). Green roofs as a tool for solving the rainwater runoff problem in the urbanized 21st century? *Landscape Urban Planning, 77*(3), 217-226. http://dx.doi.org/10.1016/j.landurbplan.2005.02.010

Nagase, A., & Dunnett, N. (2010). Drought tolerance in different vegetation types for extensive green roofs: Effects of watering and diversity. *Landscape and Urban Planning, 97*, 318-327. http://dx.doi.org/10.1016/j.landurbplan.2010.07.005

Nagase, A., & Dunnett, N. (2012). Amount of water runoff from different vegetation types on extensive green roofs: Effects of plant species, diversity and plant structure. *Landscape and Urban Planning, 104*, 356-363. http://dx.doi.org/10.1016/j.landurbplan.2011.11.001

Neila, F. J., Bedoya, C., Acha, C., Olivieri, F., & Barbero, M. (2008). Las cubiertas ecológicas de tercera generación: un nuevo material constructivo. *Informes de la Construcción, 60*, 15-24.

Oberndorfer, E., Lundholm, J., Bass, B., Coffman, R. R., Doshi, H., Dunnett, N., ... Rowe, B. (2007). Green roofs as urban ecosystems: Ecological structures, functions, and services. *BioScience, 57*(10), 823-834. http://dx.doi.org/10.1641/B571005

Olivieri, F., Di Perna, C., D'Orazio, M., Olivieri, L., & Neila, J. (2013) Experimental measurements and numerical model for the summer performance assessment of extensive green roofs in a Mediterranean coastal climate. *Energy and Buildings, 63*, 1-14. http://dx.doi.org/10.1016/j.enbuild.2013.03.054

Papafotiou, M., Pergialioti, N., Tassoula, L., Massas, I., & Kargas, G. (2013). Growth of Native Aromatic Xerophytesin an Extensive Mediterranean GreenRoof as Affected by Substrate Typeand Depth and Irrigation Frequency. *Hortscience, 48*(10), 1327-1333.

Provenzano, M. E., Cardarelli, M., Crasso, M., & Colla, G. (2010). *Evaluation of potential green roof plant species in a Mediterranean climate*. Proceedings of World Green Roof Congress: Green roofs for a changing climate, 15-16 September 2010, London, UK.

Samangooei, M. (2006). *Green spaces in the sky. What role do green roofs play in a Twenty-first century city?* (Dissertation). Department of Architecture. Oxford Brookes University. Oxford.

Scott MacIvor, J., Ranalli, M. A., & Lundholm, J. (2011). Performance of dryland and wetland plant species on extensive green roofs. *Annals of Botany, 107*, 671-679. http://dx.doi.org/ 10.1093/aob/mcr007

Spala, A., Bagiorgas, H. S., Assimakopoulos, M. N., Kalavrouziotis, J., Matthopoulos, D., & Mihalakakou, G. (2008). On the green roof system. Selection, state of the art and energy potential investigation of a system installed in an office building in Athens, Greece. *Renewable Energy, 33*(1), 173-177. http://dx.doi.org/10.1016/j.renene.2007.03.022

So-Young, P., Jin-Su, S., Hyoung-Deug, K., Kenji, Y., & Ki-Cheol, S. (2008). Effects of Interior Plantscapes on Indoor Environments and Stress Level of High School Students. *Journal of the Japanese Society for Horticultural Science, 77*(4), 447-454. http://dx.doi.org/10.2503/jjshs1.77.447

Takahiro, S., Michio, K., Yuichi, U., & Noboru, I. (2010). Evaluation of Growth and Green Coverage of Ten Ornamental Species for Planting as Urban Rooftop Greening. *Journal of the Japanese Society for Horticultural Science, 79*(1), 69-76. http://dx.doi.org/10.2503/jjshs1.79.69

Van Mechelen, C., Dutoit, T., & Hermy, M. (2014). Mediterranean open habitat vegetation offers great potential for extensive green roof design. *Landscape and Urban Planning, 121*, 81-91. http://dx.doi.org/10.1016/j.landurbplan.2013.09.010

Williams, N. S. G., Hughes, R. E., Jones, N. M., Bradbury, D. A., & Rayner, J. P. (2010). In G. P. Gianquinto & F. Orsini (Ed.), *The Performance of Native and Exotic Species for Extensive Green Roofs in Melbourne, Australia (Acta Hort. 881)*. Procedings of the IInd International Conf. on Landscape and Urban Hort.

ZinCo. (2010). Retrieved January 22, 2010, from http://www.zincogreenroof.com/EN/downloads/pdfs/ZinCo_Product_List.pdf

6

Characterization and Recycling of Organic Waste after Co-Composting – A Review

Z. Anwar[1], M. Irshad[1], I. Fareed[2] & A. Saleem[1]

[1] Department of Environmental Sciences, COMSATS Institute of Information Technology, Abbottabad, Pakistan

[2] Resources Engineering and Management, University of Kurdistan, Hewler, Iraq

Correspondence: Z. Anwar, Department of Environmental Sciences, COMSATS Institute of Information Technology, Abbottabad, Pakistan. E-mail: zobiaanwar@ciit.net.pk

Abstract

Co-composting produces a valuable compost material that can be used as valuable soil amendment. The process of the co-composting and control of the composting factors are the current challenges for the researchers. There are different factors that govern the quality, stability and the maturity of the co-compost in terms of amount of plant nutrients and reduction of heavy metals. Among these, C:N ratio is a parameter that can affect the loss of plant nutrients. Different studies showed wide ranges of C:N ratios (14-40) for maturity of quality compost. Temperature, aeration and types of the bulking agents also regulate the process of co-composting. Most widely used co-composted materials are animal manures with agro-wastes (sawdust, wheat straw, rice straw, corn stalks etc.). This practice brought substantial loss of heavy metals and maximum retention of plant nutrients. Higher nutrients contents of the compost and favourable soil properties as a result of co-composting of the saw dust, cow dung and egg shells have been reported. The application of co-composted dairy manure with wheat straw and sawdust produced higher plant biomass. Co-compost of cattle manure with rice straw produced an organic matter, total N and C:N ratio contents suitable for soil amendment. Therefore, this review focuses on the characteristics and utilization of organic waste after a reasonable co-composting process.

Keywords: co-composting, livestock manures, sewage sludge, nutrient transformation, chemical characterization

1. Introduction

Cattle farms are producing a huge amount of manure throughout the world which must be managed through proper disposal practices to avoid the adverse impacts on environment (Burton & Turner, 2003). This has become a significant management problem in several countries. The issue of waste management is increasing due to the current environmental awareness. It has stimulated the researchers to identify economically feasible and environmental friendly technologies for animal manure to be used as a source of soil fertilizer (Zhang & He, 2006). Composting is microbial based aerobic process considered to be an environmentally sound way to minimize organic waste and produce organic fertilizer or soil conditioner (Gajdos, 1992). The process is able to transform unstable organic waste (such as sewage sludge, municipal solid waste, tannery waste, animal manure, poultry manure etc) into stable ones by converting them into a humus like structure called compost which results into a valuable agronomic by-product (Kashmanian et al., 2000).

Composting presents a low-technology and low-investment process to add value to organic solid waste through conversion into an organic fertilizer known as compost (Neves et al., 2009). It also considerably reduces odour emissions, biodegradable hydrocarbons, and dries up the waste making it unattractive to insects (Barrington et al., 2002). Co-composting is an important bio-waste treatment for achieving sustainable process and zero waste. Waste co-composting also enhances the compost quality by the comprehensive usage of diversified material properties (Fang & Wong, 1999). This type of waste management is well accepted when compared with other types, such as incineration or landfill. Certain physical and chemical characteristics of animals manure are not adequate for composting and may limit the efficiency of composting process. Several modifications in the process of composting has been suggested such as the addition of biodegradable waste, bulking agents etc to reach the optimal C:N ratio, degradation rate and compost quality (Herwijnen et al., 2007).The selection of appropriate bulking agent would modify manure composting substrate properties (moisture, porosity, C:N ratio, pH) (Petric et al., 2009).

The performance of the co-composting process and the quality of the end-product (the compost) are governed by the composition of the waste material, bulking agents and the management strategies. Composting and co-composting are among the few natural process available for the disposal of solid pollution combined with resource regeneration via stabilizing different types of organic waste. Co-composting has received increasing interest as a method of handling various types of livestock manures (Ogunwande et al., 2008) and presented a low investment technology. As in the most production processes, the implementation of large-scale composting brings greater economic benefits. Co-composting of more than one type of waste can be a more sustainable management process of solid waste (Neves et al., 2009). The addition of a bulking agent for composting optimizes substrate properties such as air space, moisture content, C:N ratio, particle density, pH and mechanical structure, affecting positively the decomposition rate (Neves et al., 2009).

The majority of research studies have mainly focused on monitoring the influence of bulking agent types on the composting of manure (Petric et al., 2009), but the research articles describing the effect of bulking agents on the physico-chemical characteristics of composting manures are deficient. Therefore, efforts have been made to review the co-composting of different materials and its effects on plant nutrient transformations under varied conditions.

2. Factors Affecting Co-Composting

Compost maturity depends on several factors such as characteristics of initial wastes (C:N ratio, pH, organic matter, moisture and porosity) and process conditions (time and process) (Khalil et al., 2008). Temperature is another parameter to evaluate evolution of the composting process, since it determines the biological reactions rate, as well as the sanitation capacity of the process. Carbon to N proportion is a most widely used parameter in composting and cocomposting (Gloueke, 1991). Generally, composting could be carried out under a wide range of initial C/N ratios, namely, 11 to 105, depending on the starting materials (Eiland et al., 2001; Ghosh et al, 2007). Tuomela et al. (2000) indicated that initially higher C:N ratio will cause a slower beginning of the process and the required composting time to be longer than usual while Tiquia and Tam (2000) reported that low initial C:N ratio results in higher emission of NH_3. Huang et al. (2004), by co-composting of pig manure and sawdust at the initial C:N ratios of 30 and 15, reported that the pile with initial C:N ratio of 30 reached maturity within 49 days. Recent studies have shown that composting can be carried out effectively at a lower C:N ratio of 15 (Zhu, 2007). Eiland et al. (2001) composted miscanthus straw with pig slurry at the initial C:N ratio of 11 to 35 with the degradation rate of 40 to 80 %. Kumar et al. (2010) found that composting of food waste together with green waste at the initial C/N ratio ranging from 14 to 20 was effective with the C:N ratio of 19.6. A decrease in the C:N ratio implies an increase in the degree of humification of the organic matter.

Composting is a biotechnological process characterized by an initial phase in which different microbial communities degrade organic matter into simpler nutrients and thermophilic temperatures are developed owing to the heat generation. The higher temperatures achieved in this phase appear to be the most important factor in killing heat-sensitive microorganisms. In a second stage, complex organic macromolecules such as humic acids are formed and temperature decreases to ambient levels, leading to a product that may contribute to soil conditioning and fertility (Hsu & Lo, 1999). Huang et al. (2004) evaluated that composting piles of higher C:N ratio would show a slower rise in temperature, lower maximum temperature and shorter thermophilic phase. Temperature is a simple and excellent indicator of how well the composting process is progressing and how much oxygen is being used (Walker, 2004). Misra et al. (2003) reported that higher temperatures during composting contribute to the killing of weed seeds and pathogens.

The aeration rate (AR) is considered to be the most important factor influencing successful composting (Diaz et al., 2002). Insufficient aeration can lead to anaerobic conditions due to the lack of oxygen, while excessive aeration can increase costs and slow down the composting process via heat, water and ammonia losses. The optimal AR depends on the composition of the raw materials and ventilation methods (Guo et al., 2012). Turning is often cited as the primary mechanism of aeration and temperature control during windrow composting (Michel et al., 1996; Tiquia, 1996), while turning frequency is commonly believed to be a factor which affects the rate of composting as well as compost quality (Tiquia, 1996).

Nitrogen losses can be reduced firstly by adjusting C:N ratio (Lhadi et al., 2004; Meijide et al., 2007); secondly, reducing N losses by manipulating the temperature changes (Hiraku et al., 2004). Substantial loss of nitrogen results in the reduction of nutrient value of compost products. Among all practicable N conservation measures, an amendment was thought to be the most efficient method to reduce N loss during composting. For example, alum, peat, and zeolite lime and coal fly ash amendments have been used to reduce ammonia volatilization and heavy metals mobility during composting (Bernal et al., 1993; Kithome et al., 1999; Fang & Wong, 1999; Qiao & Ho, 1997). The type of bulking agent had a significant effect. There are a lot of bulking agents that could be used as a carbon source. Farmers make their choice on the basis of the availability of specific bulking agent. To-date, a majority of conducted studies have mainly focused on monitoring the influence of bulking agents on the composting of different types of manure (Petric et al., 2009). Yeoh et al. (2011) reported the selection of an inoculation of bacteria or fungus which may shorten the composting time, improve the compost quality and inhibit the growth of some plant pathogens.

3. Co-Composting Animal Manure

Understanding the quantity and quality of waste is an essential step for the productive management of a co-composting process (Yeoh et al., 2011). The factors that affect composting are essential in evaluating the performance of an overall composting process (Yeoh et al., 2011). Karnchanawong and Tadkaew (2013) reported higher nutrient content of the compost and the higher porosity of the soil mixture, as a result of the co-composting of the saw dust, cow dung and egg shells. Dehshan et al. (2013) reported co-compost of cattle manure with rice straw produced an organic matter, total N and C:N ratio contents suitable for soil amendment. Huang et al. (2006) studied the organic matter transformation during co-composting of pig manure with saw dust using chemical and spectroscopic methods and concluded with a stable organic matter in the mature compost. Zhang and He (2005) investigated the physicochemical properties in co-composting of swine manure with pine saw dust and found that 30% swine manure with initial C:N ratio of about 40 was desirable for composting with organic substrate. The importance of different factors affecting the quality, the stability and the maturation of the co-compost final product by composting pig feces and corn stalks under different aeration rates, C:N ratio and moisture content reported that the lowest initial C:N ratio had the lowest germination index. Aeration considered to be the main factor affected the stability of the final co-compost product while C/N ratio contributed to maturity and moisture content had a significant effect on compost quality (Guo et al., 2012).

Frederick et al. (2003) amended dairy manure of 83% of moisture with either straw dust or straw and reported that an initial C/N ratio more than 40 to minimize N loss. Li-li et al. (2013) compared the co-composting process of rabbit manure mixed with mushroom residue and rice straw and reported that both compost could help seed growth and a much more efficient way to utilize the material. On comparison, the co-composted manure with rice straw showed significant effects on improving seed growth than with mushroom residue. Dahshan et al. (2013) reported the addition of a bulking agent for composting cattle manure, especially, rice straw which produced compost with an organic matter, total nitrogen, and C:N ratio suitable for soil amendment.

Huang et al. (2006) examined the transformation of organic matter during the co-composting process of pig manure with saw dust. Organic C, C:N ratio and elemental analysis indicated the presences of more stable organic matter in the matured compost. Hao et al. (2004) investigated greenhouse gases emission during composting of straw bedded manure and wood chip bedded manure and concluded that the net contribution of greenhouse gas emission was greater for CH_4, since it is 21 times more effective at trapping heat than CO_2 and suggested further study to evaluate the impacts of composting on overall greenhouse gases emission and C sequestration. Wang et al. (2004) investigated the comparison of the co-composting process of dairy manure with wheat straw and sawdust which showed the positive initial plant growth response without fertilization.

The co-composting process of pulp and paper industry sludge with sawdust under aerobic conditions with two different treatments of cow dung only and the one with cow dung with effective microorganisms showed that the pulp and paper industry sludge co-composted with sawdust at the ratio of 3:1 by mixing cow dung and this could have successfully converted into valuable compost (Lakshmi et al., 2010). Eneji et al. (2003) reported that composting farm wastes before soil application could reduce the vulnerable P fraction, while increasing more resistant P fractions.

Neves et al. (2009) investigated the influence of the lipids content by co-composting cow manure with food waste and concluded with a positive effect due to the oily waste addition. Due to the hydrophobic properties of this specific waste, which may imply reduce need of a bulking agent. Tripetchkul et al. (2012) reported co-composting of coir pith and cow manure to evaluate the influence of C:N ratios i.e. 30, 25 and 20 on the physicochemical changes and found that simple pile turning yielded effective co-composting of coir pith comparatively with low C:N ratio. Considering the composting performance and the amount of coir pith to be utilized, the initial C:N ratio of 30 was considered suitable for coir pith composting. Karnchanawong and Tadkaew (2013) reported higher nutrient contents and vegetable growth using co-compost of saw dust, cow dung and egg-shell and the soil applied with this material was found with higher porosity. Zhang and He et al. (2006) reported total P loss during mineralization of organic P and consumption by microbes after co-composting of swine manure with pine sawdust. Eneji et al. (2003) reported favorable changes in the organic matter and elemental composition during composting of livestock manure for the use as soil amendment.

4. Co-Composting of Poultry Manure

Ogunwande et al. (2008) reported that moisture loss increased as C:N ratio and turning frequency increased when co-composted raw chicken manure with saw-dust. A significant loss of total N through ammonia volatilization and total C losses were attributed to the degradation of organic matter. The study indicated that C:N ratio significantly affected total N and P while C:N and turning frequency has a significant effect on total C and K. Tiquia et al. (2001) reported that co-composting of poultry manure and yard trimmings showed physicochemical and biochemical changes in compost mass, increase in enzymatic activities, decrease in the water soluble components (i.e., water extractable C, inorganic N and heavy metals) and removal of phytotoxicity. Steiner et al. (2010) focused on reducing N losses by using poultry litter co composted with biochar which acted as absorber of ammonia and total N losses were reduced by up to 25% by lowering ammonia concentration by up to 64%. Salma et al. (2006) co-composted poultry manure, olive mill waste and mineral rich wastewater and reported higher levels of nutrients, low C:N ratio of 15-17, free of phytotoxicity in co-compost and no negative impacts on soils. Tiquia and Tam (2000) examined the changes of N during co-composting of chicken litter i.e. the mixture of chicken manure, wood shaving, waste feed and feathers, reduced the value of chicken litter as N fertilizer and contained more stable organic matter in composted chicken litter. Huck et al. (2013) co-composted pine apple leaves and chicken slurry and reported no foul odor, low heavy metals content and considerable amount of nutrients with phytotoxic free seed germination.

5. Co-Composting of Municipal Solid Waste

Marco et al. (2013) used disposed white rot fungus Trametesversicolor after its application in the biotechnological process and co-composted with organic fraction of municipal solid waste and suggested that the final compost obtained through co-composting process ensured the higher stable final product than that obtained through organic fraction of municipal solid waste composting. Municipal solid waste with cow manure, poultry manure and yard trimmings etc helped to maintain soil fertility and improved moisture holding capacity of soils and found an acceptable release of plant nutrients (Gautam et al., 2010). Fourti et al. (2008) assessed co-composting of municipal solid waste and sewage sludge in semi-arid pedo-climatic conditions and finally confirmed amongst others that temperature is the main parameter to consider in composting since it acclimatizes the functioning of all the others parameters, specifically the microbial activity and diversity.

6. Co-Composting of Sludge

The effects of bamboo charcoal on N conservation and Cu, Zn mobility during sludge composting concluded that N loss and Cu and Zn mobility could significantly lessened through bamboo charcoal capability of ammonia adsorption (Hua et al., 2009). Wu et al. (2010) carried out N transformation during co-composting of herbal residue, spent mushrooms and sewage sludge at different ratios of 3:1:1 (sewage sludge:sawdust:spent mushrooms or herbal residue) and 3:1:2 (sewage sludge:sawdust:spent mushrooms or herbal residue) and concluded that the germination index values for all samples at maturity were above 80%, demonstrating optimal maturity. The addition of spent mushrooms or herbal residues augmented sewage composting.

The microbial and enzymatic activities during co-composting process of empty fruit bunches and partially treated with palm oil mill effluent showed that different species of uncultured bacteria dominated the process from the phylum proteobacteria in the thermophilic stage to the phylum chloroflexi in the maturing stage (Baharuddin et al., 2009). Hock et al. (2009) investigated the characteristics and physicochemical changes in the windrow in co-composting process of oil palm mesocarp fiber and palm oil mill effluent with anaerobic sludge. The final co composted product was achieved within 50 days with C:N ratio of 12.6 with considerable amount of nutrients and low levels of heavy metals.

7. Use of Co-Composted Waste Material

7.1 Nitrogen/Carbon Losses during Co-Composting

Composting is a bio-oxidative process in which organic matter is transformed by N mineralization, ammonia volatilization, nitrification, denitrification and partial humification into compost which is free of phyto-toxicity, pathogens and with certain humic properties by aerobic microorganisms. It comprises three major phases: mesophilic, thermophilic and cooling phase (the compost stabilization phase) (Tiquia & Tam, 2000; Singh & Kalamdhad, 2012; Neves et al., 2009). Nitrogen mineralization is of extreme importance because it converts organic N into ammonium (NH_4). Ammonia (NH_3) volatilization and de-nitrification may lead to significant losses of N (Martins & Dewes, 1992; Bernal et al., 1996). One of the most negative effects of composting animal manures is the loss of N through NH_3 volatilization which reduces the fertilizer value of the manure and constitutes an important economic loss (Tiquia & Tam, 2000). Ammonia emissions result from aerobic or anaerobic bacteria in manure (Zhang et al., 1991). Several studies have shown that NH_3 volatilization increased with an increase in pH, moisture content, aeration, NH_3 concentration or temperature (Bishop & Godfrey, 1983). These losses during composting of animal manure range from 21% upward to 77% (Martins & Dewes, 1992; Rao Bhamidimarri & Pandey, 1996). Nitrogen losses varied depending on several environmental factors such as aeration, moisture content, and temperature (Bishop & Godfrey, 1983). Ogunwande et al. (2008) reported that C:N ratio had significant ($P \leq 0.05$) effect on pile temperature, total N, total C, C: N ratio, dry matter, P and K while turning frequency had significant ($P \leq 0.05$) effect on pile temperature, pH, TC, C:N ratio and K. The decreasing level of C:N ratio is correlated by the loss of TOC to produce carbon dioxide and smaller increase of nitrogen as the decomposition progressed. The decreasing pattern of the C:N ratio and TOC, and the increasing pattern of nitrogen content were reported by Abdullah et al. (2013) and Goyal et al. (2005) on sugarcane trash, cattle dung, pressmud, poultry waste and water hyacinth composting. The key to the development of compost technology is to control the changes and N losses, and this has attracted the attention of many compost researchers (Cordovil et al., 2007; Meijide et al., 2007; Wu et al., 2010; Huang et al., 2005; Hua et al., 2009).

Apart from N, C is another element that is most likely to be lost during the composting process. Zhang and He (2006) reported that N and P decomposition primarily occurred in the mesophilic phase, while organic carbon decomposed in the thermophilic phase during co-composting of solid swine manure with pine sawdust and 30% swine manure with initial C:N ratio of about 40 was more desirable for composting organic substrates. The nutrient loss especially nitrogen is an unavoidable problem during the composting of organic waste (Cooperband & Middleton, 1996; Rao & Pandey, 1996). Carbon may be lost due to either bio-oxidation, in which carbonaceous materials are lost as CO_2 (Bishop & Godfrey, 1983; Eghball et al., 1997) or mineralization of C, in which inorganic C is converted to organic C (Bernal et al., 1998). The compost quality could further be improved by adding cow manure, poultry manure or yard waste etc (Gautam et al., 2010). The adjustment of the C:N ratio of lingo-cellulosic composts by the addition of N source from animal residues proved to a reliable composting process and the end products characterized by a good degree of stability, and compatible with safe and effective soil utilization (Cayuela et al., 2009).

7.2 Effects of Co-Composting on Soil Quality

The application of stabilized co-composted manure to soil can affect soil fertility by improving physical, chemical and biological properties of the soil (Dick et al., 1993). The physical changes involve modifications of soil bulk density, structure, strength and water relations. The chemical changes include the accumulation of organic plant nutrients, the soil's ion exchange capacity, chelating activity and buffering ability. Composted animal manure can enlarge the nutritional base for soil microorganisms. Incorporated manure composts also become part of the soil humus and employ their influence on long-term basis. Such inclusions of soil amendments are particularly important where the soils are poor enough in terms of OM content (Tiquia & Tam, 2002).

The effect of compost on soil pH depends both on the initial pH of the compost and the soil pH (Tiquia & Tam, 2002). Organic amendments significantly increase the soil reducing exchangeable acidity, exchangeable Al, and exchangeable Fe that altered soil chemical properties in a way that enhanced the availability of phosphorus pH (Schulz et al., 2014; Chng et al., 2014). Manures have been proven to lower the pH of alkaline soils (Chang et al., 1991). Increasing the pH of acidic soils has also been demonstrated frequently by the use of livestock manure (Hue, 1992; Whalen et al., 2000; Baziramakenga et al., 2001). Elamin (1991) demonstrated that the decomposition of organic matter tended to improve the soil physical and chemical proprieties.

End product of co-composted plant residues and animal manure greatly influence the biochemical and structural properties of the soil. Mineralization of plant residue has been demonstrated to be the best soil amendment and animal manures as organic fertilizer (Mondini et al., 2010). The addition of plant residues tended to increase the microbial pool due to lack of N availability, the addition of animal by-products led to the rapid increase of the microbial biomass (Cayuela et al., 2008). Co-composted biochar incorporation induced soil alkalization which increasd soil nitrification (Lehmann et al., 2003; Yamato et al., 2006; Amlinger et al., 2007).

7.3 Heavy Metals during Co-Composting

Compost usually contains heavy metals based on initial raw material. Generally, these heavy metals (micronutrients) are essential for plant growth. The micronutrients present in the compost, such as Zn, Cu and Fe, are utilized by the plants during the fertilizing process. These trace elements are required for a better plant growth in trace quantities but in large quantities can cause phytotoxicity, consequently can impose hazardous risks to human and fauna (Eneji et al., 2003; Simth, 2009; Van Herwijnen et al., 2007). However, constant and extensive application of organic compost will lead to the accumulation of heavy metals in the soil and an increase in toxic levels. Heavy metals can render the growth, morphology and metabolism of soil microorganisms, consequently affect the fertility of soil reduced (Bragato et al., 1998). Uptake of heavy metals by plants and successive accumulation in human tissues and bio magnifications by the food chain can adversely affect human health and environment concerns (Wong et al., 2006; Chiroma et al., 2013). Addition of lime during composting process, natural zeolite and bamboo charcoal amendments tend to immobilize metals and reduced heavy metals availability and leachibility. (Singh & Kalamdhad, 2013) suggested that bioavailability and leachability of metals should be assessed before adding composted manure to the agricultural fields. Composting process tended to increase total P, K, Cu, and Zn (Jusoh et al., 2013) due to the losses of organic C, H, N, and O in the forced-aeration piles as demonstrated by (Tiquia & Tam, 2002) where CO_2 and H_2O during composting process, leaving Cu and Zn behind and giving a relative increase in the concentrations of these elements. The organic amendments tend to increase organic P compared with soil alone as showed by Chang et al. (2014). Moreover the high sorption capacity caused by aromaticity of the bio char (co-composted) influenced nutrient cycling (Glaser et al., 2002). Biochar applied in higher amounts, soil nutrient availability increased in highly weathered tropical soils comparable (Lehmann et al, 2002, with ~560 Mg ha^{-1}; Lehmann et al., 2003, with 67.6–135.2 Mg ha^{-1}; Steiner et al., 2007, with 11 Mg ha^{-1}). According to Kostov et al. (1995), compost had a significant impact over manure in respect of nutrient release.

7.4 Bulk Density during Co-Composting

Reduction in soil bulk density may be expected when soil is mixed with less dense organic material, but there may also be associated with alteration of soil structure. The magnitude of change for bulk density and other soil properties likely to vary with soil texture as demonstrated by Aggelides and Londra (2000). Decreased bulk density and increased water holding capacity of soil have been characterized by the addition of the organic amendments. Bulking agents acted as moisture adjuster due to their low moisture level and higher cellulose content, which can be a source of carbon (Abdullah et al., 2013). The addition of bulking agent (yard trimmings, woodchips, rice husks) to the chicken litter help to reduce the loss of N by reducing ammonia loss (Wan et al., 1999, Tiquia & Tam, 2000). The decrease bulk densities and increase soil water holding capacities of the organically amended treatments in comparison to the control is due to enhanced aggregation of soil particles by the addition of organic matter (Khaleel et al., 1981; Grandy et al., 2002; Elsharawy et al., 2003).

7.5 Agronomic Benefits of Co-Composting

Plant growth significantly increased with compost amendment (Schulz et al., 2014). Co-composting rabbit manure with rice straw and mushroom residues was found efficient to utilize the material to enhance seed production (Li-li et al., 2013). Chicken manure resulted in an increase growth attributes as well as forage yield as indicated by Ismaeil et al. (2012). Several researchers reported that nitrogen and phosphorus uptake, as a function of chicken manure application, progressively exceeded with composted manure (Magid et al., 1998; Postma et al., 1998; Hue & Sobieszczyk, 1999; Abusuwar & El Zalal, 2010). The increase in organic P with the increasing of time is essential to mineralize P in soil for crop use and influenced the crop yield and growth (Chang et al., 2014). Agronomic considerations including increase crop productivity, reduced fertilizer and pesticide use have been reported with the application of composted manure (Schulz & Glaser, 2013; Jaffery et al., 2011; Blackwell et al., 2013).

8. Conclusions

Co-composting is an effective solid waste management that helps to manage a variety of solid waste (poultry waste, animal waste, food waste agro-waste, municipal solid waste, industrial sludges and effluents etc.) and converts them into a valuable product that may enhance the fertility of the soil by acting as soil conditioner. This process is simple having lowest operational cost and an alternative of costly fertilizer used for enhancing the soil nutrients. The process of co-composting can help to promote the complexation of heavy metals whose mobility and availability tend to decrease with decreasing toxicity. Bio-degradable waste can successfully be co-composted with different types of organic wastes to get higher quality cocomposted product. There is a need to study quality characteristics and possible benefits of compost.

References

Abdullah, N., Chin, N. L., Mokhtar, M. N., & Saleena, F. T. (2013). Effects of bulking agents, load size or starter cultures in kitchen waste composting. *International Journal of Recycling of Organic Waste in Agriculture, 3*, 1-10.

Aggelides, S. M., & Londra, P. A. (2000). Effects of compost produced from town wastes and sewage sludge on the physical properties of a loamy and a clay soil. *Bioresource Technology, 71*, 253-259. http://dx.doi.org/10.1016/S0960-8524(99)00074-7

Abusuwar, A. O., & El Zalal, H. A. (2010). Effect of chicken manure on yield, quality and HCN concentration of two forage Sorghum (*Sorghum bicolor* L. (Moench) cultivars. *Agriculture and Biology Journal of North America, 1*(1), 27-31.

Amlinger, F., Peyr, S., Geszti, J., Dreher, P., Karlheinz, W., & Nortcliff, S. (2007). Beneficial Effects of Compost Application on Fertility and Productivity of Soils. *Federal Ministry for Agricultural and Forestry, Environment and Water Management*. Lebensministerium: Vienna, Austria.

Bishop, P. L., & Godfrey, C. (1983). Nitrogen variations during sludge composting. *BioCycle, 24*, 34-40.

Baharuddin, A. S., Wakisaka, M., Shirai, Y., Abd-Aziz, S., Abdul Rahman, N. A., & Hassan, M. A. (2009). Co-composting of empty fruit bunches and partially treated palm oil mill effluents in pilot scale. *International Journal of Agricultural Research, 4*, 69-78. http://dx.doi.org/10.3923/ijar.2009.69.78

Barrington, S., Choiniere, D., Trigui, M., & Knight, W. (2002). Effect of carbon source on compost nitrogen and carbon losses. *Bioresource Technology, 83*, 189-194. http://dx.doi.org/10.1016/S0960-8524(01)00229-2

Baziramakenga, R., Simard, R. R., & Lalande, R. (2001). Effect of de-inking paper sludge compost application on soil chemical and biological properties. *Canadian Journal of Soil Science, 81*, 561-575. http://dx.doi.org/10.4141/S00-063

Bernal, M. P., Lopez-Real, J. M., & Scott, K. M. (1993). Application of natural zeolites for the reduction of ammonia emission during the composting of organic waste in a laboratory composting simulators. *Bioresource Technology, 43*, 35-39. http://dx.doi.org/10.1016/0960-8524(93)90079-Q

Bernal, M. P., Paredes, C., Sánchez-Monedero, M. A., & Cegarra, J. (1998). Maturity and stability parameters of composts prepared with a wide range of organic wastes. *Bioresource Technology, 63*, 91-99. http://dx.doi.org/10.1016/S0960-8524(97)00084-9

Bernal, M., Navarro, A. F., Roig, A., Cegarra, J., & Garcia, D. (1996). Carbon and nitrogen transformation during composting of sweet sorghum bagasse. *Biology and Fertility of Soils, 22*, 141-148. http://dx.doi.org/10.1007/BF00384446

Barrington, S., Choiniere, D., Trigui, M., & Knight, W. (2002). Effect of carbon source on compost nitrogen and carbon losses. *Bioresource Technology, 83*, 189-194. http://dx.doi.org/10.1016/S0960-8524(01)00229-2

Blackwell, P., Krull, E., Butler, G., Herbert, A., & Solaiman, Z. (2013). Effect of banded biochar on dryland wheat production and fertiliser use in south-western Australia: An agronomic and economic perspective. *Soil Research, 48*, 531-545. http://dx.doi.org/10.1071/SR10014

Bragato, G., Leita, L., Figliolia, A., & Nobili, M. (1998). Effects of sewage sludge pre-treatment on microbial biomass and bioavailability of heavy metals. *Soil Tillage Research, 46*, 129-134. http://dx.doi.org/10.1016/S0167-1987(97)00084-6

Burton, H., & Turner, C. (2003). Treatment Strategies for sustainable Agriculture. *Manure management* (2nd ed.). Silsoe Institute, Lister and Durling printers. Flitwick Bedford, UK.

Cayuela, M. L., Sinnico, T., & Mondini, C. (2009). Mineralization dynamics and biochemical properties during initial decomposition of plant and animal residues in soil. *Applied Soil Ecology, 4*, 118-127. http://dx.doi.org/10.1016/j.apsoil.2008.10.001

Chang, H. Y., Ahmad, O. H., Kassim, S., & Majid, N. M. A. (2013). Co-composting of pineapple leaves and chicken slurry. *International Journal of Recycling of Organic Waste in Agriculture, 2*(23), 1-8.

Chang, H. Y., Ahmed, O. H., & Majid, N. M. A. (2014). Improving phosphorous availability in an acid soil using organic amendments produced from agro industrial wastes. *The Scientific World Journal, 10*, 1-6. http://dx.doi.org/10.1155/2014/794756

Chiroma, T. M., Ebewele, R. O., & Hymore, F. K. (2012). Levels of heavy metals (Cu, Zn, Pb, Fe and Cr) in bush green and roselle irrigated with treated and untreated urban sewage water. *International Research Journal of Environmental Sciences, 1*, 50-55.

Cordovil, C. M. D. S., Cabral, F., & Coutinho, J. (2007). Potential mineralization of nitrogen from organic wastes to ryegrass and wheat crops. *Bioresource Technology, 98*, 3265-3268.

http://dx.doi.org/10.1016/j.biortech.2006.07.014

Dahshan, H., Abd- Elall, A. M. M., Abd-El-Kader, M. A., & Megahed, A. M. (2013). Field-scale management and evaluation of recycled cattle manure solids from livestock in Nile Delta ecosystem. *African Journal of Agricultural Research, 8*, 2829-2836.

Diaz, M. J., Madejon, E., Lopez, F., Lopez, R., & Cabrera, F. (2002). Optimization of the rate vinasse/grape marc for co-composting process. *Process Biochemistry, 37*, 1143-1150. http://dx.doi.org/10.1016/S0032-9592(01)00327-2

Dick, W. A., & McCoy, E. I. (1993). Enhancing soil fertility by addition of compost. In H. A. J. Hoitink, & H. M. Keener (Eds.), *Science and engineering of composting: Design, environmental, microbiological and utilization aspects* (pp. 623-644). Worthington, Ohio: The Renaissance Publications.

Eghball, B., Power, J. F., Gilley, J. E., & Doran, J. W. (1997). Nutrient, carbon and mass loss during composting of beef cattle feedlot manure. *Journal of Environmental Quality, 26*, 189-193. http://dx.doi.org/10.2134/jeq1997.261189x

Eneji, A. E., Yamamoto, S., Honna, T., & Ishiguro, A. (2001). Physico-chemical changes in livestock feces during composting. *Communications in Soil Science and Plant Analysis, 32*, 477-489. http://dx.doi.org/10.1081/CSS-100103023

Eneji, A. E., Honna, T., Yamamoto, S., Masuda, T., & Irshad, M. (2003). Changes in humic substances and phosphorus fractions during composting. *Communications in Soil Science and Plant Analysis, 34*, 2303-2314. http://dx.doi.org/10.1081/CSS-120024065

Eneji, E., Honna, T., Yamamoto, S., Masuda, T., Endo, T., & Irshad, M. (2003). The relationship between total and available heavy metals n composted manure. *Journal of Sustainable Agriculture, 23*, 125-134. http://dx.doi.org/10.1300/J064v23n01_09

Eiland, F., Klamer, M., Lind, M. A., Leth, M., & Bååth, E. (2001). Influence of initial C/N ratio on chemical and microbial composition during long term composting of straw. *Microbiology and Ecology, 41*, 272–280.

Elamin, E. E. (1991). *Effect of organic manure decomposition on some soil properties* (M. Sc (Agric). Thesis). Faculty of Agriculture, University of Khartoum, Sudan.

Fang, M., Wong, J. W. C., Ma, K. K., & Wong, M. H. (1999). Co-composting of sewage sludge and coal ash: nutrient transformation. *Bioresource Technology, 67*, 19-24. http://dx.doi.org/10.1016/S0960-8524(99)00095-4

Fourti, O., Haydri, Y., Murano, F., Jedidi, N., & Hassen, A. (2008). A new process assessment of co composting of municipal solid waste and sewage sludge in semi arid pedo climate condition. *Annals of Microbiology, 58*, 403-409. http://dx.doi.org/10.1007/BF03175535

Gajdos, R. (1992). The use of organic waste materials as organic fertilizer – Recycling of plant nutrients. *Acta Horticulturae, 302*, 325-331.

Gautam, S. P., Bundela, P. S., Pandey, A. K., Awasthi, M. K., & Sarsaiya, S. (2010). Composting of municipal solid waste of Jabalpur city. *Global Journal of Environmental Research, 4*, 43-46.

Glaser, B., Lehmann, J., & Zech, W. (2002). Ameliorating physical and chemical properties of highly weathered soils in the tropics with charcoal- A review. *Biology and Fertility of Soils, 35*, 219-230. http://dx.doi.org/10.1007/s00374-002-0466-4

Ghosh, P. K., Sarma, U. S., Ravindranath, A. D., Radhakrishman, S., & Ghosh, P. (2007). A novel method for accelerated composting of coir pith. *Energy Fuel, 21*, 822-827. http://dx.doi.org/10.1021/ef060513c

Golueke, C. G. (1991). Principle of Composting. The staff of Biocycle Journal of Waste Recycling. *The art and science of composting* (pp. 32-55). JG Press, Pennsylvania.

Guo, R., Li, G., Jiang, T., Schuchardt, F., Chen, T., Zhao, Y., & Shen, Y. (2012). Effect of aeration rate, C/N ratio and moisture content on the stability and maturity of compost. *Bioresource Technology, 112*, 171-178. http://dx.doi.org/10.1016/j.biortech.2012.02.099

Hachicha, S., Chtourou, M., Medhioub, K., & Ammar, E. (2006). Compost of poultry manure and olive mill waste as an alternative fertilizer. *Agronomy for Sustainable Development, 26*, 135-142. http://dx.doi.org/10.1051/agro:2006005

Herwijnen, V. R. (2007). Remediation of metal contaminated soil with mineral amended compost.

Environmental Pollution, 150, 357-54.

Herwijnen, R. V., Hutchings, T. R., Tabbaa, A. A., Moffat, A. J., Johns, M. L., & Ouki, S. K. (2007). Remediation of metal contaminated soil with mineral amended composts. *Environmental Pollution, 150*, 347-354. http://dx.doi.org/10.1016/j.envpol.2007.01.023

Hiraku, S., Gen, M., Hanatsu, S., Jun, N., Msaaki, S., Takako, S., ... Yutaka, N. (2004). Distribution of ammonia assimilating bacteria in the composting process. *Compost Science and Utilization, 12*, 108-113. http://dx.doi.org/10.1080/1065657X.2004.10702168

Hock, L. S., Baharuddin, A. S., Ahmad, M. N., & Shah, U. K. (2009). Physicochemical changes in windrow co-composting process of oil palm mesocrap fiber and palm oil mill effluent anaerobic sludge. *Australian Journal of Basic and Applied Sciences, 3*(3), 2809-2816.

Hua, L., Wu, W., & Liu, Y. (2009). Reduction of nitrogen loss and Cu and Zn mobility during sludge composting with bamboo charcoal amendment. *Environmental Science and Pollution Research, 16*, 1-9. http://dx.doi.org/10.1007/s11356-008-0041-0

Huang, G. F., Wu, Q. T., Wong, J. W. C., & Nagar, B. B. (2006). Transformation of organic matter during co-composting of pig manure with saw dust. *Bioresource Technology, 97*, 1834-1842. http://dx.doi.org/10.1016/j.biortech.2005.08.024

Huang, G. F., Wong, J. W. C., Wu, Q. T., & Nagar, B. B. (2004). Effect of C/N on composting of pig manure with sawdust. *Waste Management, 24*(8), 805-813. http://dx.doi.org/10.1016/j.wasman.2004.03.011

Hue, N. V., & Sobieszczyk, B. A. (1999). Nutritional values of some bio-wastes as soil amendments. *Compost Science and Utilization, 7*, 34-41. CAB Abstracts, 98/2000. http://dx.doi.org/10.1080/1065657X.1999.10701950

Hue, N. V. (1992). Correcting soil acidity of a highly weathered Ultisol with chicken manure and sewage sludge. *Communications in Soil Science and Plant Analysis, 23*, 241-264. http://dx.doi.org/10.1080/00103629209368586

Hsu, J. H., & Lo, S. L. (1999). Chemical and spectroscopic analysis of organic matter transformations during composting of pig manure. *Environmental Pollution, 104*, 189-196. http://dx.doi.org/10.1016/S0269-7491(98)00193-6

Ismaeil, F. M., Abusuwar, A. O., & Naim, A. M. E. (2012). Influence of chicken manure on growth anf yield of Sorghum (*Sorghum biocolor* L. Moench). *International Journal of Agriculture and Forestry, 2*(2), 56-60. http://dx.doi.org/10.5923/j.ijaf.20120202.09

Jeffery, S., Verheijen, F. G. A., Van der Velde, M., & Bastos, A. C. (2011). A quantitative review of the effects of biochar application to soils on crop productivity using meta-analysis. *Agriculture, Ecosystems and Environment, 144*, 175-187. http://dx.doi.org/10.1016/j.agee.2011.08.015

Jusoh, M. L. C., Manaf, L. A., & Latiff, P. A. (2013). Composting of rice straw with effective microorganisms (EM) and its influence on compost quality. *Journal of Environmental Health Science and Engineering, 10*, 17-26. http://dx.doi.org/10.1186/1735-2746-10-17

Kashmanian, R. M. (2000). Quantities, Characteristics, Barriers and Incentives for use of Organic Municipal By products, Land Application of Agricultural, Industrial and Municipal By-products. *Science Society of America Books Series, 6*, 128-167.

Karnchanawong, S., & Tadkaew, N. (2013). Effects of compost from co composting of sawdust, cow dung and eggshell on soil acidity neutralization and growth of vegetable. *Digital proceeding of the ICOEST' 2013*. Nevsehir, Turkey. June 18-21, 2013.

Kithome, M., Paul, J. W., & Bomk, A. A. (1999). Reducing nitrogen losses during simulated composting of poultry manure using absorbents or chemical amendments. *Journal of Environmental Quality, 28*, 194-201. http://dx.doi.org/10.2134/jeq1999.00472425002800010023x

Kostov, O., Tzvetkov, Y., Kaloianova, N., & Cleemput, O. (1995). Cucumber cultivation on some waste during their aerobic composting. *Bioresource Technology, 53*, 237-242. http://dx.doi.org/10.1016/0960-8524(95)00076-Q

Kumar, M., Ou, Y. L., & Lin, J. G. (2010). Co-composting of green waste and food waste at low C/N ratio. *Waste Management, 30*, 602-609. http://dx.doi.org/10.1016/j.wasman.2009.11.023

Lehmann, J., Pereira da Silva, J., Steiner, C., Nehls, T., Zech, W., & Glaser, B. (2003). Nutrient availability and leaching in an archaeological Anthrosol and a Ferralsol of the Central Amazon basin: Fertilizer, manure and charcoal amendments. *Plant and Soil, 249*, 343-357. http://dx.doi.org/10.1023/A:1022833116184

Lehmann, J., Da Silva, J. P. Jr., Rondon, M., Cravo, M. S., Greenwood, J., Nehls, T., & Steiner, C. (2002). Slash-and-char- A feasible alternative for soil fertility management in the Central Amazon. *Proceedings of the 17th World Congress of Soil Science*. Bangkok, Thailand, August 14-21, 2002.

Lhadi, E. K., Tazi, H., Aylaj, M., Tambone, F., & Adani, F. (2004). Co-composting separated MSW and poultry manure in Morocco. *Compost Science and Utilization, 121*, 37-144.

Li-Li, B., Tie-Jun, Y., Bin, W., Lin, B., De-gui, T., & Xiang-Chao, F. (2013). Evaluation and comparison of composting rabbit manure mixed with mushroom residue and rice straw. *Journal of Agricultural Science and Technology, 15*, 1069-1081.

Lokman, M., Jusoh, C., Manaf, L. A., & Latiff, A. (2013). Composting of rice straw with effective microorganisms (EM) and its influence on compost quality. *Journal of Environmental Health Science and Engineering, 10*, 1-9.

Meijide, A., Diez, J. A., Sanchez-Martin, L., Lopez-Femandez, S., & Vakkejo, A. (2007). Nitrogen oxide emissions from an irrigated maize crop amended with treated pig slurries and composts in a Mediterranean climate. *Agriculture, Ecosystem and Environment, 121*(4), 383-394. http://dx.doi.org/10.1016/j.agee.2006.11.020

Magid, H. M., Abdel Al, S. I., Rabie, R. K., & Sabra R. E. (1998). Chicken manure as a bio-fertilizer for wheat gown in sandy soils of Saudi Arabia. *Egyptian Journal of Soil Science, 38*, 329-338.

Martins, O., & Dewes, T. (1992). Loss of nitrogenous compounds during composting of animal wastes. *Bioresource Technology, 42*, 103-111. http://dx.doi.org/10.1016/0960-8524(92)90068-9

Michel, F. C., Forney, L. J., Huang, A. J. F., Drew, S., Czuprendski, M., Lindeberg, J. D., & Reddy, C. A. (1996). Effects of turning frequency, leaves to grass mix ratio and windrow vs. pile configuration on the composting of yard trimmings. *Compost Science and Utilization, 4*, 126-143. http://dx.doi.org/10.1080/1065657X.1996.10701816

Michel, F. C. J., Pecchia, J. A., Rigot, J., & Keener, H. M. (2003). Mass and nutrient losses during composting of dairy manure with sawdust versus straw amendment. *Compost Science and Utilization, 23*, 1-33.

Misra, R. V., Roy, R. N., & Hiraoka, H. (2003). *On-farm composting methods*. FAO Corporate Document Repository. Retrieved from http://www.fao.org/docrep/007/y5104e/y5104e05.htm

Mondini, C., Sinicco, T., & Cayucla, M. L. (2010). Mineralization dynamics and biochemical properties following application of organic residue to soil. *19th World Congress of Soil Sc. soil solutions for a changing World*. August 1-6, 2010, Brisbane, Australia.

Neves, L., Ferreira, V., & Oliveira, R. (2009). Co-composting cow manure with food waste: The influence of lipid content. *World Academy of Science, Engineering and Technology, 58*, 986-991.

Ogunwande, G. A., Osunade, J. A., & Ogunjimi, L. A. O. (2008). Effects of carbon to nitrogen ratio and turning frequency on composting chicken litter in turned windrow piles. *CIGR E-journal, X*, 1-16.

Postma, R., Erp, P. V., & Van, E. P. (1998). Nutrient management in arable farming and outdoor vegetable growing. *Mestoto Hen*, 13-19. CAB Abstract 1998/99.

Petric, I., Sestan, A., & Sestan, I. (2009). Influence of wheat straw addition on composting of poultry manure. *Process Safety and Environmental Protection, 78*, 206-212. http://dx.doi.org/10.1016/j.psep.2009.02.002

Petrica, I., Šestan, A., & Šestan, I. (2009). Influence of initial moisture content on the composting of poultry manure with wheat straw. *Biosystems Engineering, 10*, 125-134. http://dx.doi.org/10.1016/j.biosystemseng.2009.06.007

Qiao, L., & Ho, G. (1997). The effects of clay amendment and composting on metal speciation in digested sludge. *Water Research, 31*, 951-964. http://dx.doi.org/10.1016/S0043-1354(96)00289-8

Rao Bhamidimarri, S. M., & Pandey, S. P. (1996) Aerobic thermophilic composting of piggery solid wastes. *Water Science and Technology, 33*(8), 89-94. http://dx.doi.org/10.1016/0273-1223(96)00265-X

Schulz, H., Dunst, G., & Glaser, B. (2013). Positive effects of composted biochar on plant growth and soil fertility. *Agronomy for Sustainable Development, 33*, 817-827.

http://dx.doi.org/10.1007/s13593-013-0150-0

Schulz, H., Dunst, G., & Glasser, B. (2014). No effect of co-composted biochar on plant growth and soil properties in a greenhouse experiment. *Agronomy, 4*, 34-51. http://dx.doi.org/10.3390/agronomy4010034

Smith, S. R. (2009). A critical review of the bioavailability and impacts of heavy metals in municipal solid waste composts compared to sewage sludge. *Environment International, 35*, 142-156. http://dx.doi.org/10.1016/j.envint.2008.06.009

Steiner, C., Das, K. C., Melear, N., & Lakly, D. (2010). Reducing nitrogen loss during poultry litter composting using biochar. *Journal of Environmental Quality, 39*. http://dx.doi.org/10.2134/jeq2009.0337

Steiner, C., Teixeira, W., Lehmann, J., Nehls, T., de Macêdo, J., Blum, W., & Zech, W. (2007). Long term effects of manure, charcoal and mineral fertilization on crop production and fertility on a highly weathered Central Amazonian upland soil. *Plant and Soil, 291*, 275-290. http://dx.doi.org/10.1007/s11104-007-9193-9

Singh, J., & Kalamdhad, A .S. (2013). Bioavailability and leachability of heavy metals during composting. A review. *International Research Journal of Environmental Science, 2*(4), 59-64.

Thyagarajan, L. P., Meenambal, T., Mangaleshwaran, L., & Lashminarasimaiah, N. (2010). Recycling of pulp and paper industry sludge with sawdust by aerobic composting method. *Nature Environment and Pollution Technology, 1*, 149-154.

Tiquia, S. M., Richard, T. L., & Honeyman, M. S. (2000). Effects of windrow turning and seasonal temperatures on composting of hog manure from hoop structures. *Environmental Technology, 21*, 1037-1046. http://dx.doi.org/10.1080/09593332108618048

Tiquia, S. M., Tam, N. F. Y., & Hodgkiss, I. J. (1996). Microbial activities during composting of spent pig-manure sawdust litter at different moisture contents. *Bioresource Technology, 55*, 201-206. http://dx.doi.org/10.1016/0960-8524(95)00195-6

Tiquia, S. M., & Tam, N. F. Y. (2000). Fate of nitrogen during composting of chicken litter. *Environmental Pollution, 110*, 535-541. http://dx.doi.org/10.1016/S0269-7491(99)00319-X

Tiquia, S. M., & Tam, N. F. Y. (2002). Characterization and composting of poultry litter in forced aeration piles. *Process Biochemistry, 37*, 869-880. http://dx.doi.org/10.1016/S0032-9592(01)00274-6

Tripetchkul, S., Pundee, K., Koonsrisuk, S., & Akeprathumchai, S. (2012). Co-composting of coir pith and cow manure, initial C/N ratio Vs physicochemical changes. *International Journal of Recycling of Organic Waste in Agriculture, 1*(15). http://dx.doi.org/10.1186/2251-7715-1-15

Tuomela, M., Vikman, M., Hatakka, A., & Itavaara, M. (2000). Biodegradation of lignin in a compost environment: A review. *Bioresource Technology, 72*(2), 169-183. http://dx.doi.org/10.1016/S0960-8524(99)00104-2

Vallini, M. G., & Pera, A. (1983). The biology of cornposting: A review. *Waste Management Research, 1*, 157-176. http://dx.doi.org/10.1177/0734242X8300100118

Whalen, J. K., Chang, C., Clayton, G. W., & Carefoot, J. P. (2000). Cattle manure amendments can increase the pH of acid soils. *Soil Science Society of America Journal, 64*, 962-966. http://dx.doi.org/10.2136/sssaj2000.643962x

Walker, F. (2004). On-farm composting of poultry litter. The Agricultural Extension Service, The University of Tennessee Institute of Agriculture.

Wang, P., Changa, C. M., Watson, M. E., Dick, W. A., Chen, Y., & Hoitink, H. A. J. (2004). Maturity indices for composted dairy and pig manure. *Soil Biology and Biochemistry, 36*, 767-776. http://dx.doi.org/10.1016/j.soilbio.2003.12.012

Wong, J. W. C., & Selvam, A. (2006). Speciation of heavy metals during co-composting of sewage sludge with lime. *Chemosphere, 63*, 980-986. http://dx.doi.org/10.1016/j.chemosphere.2005.08.045

Wu, D. L., Liu, P., Luo, Y. Z., Tian, G. M., & Mahmood, Q. (2010). Nitrogen transformation during co-composting of herbal residues, spent mushrooms and sludge. *Journal of Zhejiang University Science, B*, 1862-1783.

Yamato, M., Okimori, Y., Wibowo, I., Anshori, S., & Ogawa, M. (2006). Effects of the application of charred bark of Acacia mangium on the yield of maize, cowpea and peanut, and soil chemical properties in South Sumatra, Indonesia. *Journal of Soil Science and Plant Nutrition, 52*, 489-495.

http://dx.doi.org/10.1111/j.1747-0765.2006.00065.x

Yeoh, C. Y., Chin, N. L., & Tan, C. S. (2011). Co-composting of palm oil mill wastes. *Journal of Food Agriculture and Environment, 9*(3-4), 880-885.

Zhang, Y., & He, Y. (2006). Co-composting solid swine manure with pine sawdust as organic substrate. *Bioresource Technology, 97*, 2024-2031. http://dx.doi.org/10.1016/j.biortech.2005.10.004

Zhang, R., Ishibashi, K., & Day, D. L. (1991). Experimental study of microbial decomposition in liquid swine manure, and generation rates of ammonia. *Proc. Livestock Waste Management Conference*. ASAE St. Joseph, Michigan.

Zhu, N. (2007). Effect of low initial C/N ratio on aerobic composting of swine manure with rice straw. *Bioresource Technology, 9*, 89-93.

Zorpas, A. A., Constantinides, T., Vlyssides, A. G., Haralambous, I., & Loizidou, M. (2000). Heavy metal uptake by natural zeolite and metals partitioning in sewage sludge compost. *Bioresource Technology, 72*, 113-119. http://dx.doi.org/10.1016/S0960-8524(99)00110-8

7

Biodegradation of Petroleum Hydrocarbons in a Tropical Ultisol Using Legume Plants and Organic Manure

B. E. Udom[1] & B. O. Nuga[1,2]

[1] Faculty of Agriculture, Department of Crop and Soil Science, University of Port Harcourt, Port Harcourt, Nigeria

[2] Nukot Consulting Services Inc., Winnipeg, Manitola, Canada

Correspondence: B. E. Udom, Faculty of Agriculture, Department of Crop and Soil Science, University of Port Harcourt, P.M.B. 2353, Port Harcourt, Nigeria. E-mail: ebassidy@yahoo.com

Abstract

Global emphasis on food security and soil health should consider rehabilitation of degraded lands, especially where oil contamination limits the use of such lands. Three legume plants (*Gliricidia sepium*, *Leucaena leucocephala* and *Calapoconium caerulean*) alone or along with 0.5% (w/w) (equivalent of 10 tons ha^{-1}) poultry manure, were used to treat the soil, in which the oil residuals were monitored for three years. Results showed that significantly high levels of residual total petroleum hydrocarbon content (RTPHs) persisted in the non-amended soil after 36 months. At 3 months, 43% of RTPHs was removed by *Gliricidia sepium* and *Leucaena* luecocephala along with poultry manure. Net loss of RTPHs after 12 months was 69% for *Gliricidia sepium* and *Leucaena luecocephala* combined with poultry manure and only 38% for A_5, explaining that, degradation of petroleum hydrocarbon by indigenous soil micro-organisms was very low. At 18 months when additional load of oil was applied, the degradation rate increased from 71.7 mg kg^{-1} day^{-1} to 142 mg kg day^{-1} within 6 months for all the legume plants along with poultry manure. This explained increased in number of hydrocarbon-degrading micro-organisms due to increase in oil load. The oil inhibited germination and yield of maize crop completely. Treatments with legume plant along with poultry manure significantly ($p < 0.05$) increased both germination and yield of maize crop. The effects of legume plants combined with poultry manure were the better treatment than legume or poultry manure alone in improving the soil properties for growth and performance of maize crop.

Keywords: contaminated soil, residual hydrocarbons, degradation rate, poultry manure, legume plants

1. Introduction

Biodegradation of petroleum hydrocarbons is presumed to be based on the stimulation of microbial degradation in the rhizosphere. Effective soil treatment technologies require or suggest that residual total petroleum hydrocarbons (RTPHs) concentration in soils be reduced to 1000 mg kg^{-1} or in some areas, below 100 mg kg^{-1} (USEPA, 1991; Ram et al., 1993). The treatment technologies are often developed and evaluated in order to conform to the regulatory demands and International standards (Lee & Banks, 1993). However, many of the standard treatment methods available for cleaning-up soils contaminated with petroleum hydrocarbons have been limited in their application, are prohibitively expensive, or may be only partially effective (Merkl et al., 2005). The high cost associated with a number of these methods, has also been a major limitation to the sustainability of such methods (Kelly et al., 1992; Van Gestel et al., 1992). The excavation and soil washing methods are often extremely costly and insufficient risk-reducing (Van Gestel et al., 1992; Alloway & Ayres, 1997). Literature reports many examples in which bacterial strains and microbial systems have been successfully utilized to reduce and/or transform selected pollutants in petroleum-contaminated soils under laboratory conditions (Gallizia et al., 2003; Harayama et al., 2004).

Naturally occurring micro-organisms may be able to biodegrade hydrocarbons and other organic compounds in unsaturated soil if the level of contaminations is low and does not produce toxicity for the active bacteria (Molina-Barahona et al., 2004). Biodegradation of petroleum hydrocarbons in soil can be limited by many factors, for example: type of micro-organisms, nutrients, pH, temperature, moisture, oxygen and soil properties. The use of organic wastes, such as cow dung and pig droppings (Okurumeh & Okiemen, 1998) and rubber processing sludge (C. O. Okiemen & F. E. Okiemen, 2002), have also been reported to give positive results in *in situ* remediation of oil-contaminated soils.

The indiscriminate discharge of petrol oils and grease into open vacant plots and farm lands is becoming an acute environmental problem in Nigeria, particularly, when large areas of agricultural lands are affected (Anoliefo & Vwioko, 1995). In most cases, the soils may remain unsuitable for crop growth for months or years, until oil is degraded to a tolerable level. Depletion in nutrient status, inhibition of microbial activities and reduced oxygen has been reported in spent-oil-contaminated soils (Kirk et al., 2005). Formation of waxy texture, reduced water affinity to soil aggregates and formation of hydrophobic micro-aggregates with clay surfaces are a few other problems associated with soils contaminated with petroleum hydrocarbons (Anoliefo & Vwioko, 1995). The oil succeeding water in the competition for pore spaces and reduction of water film thickness around macro-aggregates decreased both saturated and unsaturated hydraulic conductivity Rasiah et al., 1990; Amadi et al., 1993).

Major report on the use of plants to clean-up soils contaminated with petrol oils and grease has been concentrated on heavy metal removal (Gallizia et al., 2003; Harayama et al., 2004). Whereas, the use of legume plants, or along with organic manure to remove petroleum hydrocarbons in soils, is not well known (Mager & Hernandez-Valencia, 2003; Revera-Cruz et al., 2004). The method is effective, economical and ecologically acceptable. It is also rapidly deployable in a wide range of physical settings (Catallo & Portier, 1992; Kelly et al., 1992; Ram et al., 1993). Plants can enhance microbial degradation of hydrocarbons by providing oxygen in the root area along root channels and loosened soil aggregates. Legumes plants are especially promising in this method, because they are nitrogen independence, which is of significance in oil-contaminated soils (Yeung et al., 1997; Revera-Cruz et al., 2004). The use of some tropical legume plants to remove petroleum hydrocarbon from contaminated sites with a view to making it available for crop production is of research interest. Therefore, the concern in this study was to bridge the gap in knowledge regarding the use of tropical legume plant species and poultry manure for *in situ* degradation of petroleum hydrocarbons. It will provide valuable data on the effectiveness of *Gliricidia sepium* and *Leucaena leucocephala* and *Galapogonium caerulean* along with or without poultry manure in cleaning up soil contaminated with petroleum hydrocarbons.

2. Materials and Methods

2.1 Site Description and Application of Treatments

The study was carried out on 45 plots (each measuring 2.5 × 1.5 m) at the University of Nigeria, Nsukka, Research Farm (Lat. 06°52′N, and Long. 07°24′E). The soil is sandy loam (*Typic Kandiustult*) derived from False-Bedded sand stone (Nwadialo, 1989). The average slope of the site is < 5%. Mean annual rainfall is more than1700 mm with maximum temperature of 32 °C (Inyang, 1978).

2.2 Application of Oil and Treatments

The plots were impacted with 5% (w/w) (equivalent of 50,000 mg kg^{-1}) mono- and multi-grade crankcase oils from petrol and diesel engines, together with gear oils and transmission fluids in a single dose, each for two years. By the second year, oil contaminated plots had spent oil application load of 100,000 mg kg^{-1}, representing a total oil load of 10% (w/w). Three (3) legumes: Biodegradation was enhanced using three (3) legume plants: *Gliricidia sepium*, *Leucaena leucocephala*, and *Galapogonim caerulean* alone or combinations with 0.5% (w/w) (equivalent of 10 tons ha^{-1}) of poultry manure. The legume plants used in the experiment have the ability to grow fast, generate high biomass, nitrogen-independent and encourage high population of hydrocarbon-degrading micro-organisms in the rhizosphere (Anderson et al., 1993). The experiment was arranged as a Randomized complete block design (RCBD), with nine (9) treatments in five replications viz:

Treatments:	Designation:
- Uncontaminated soil (Control)	C
- Spent oil only	A_5
- Spent oil + *Calapogonium spp*	A_5 +Ca
- Spent oil + *Gliricidia spp*	A_5 + Gl
- Spent oil + *Leucaena spp*	A_5 + Le
- Spent oil + poultry manure	A_5 + Pm
- Spent oil+ *Calapogonium* + poultry manure	A_5 + Ca +Pm
- Spent oil + *Gliricidia* + poultry manure	A_5 + Gl +Pm

- Spent oil+*Leucaena* +poultry manure $A_5 + Le + Pm$

The legume plants along with poultry manure were introduced to the plots seven days after oil contamination during the early rains to allow for incubation. The second applications of 5% (w/w) spent oil was carried out at 360 days after the first application.

2.3 Planting

The *Calapogonium caerulean* was planted at 30 × 90 cm spacing, (density of 37,000 plants ha^{-1}), *Gliricidia sepium* and *Leucaena leucocephala* was planted at 1 m × 90 cm spacing (density of 11,100 plants ha^{-1}). Maize variety, (FASR-W) used as test crop was planted at 30 cm × 75 cm spacing.

2.4 Sampling

Disturbed soil samples were collected from the 0-30 cm depth using a soil auger at 3, 6, 12, 18, 24 and 36 months after oil-contamination for laboratory studies. The implications of the oil and treatments on maize performance were evaluated using germination index at 2-4 weeks after planting and grain yield at maturity

2.5 Determination of total Hydrocarbon

Total Hydrocarbon (TH) at each sampling period was determined gravimetrically by toluene extraction (cold extraction) method described by Odu et al. (1989). The liquid phase of the cold extract was measured with a spectrophotometer and fitted into a standard curve derived from fresh spent oil treated with toluene.

2.5 Biodegradation Rate of Total Hydrocarbon Loss

Average biodegradation rates (mg kg^{-1} day^{-1}) of hydrocarbons were calculated according to the method of Yeung et al. (1997) as:

$$\Delta HL = HC_{ini} - HC_{end} / Time_{inc} \quad (1)$$

Where: ΔHL is the average hydrocarbon content in the soil (mg kg^{-1}), HC_{ini} is the initial hydrocarbon content in the soil (mg kg^{-1}), HC_{end} is the hydrocarbon content when the experiment ended (mg kg^{-1}), and $Time_{inc}$ is the degradation time (d).

2.6 Measurement of Unsaturated Hydraulic Conductivity

Unsaturated hydraulic conductivity, $K_{(o)}$ was predicted from saturated hydraulic conductivity, (K_{sat}) and soil moisture retention characteristics data as proposed by Campbell (1974), and confirmed to be reliable by Rasiah et al. (1990), in this procedure, the pressure potential H_o is related to the relative saturation water content (Q_v/Q_s) by a power function equation:

$$H_o = c\,[Q_v/Q_s]^{-b} \quad (2)$$

and

$$K_{(o)} = K_{sat}\,[Q_v/Q_s]^{(2b+3)} \quad (3)$$

Where: b and c are fitting parameters, and H_o, $K_{(o)}$, Q_v, Q_s and K_{sat} are the pressure potential, unsaturated hydraulic conductivity (cm hr^{-1}), volumetric moisture content at any specific matrix potential (cm^3 cm^{-3}), volumetric moisture content at saturation (cm^3 cm^{-3}), and saturated hydraulic conductivity (cm hr^{-1}) respectively. Soil water matrix potential of -6 Kpa (60 cm water suction or tension), representing field capacity which drained pores > 50 μm equivalent cylindrical diameter was used. The b estimate obtained in Equation (2) was used in Equation (3) to predict $K_{(o)}$ for the soil. Soil moisture retention was measured by the hanging column method as described by Galganov et al. (1993). Saturated hydraulic conductivity (K_{sat}) was determined by the constant-head permeameter technique (Klute & Dirksen, 1986). Volume of water draining out was measured overtime, until flow was constant, at which the final flow rate was determined from the equation

$$K_{sat} = Q/AT \times L/\Delta H \quad (4)$$

Where Q is the volume of water (cm^3) that flows through across-sectional area A (cm^2) in time T (sec.), ΔH is the hydraulic head difference, and L is the length of core sample.

2.7 Data Analysis

Statistical analyses were carried out using the SAS Software (SAS Institute, 2001). Means were separated using the LSD (Fishers protected test) (K. A. Gomez & A. A. Gomez, 1984).

3. Results and Discussions

The soil data before the experiment showed that the soil is sandy loam. The sand, silt and clay contents of the top soil are 820, 60 and 120 g kg^{-1} respectively (Table 1). The low clay and silt content is an indication of the highly weathered, low fertility soils of the southern Nigeria with characteristic low buffering capacity to attenuate contaminants (Alloway & Ayres, 1997). The soil also has low C:N ratio and low water holding capacity (Table 1). The spent oil and poultry manure (Pm) have Organic carbon content in spent oil and poultry manure was 31.5 and 28.6 g kg^{-1} respectively. The pH of the soil was 4.7 and that of poultry manure was 6.5.

Table 1. Some characteristics of the site, poultry manure and spent oil used in the experiment

Parameter	**Unit**	**Soil**	**Poultry Manure**	**Spent Oil**
Sand (2000-50 μm)	g kg^{-1}	820	-	-
Silt (50-2 μm)	g kg^{-1}	60	-	-
Clay (< 2μm)	-	120	-	-
Texture	g kg^{-1}	Sandy loam	-	-
Organic carbon	g kg^{-1}	6.84	28.6	31.5
Total N	g kg^{-1}	0.76	4.5	2.79
C:N	-	9	6	11
p^H (H_2O)	-	4.7	6.5	-
Na	C mol kg^{-1}	0.10	1.94	-
Exchangeable acidity	C mol kg^{-1}	2.6	-	-
Saturated hydraulic conductivity	cm hr^{-1}	20.44	-	-
Bulk density	g cm^{-3}	1.52	-	-
Water holding capacity	cm^3 cm^{-3}	0.31	-	-
Total porosity	%	51.	-	-

3.1 Residual Total Hydrocarbons

The distributions of residual total hydrocarbons in the soil as modified by the treatments are shown in Tables 2 and 3. Very high amount of residual total hydrocarbon (RTHs) (35064 mg kg^{-1}) persisted in the contaminated soil after 36 months (Table 2). In 12 months, the RTH due to the effects of A_5 + Gl + Pm, A_5+ Le + Pm and A_5 + Ca + Pm, were 15471, 15549 and 15816 mg kg^{-1} respectively, representing 69%, 69% and 68% reductions in compared to high amount of 30648 mg kg^{-1} found in the A_5 soil. When additional 5% (w/w) load of spent oil was added to the soil after 12 months, residual RTPHs in the soil at 24 month were significantly low ($p < 0.05$), in A_5 + Gl + Pm, A_5 + Le + Pm and A_5 + Ca + Pm soils. Treatment with only poultry manure (Pm) or legume plants alone did not show significant reductions in RTPHs (Table 2). This implied that biodegradation of RTPHs was higher when the *Gliricidia spp*, *Leucaena spp* and *Calapogonium spp* were used along with 10 tons ha^{-1} of poultry manure. It could be inferred that the significant reductions in total hydrocarbon found in soils amended with the legume plants along with poultry manure may have been possible due to enhanced positive changes in the physico-chemical conditions of the soils by the legume plants. It is also believed that the plants may have participated in biodegradation of hydrocarbons via support of symbiotic root-associated micro-organisms which was earlier reported by Stamps et al. (1994) and Ensley et al. (1997). The *Gliricidia* and *Leucaena* showed better promise than *Galapogonium* in reducing RTPHs in the soil, indicating that different species of plants have varying effects on rhizosphere micro-organisms and on hydrocarbon degradation activities in the soil (Merkl et al., 2005). As low as 14,936 mg kg^{-1} total hydrocarbons, representing 30% was degraded naturally by the native soil microorganisms after 36 months (Table 2).

Table 2. Changes in total hydrocarbon content (THC) of the soil as affected by treatment after 36 months

Treatments	THC (mg kg^{-1}) Months after oil application					
	3	12	18	24	36	Mean
A_5	35492	30648	41033	36416	31731	35064
A_5 +Gl	34784	17742	36617	29011	20619	27755
A_5 +Le	34652	17886	36214	29930	21174	27971
A_5 +Ca	33964	17421	36347	30662	24366	28552
A_5 +Pm	34011	16638	39118	31457	28694	29984
A_5 +Gl + Pm	28413	15471	35473	21974	20416	24349
A_5 +Le + Pm	28519	15549	35718	22603	20544	24587
A_5 +Ca + Pm	28944	15816	35736	23146	20712	24871
C	2390	2075	1964	19103	19004	2048
Mean	29018	16583	33136	25224	21128	-

Note. Total loss (%) = [Spent oil loading - Residual THC (Treatment)/Spent oil loading] × 100; LSD (0.05): Treatment = 9646, Month = 125.318, T × M = 594.437.

3.2 Quantification of Total Hydrocarbon Degradation Rate

The net loss and degradation rate of the RTPHs is shown in Table 3. At 3 months, net loss of RTPHs was 14% and degradation rate of 240 mg kg^{-1} day^{-1} for *Gliricidia* and *Leucaena spp* along with 10 tons ha^{-1} poultry manure. When net degradation rate of RTPH was 240 mg kg^{-1} day^{-1}, then by calculations, 50,000 mg kg^{-1} total petroleum hydrocarbon would have been completed removed from the soil in 208 days, if the soil is to be treated with *Gliricidia* or *Leucaena spp* supplemented with 10 tons ha^{-1} poultry manure. There were progressive reductions in residual total hydrocarbons, particularly in the amended soils. Degradation rates decreased during 12 months, because, the numbers of hydrocarbon utilizing micro-organisms have usually been found to be high in soil immediately and a few months after oil application (Molina-Barahona et al., 2004). Degradation rate under natural conditions (A_5), was 161 mg kg^{-1} day^{-1} in the first 3 months and drop to 72 mg kg^{-1} day^{-1} at 12 months (Table 3). This implies by calculations that it will take more than 674 days for 50,000 mg kg^{-1} total petroleum hydrocarbon to be degraded naturally from the top soil. This further confirmed that some naturally occurring indigenous micro-organisms which degrade petroleum hydrocarbons by nature exist in some tropical soils environments as reported earlier by Wang and Bartha (1990); Wilson and Jones (1992).

Treatment with *Gliricidia sepium* along with poultry manure increased the RTPHs degradation rates from 127.9 mg kg^{-1} in 12 months to 443.5 mg kg^{-1} day^{-1} in 24 months and 442.1 mg kg^{-1} in 36 months. This was not surprising, because the legume plant residues may have acted as bacterial biomass suppliers, which supported the high removal petroleum hydrocarbons from the soil. Improvement in intrinsic properties of the soil, such as the unsaturated hydraulic conductivity (Table 4), by the legumes and poultry may have enhanced rapid biodegradation of the RTPHs.

3.3 Unsaturated Hydraulic Conductivity

The predicted unsaturated hydraulic conductivity in the A_5, at the vicinity of saturated water content was in the order of magnitude less than in the uncontaminated plots (C) (Table 4). The decrease in unsaturated hydraulic conductivity may have been due to the formation of oily scum or coatings on soil aggregates which acted as a barrier to water flow in the soil. Reduction of water film thickness around the macro-aggregates which lead to retarded movement of water into and out of macro-aggregates may led to low unsaturated hydraulic conductivity observed in the A_5 soil. The A_5 + Gl + Pm treatment significantly ($P < 0.05$) increased the unsaturated hydraulic conductivity $K_{(0)}$ of the soils from 71% in 3 months, to 602% in 36 months after oil contamination compared with the (A_5). Generally, treatments improved the unsaturated hydraulic conductivity of the soil in the order of A_5 + Gl + Pm > A_5 + Ca + Pm > A_5 + Le + Pm > A_5 + Gl > A_5 + Pm > A_5 + Ca > A_5 + Le.

Table 3. Degradation rate of residual total petroleum hydrocarbons ($RTPH_s$) soil as influenced by treatments

Treatments	Net Loss of TPHs (%) Months after application					TPHs Degradation Rate (mg kg^{-1} day^{-1}) Months after application				
	3	12	15	24	36	3	12	18	24	36
A_5	-	-	-	-	-	131.2	71.7	327.6	353.2	379.3
A_5 + Gl	1.4	26.1	4.4	7.4	7.1	169.1	109.5	352.1	394.4	441.0
A_5 + Le	1.7	25.8	4.8	6.5	10.5	170.5	118.9	354.4	389.3	437.9
A_5 + Ca	3.1	26.8	4.7	5.7	7.3	178.5	120.7	353.6	385.2	420.2
A_5 + Pm	3.0	28.3	1.9	4.9	3.0	177.7	123.6	338.2	380.8	396.2
A_5 + Gl + Pm	14.2	30.7	5.5	14.4	11.3	239.9	127.9	358.5	443.5	442.1
A_5 + Le + Pm	14.0	30.5	5.3	13.8	11.2	238.7	127.6	357.1	430.0	441.4
A_5 + Ca + Pm	13.1	30.0	5.3	13.3	11.0	234.0	126.6	357.0	427.0	440.5

Note. LSD (0.05): Treatment = 1.642, Months = 19.971, T × M = 10.442; Net loss of TPHs = % loss in TPHs (Treatment) - % loss in oil (Control); TPHs Degradation = initial TPHs – TPHs at the end; TPHs = total petroleum hydrocarbons.

Table 4. Unsaturated hydraulic conductivity of the oil contaminated soil as influenced by the treatments

Treatment	**$K_{(o)}$ (cm hr^{-1}) Months after oil application**						
	3	**6**	**12**	**18**	**24**	**30**	**36**
A_5	9.38	4.32	2.51	2.19	3.76	2.83	2.87
A_5 + Gl	12.09	10.49	17.47	14.23	17.93	18.82	19.56
A_5 + Le	9.11	13.05	12.48	12.67	15.17	16.92	17.12
A_5 + Ca	10.28	12.00	14.66	16.18	17.06	19.06	17.69
A_5 +Pm	12.88	16.18	17.55	20.78	18.07	19.29	19.77
A_5 +Gl + Pm	15.99	15.58	18.85	17.06	19.55	20.57	20.16
A_5 +Le + Pm	14.81	16.33	18.74	15.26	19.44	19.72	19.88
A_5 +Ca + Pm	16.05	16.97	16.63	18.62	20.16	18.09	20.41
C	17.13	10.69	10.49	14.12	9.87	11.18	11.26
LSD (0.05)	1.98	1.88	1.97	1.76	1.99	1.64	1.86

Note. $K_{(o)}$ = Unsaturated hydraulic conductivity.

The legume plants along with poultry manure significantly ($P < 0.05$), increased unsaturated hydraulic conductivity, an indication that soil water characteristics and availability to plant was enhanced. This way, water stress to plants is usually associated with oil contaminated soil was reduced. This further supports the assertion of Joner and Leyval (2004), that micro-organisms in the rhizosphere may have reduced plant stress through an increase in water availability which further enhanced oxidative enzyme production, and increased the volume of soil being remediated.

3.4 Effects on Germination Index and Yield of Maize

The germination index was adversely inhibited in the non-amended soil by the oil through the 3-year planting (Table 5). The few maize plants that germinated died after a few weeks. Hence, no yield was obtained in A_5. Several factors may have contributed to the death of the maize plants, among which may include: lack of adequate oxygen, decreased in water availability, crusting and other unsatisfactorily soil conditions due to the oil. Similar observations were reported by Molina-Barahoma et al. (2004) on maize and sugar cane. Significantly higher percent germination were found in the plots treated with the legume plants along with poultry manure. At 36 months, (corresponding to third planting season), germination counts rose to 93% and 90% for *Gliricidia* and

Leuceana with poultry mamure respectively. Maize yield were 4.91, 8.25 and 6.46 tons ha^{-1} respectively during the first, second and third planting seasons for A_5 + Gl + Pm. There was no yield in A_5 and Control soil in the three planting seasons, most probably caused by either nutrient deficiencies or adverse effects of the hydrocarbons on the soil physical properties.

Table 5. Effects of treatments on germination and grain yield of maize

Treatment	Maize grain yield (tons ha^{-1})			Germination count (%)		
	1st Planting	2nd Planting	3rd Planting	1st Planting	2nd Planting	3rd Planting
A_5	0.0^a	0.0^a	0.0^b	41^a	36^b	34^a
A_5+Gl	3.06^c	5.16^b	4.18^a	57^b	53^a	67^a
A_5+Le	2.32^a	5.05^b	4.12^a	55^b	50^d	65^a
A_5+Ca	1.87^a	4.76^b	3.47^a	60^b	63^c	89^b
A_5+Pm	4.01^b	6.23^c	4.50^b	58^b	63^c	63^a
A_5+Gl+Pm	4.91^c	8.25^b	6.46^a	62^b	80^d	93^b
A_5+Le+Pm	3.91^b	6.90^c	5.03^b	63^b	81^d	90^b
A_5+Ca+Pm	4.16^b	7.02^a	6.21^a	66^b	79^d	88^b
C	0.0^a	0.0^a	0.0^b	78^c	70^d	78^c

Note. Yield and germination count followed by different letters within the years are significantly different at $P < 0.05$.

4. Conclusion

Conclusions drawn from this study are that: Contamination of soil with spent oils and grease increased the total and residual hydrocarbons in soil. Hydraulic conductivity and water affinity to the soil aggregates were reduced which adversely affected maize germination and growth. High amounts of petroleum hydrocarbons were found in the non-amended soils after 36 months of oil contamination. By comparing the degradation effect, the legume plants combined with poultry manure were the better treatment than legume or poultry manure alone. Degradation effect by indigenous micro-organism was negligible. The *Gliricidia sepium* and *Leucaena leucocephala* are more promising species than *Calapogonium caerulean* in bioremediation of soils contaminated with petroleum products.

References

Alloway, B. J., & Ayres, A. C. (1997). *Chemical Principles of Environmental Pollut* (p. 395). Champman and Hall Publication.

Amadi, A., Dickson, A. A., & Maate, G. O. (1993). Remediation of oil pollution soil I: Effects of organic and inorganic nutrient supplement on the performance of maize (*Zea mays*). *Water, Air, Soil Pollution, 66*, 54-76. http://dx.doi.org/10.1007/BF00477060

Anderson, T. N., Guthrie, E. A., & Walton, B. T. (1993). Bioremediation in the rhizosphere. *Environ. Sci. Technol., 22*, 3620-3636.

Anoliefo, G. O., & Vwioko, D. E. (1995). Effects of spent lubricating oil on the growth of *Capsicum annum* L. and *Lycopersicon esculentum* Miller. *Environ. Pollut., 99*, 361-364. http://dx.doi.org/10.1016/0269-7491(95)93451-5

Campbell, G. S. (1974). A simple method of determining unsaturated conducting from moisture retention data. *Soil Sci., 117*, 311-314. http://dx.doi.org/10.1097/00010694-197406000-00001

Catallo, W. J., & Portier, R. J. (1992). Use of Indigenous and adapted microbial assemblages in the removal of organic chemicals from soils and sediments. *Water Sci. Technol., 25*, 229-237.

Ensley, B. O., Raskin, I., & Salt, D. E. (1997). Phytoremediation applications for removing heavy metal contamination from soil and water. In Sayler (Ed.), *Biotechnology in the sustainable environment* (pp. 59-64). New York: Plenum Press.

Gallizia, I., Mcklean, S., & Banat, I. M. (2003). Bacterial degradation of phenol and 2,4 – dichlorohenol. *J. Chem. Technol. Biotechnol., 78*, 959-963. http://dx.doi.org/10.1002/jctb.890

Galganov, Y. T., Toop, G. C., Ball, B. C., & Carter, N. R. (1993). Soil water desorption curves. Soil Sampling and Methods of Analysis. In M. R. Carter (Ed.), *Canadian Soc. Soil Sci.* (pp. 569-579). Lewis Publ. U.S.A.

Gomez, K. A., & Gumez, A. A. (1984). *Statistical procedures for Agricultural Research* (2nd ed., p. 680). Wiley Interscience publication, Singapore.

Harayama, S., Kasai, Y., & Hara, A. (2004). Microbial communities in oil-contaminated sea water. *Curr. Opin. Biotechnol., 15*, 205-214. http://dx.doi.org/10.1016/j.copbio.2004.04.002

Inyang, P. E. B. (1978). Climate regime. In G. R. K. Ofomata (Ed.), *Nigeria in Maps. Eastern States* (pp. 2-27). Ethiopia Publ. House. Benin City, Nigeria.

Joner, E. J., & Leyval, C. (2004). Influence of arbuscular mycorrhizal on clover and ryegrass grown together in a soil spiked with polycyclic aromatic hydrocarbons. *Mycorrhiza, 10*, 155-159. http://dx.doi.org/10.1007/s005720000071

Kelly, A., Pemock, S. J., Bohn, M. K., & White, M. K. (1992). *Expert software that matches remediation site and sites and strategy remediation* (pp. 183-198). U.S. department of Energy. Pacific Northwest Laboratories.

Kirk, J. L., Moutoglis, P., Klironomos, J., Lee, H., & Trevors, J. T. (2005). Toxicity of diesel fuel to germination, growth, and colonization of *Glomus intraradices* in soil and in vitro transformed carrot cultures. *Plant and Soil, 270*, 23-30. http://dx.doi.org/10.1007/s11104-004-1013-x

Klute, A., & Dirksen, C. (1986). Hydraulic conductivity and diffusivity. Laboratory methods. In A. Klute (Ed.), *Methods of Soil Analysis* (Part 1, 2nd ed., pp. 687-734). ASA-SSSA, Madison, WI.

Lee, E., & Banks, M. K. (1993). Bioremediation of Petroleum Contaminated soil using vegetation: A microbial soil. *J. Environ. Sci. Health, Part A., A 28*(10), 2187-2198.

Mager, D., & Hernandez-Valencia, I. (2003). *Use of Brachiana brizantha and panicum with crude oil* (p. 49). Paper presented at the 15th congress of Venezuela ecology Maryazine, Venezuela.

Merkl, N., Schultze-Kraft, R., & Infante, C. (2005). Assessment of tropical grasses and legumes for phytoremediation of petroleum–contaminated soils. *Water, Air and Soil Pollution, 165*, 195-209. http://dx.doi.org/10.1007/s11270-005-4979-y

Molina-Barahona, L., Rodriqueze-Vazquez, R., Hernandez-Perrez, O., Mendoza-Canty, A., & Albores, A. (2004). Diesel removal from contaminated soils by biostimulation and supplementation with crop residues. *Appl. Soil Ecology, 27*, 165-175. http://dx.doi.org/10.1016/j.apsoil.2004.04.002

Nwadialo, B. E. (1989). Soil-landscape relationship in Udi-Nsukka Plateau Nigeria, *Catena Verlag* (pp. 111-120).

Odu, C. T. L., Nwoboshi, L. C., Fagage, S. O., & Awani, P. E. (1989). *Post – Impact study of SPDC's Nun River 8 Delivering line oil spillage* (pp. 5-9). Final Report, SPDC. Nig.

Okieimen, C. O., & Okieimen, F. E. (2002). Effects of natural rubber processing sludge on the degradation of crude oil hydrocarbons in soil. *Bioresour. Technol., 82*, 95-97. http://dx.doi.org/10.1016/S0960-8524(01)00157-2

Okurumeh, O. A., & Okieimen, F. E. (1998). Effect of cow during and poultry dropping application on petroleum hydrocarbon degradation on soil. *Bioresor. Technol., 82*, 117-126.

Ram, N. M., Bass, D. H., Falotico, R., & Leahy, M. (1993). A Decision Framework for Selecting Remediation Technologies at Hydrocarbon Contaminated sites. *J. Soil Contam., 2*(2), 167-189. http://dx.doi.org/10.1080/15320389309383436

Rasiah, V., Voroney, R. P., Groenevelt, P. H., & Kachanoski, R. G. (1990). Modification in soil water retention and hydraulic conductivity by an oily waste. *Soil Technol., 3*, 367-372. http://dx.doi.org/10.1016/0933-3630(90)90017-W

Revera-Cruz, M. D., Ferrera-Corrato, R., Sanchez-Corrcia, P., Volke-Haller, V., Farandez-Linares, L., & Rodrigue-Vazquez, R. (2004). Decontamination of soils polluted with crude petroleum using indigenous microorganisms and aleman grass (Chinocloa Polyslachya Hitche). *Agrociencia, 38*, 1-12.

S.A.S. (statistical Analysis System). (2001). *Institute SAS/STAT User's Guide* (4th ed., Vol. 5). SAS Inst. Cary NC, USA.

Stamps, A. M., Han, K. H., Wilbert, S., Gordon, M. P., & Cunningham, S. D. (1994). Genetic strategies for enhancing phytoremediation. *Ann. New York Acad. Sci., 721*, 481-491. http://dx.doi.org/10.1111/j.1749-6632.1994.tb47418.x

USEPA (United States Environmental Protection Agency). (1991). *Research and Development (RD-681). EPA/600/M-91/049* (p. 49). Alternative Treament Tecnology Information Centre (ATTIC).

Van Gestel, C. A. M., Dirven-Van Breemen, E. M., & Kamerman, J. W. (1992). Evaluation of Decontaminated soils. *No. 216402005*. National Institute of Puplic Health and Environmental Protection. Bilthoven.

Wang, X., & Bartha, R. (1990). Effect of bioremediation of residues, activity and toxicity of soil contaminated by fuel spills. *Soil Biol. Biochem., 22*, 290-295. http://dx.doi.org/10.1016/0038-0717(90)90185-3

Wilson, S. C., & Jones, K. C. (1992). Bioremediation of soil contaminated with Polynuclear aromatic hydrocarbons, (PAHs): Review. *Environ. Pollut., 81*, 229-249. http://dx.doi.org/10.1016/0269-7491(93)90206-4

Yeung, P. Y., Johnson, R. L., & Xu, J. G. (1997). Bioremediation of petroleum hydrocarbons in soil as affected by heating and forced generation. *J. Environ. Qual., 26*, 1511-1516. http://dx.doi.org/10.2134/jeq1997.2661511x

8

Potential Impact of Climate Change Trends on Wheat Production and Mitigation Strategies in Afghanistan

Rajiv Sharma[1], Kai Sonder[2] & G. Sika[3]

[1] International Maize & Wheat Improvement Centre (CIMMYT), Kabul, Afghanistan

[2] International Maize & Wheat Improvement Centre (CIMMYT), Mexico

[3] International Maize & Wheat Improvement Centre (CIMMYT), Nairobi, Kenya

Correspondence: Rajiv Sharma, CIMMYT-Afghanistan, #157, Karte Parwan, Baharistan, Kabul, Afghanistan. E-mail: rk.sharma@cgiar.org

Abstract

Climate change is here and continues unabated. It is expected to affect different biological processes differently. Afghanistan has a varied climate profile and depends heavily on wheat, the staple food of its 30 million people. The irrigated wheat occupying about 1.35 million hectare and contributes major chunk of wheat to national granaries compared to rainfed wheat which is more opportunistic with an unpredictable and uncertain share in national harvest. Previous decades have already seen a reduction of up to 100 mm wheat season rainfall and more reductions are estimated by 2050. Also, wheat which is a cool season crop and is also estimated to face higher temperatures by up to 5 degree Celsius across several provinces in the country. The study predicts variable impact of the estimated changes on the productivity of irrigated and rainfed wheat in Afghanistan.

Keywords: wheat, Afghanistan, climate change, production, temperature, rainfall

1. Introduction

Climate is defined as a change in the average pattern of temperature, humidity, precipitation and other meteorological events in any given area over an extended period of time – typically decades or longer. Climate change may be due to natural internal processes or external forces such as modulations of the solar cycles, volcanic eruptions, and persistent changes caused by human activity in the composition of the atmosphere or in land use (IPCC, 2014).

Temperature has a profound impact on soil-plant-water relationships in agriculture, influencing the rate of photosynthesis, respiration and plant growth. For wheat farmers, it is the main determinant of the potential health of crops and plays a major role in decision-making about which wheat variety to plant and when, harvest times and yield potential.

2. Agriculture in Afghanistan

The importance of agriculture to Afghanistan's economy cannot be over emphasized. The sector employs 60 percent of the working population and in 2013 contributed more than 20 percent to the country's overall gross domestic product, which was valued at $20.7, billion according to the World Bank (World Bank, 2013). Statistics from the U.N. World Food Programme show that 7.6 million of its total 30.5 million people (WFP, 2014) are food insecure and 14 percent are considered to be borderline food-insecure.

However, Afghanistan, which has been embroiled in conflict for most of the past over three decades, has the potential to expand self-sufficiency and combat food insecurity notably by tackling the sector's low productivity. Exploiting this growth potential requires improving the sub-optimal productivity plaguing Afghan agriculture and further promoting the entrepreneurial capacity of semi-commercial farmers. To achieve increased production in agriculture to make a significant impact on rural areas, agriculture production needs to grow at a steady average rate of at least 5 percent per annum over the next decade. Therefore, improving production and farm income is vital to strengthening the contribution of the agriculture sector to aid in achieving economic growth and poverty reduction in Afghanistan.

3. Food Security

Afghanistan ranks among the top wheat consuming countries of the world. The wheat grain is the basic food staple in the country accounting for about 80 percent of the 3.2 million hectares of land under cereals' production in the country (MAIL, 2012). It also contributes a proportionate amount to overall cereal production. Food production capacity affects its availability in remote areas and is the basis for income generation for farmers in these areas. Afghanistan's wheat production levels have varied widely over the years, and although it came close to self-sufficiency by producing 96.6 percent of wheat consumed in 2009 due to exceptionally good weather, it has never been self-sufficient in wheat. In 2004, during a particularly bad year, Afghanistan produced only 53 percent of wheat consumed in the country (MAIL, 2009).

Out of 30.5 million, almost 11 million Afghans (36 percent) live below the national poverty line (World Bank, 2013), unable to meet basic needs and dietary requirements. Food insecurity is found also in urban areas, accompanied by malnutrition. Afghanistan was more severely affected by the food price crisis of 2007-2008 than other countries due to its high level of poverty. Overall, 27.7 percent of people face a crisis situation as far as food security is concerned (IPC, 2013). Wheat is synonymous with food security as it is the staple food crop supplying about 60 percent of caloric intake to an average Afghan with an average annual per capita consumption of about 200 Kilograms (440 pounds) (Persaud, 2012).

4. Boosting Yields

The concerted integrated efforts of research and development agencies have resulted in the positive production and productivity growth of this commodity (Figure 1). Afghanistan's wheat production trend is highly erratic, ranging from a high of more than 5 million metric tonnes in 2009 and 2012 to a low of 2.8 in 2008 and 3.3 in 2011 (MAIL, 2012). However, Afghanistan's population has increased by 10 million over the past decade (FAO, 2013). Afghanistan is reliant on such neighboring countries as Pakistan, Uzbekistan and Iran to meet its wheat needs. The Ministry of Agriculture, Irrigation and Livestock (MAIL) has estimated that Afghanistan would need to produce about 7 million metric tons of wheat by 2022 (Waziri et al., 2013) to achieve self-sufficiency. The prospects of achieving 2 million tonnes jump in wheat production looks bleak given that only 45 percent of wheat acreage is irrigated.

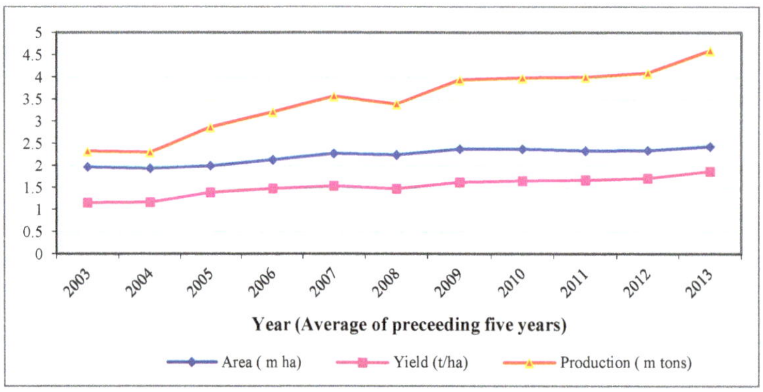

Figure 1. Five year rolling averages

However, rainfed wheat accounts for about 55% of wheat acreage in Afghanistan. It added substantially to the country's wheat harvest during such plentiful rainfall years as 2009 and 2012. Being opportunistic, rainfed cannot be a reliable contributor to Afghan wheat production, but needs to be supported with all available technology and policy support so as to become a more consistent contributor to wheat coffers of Afghanistan.

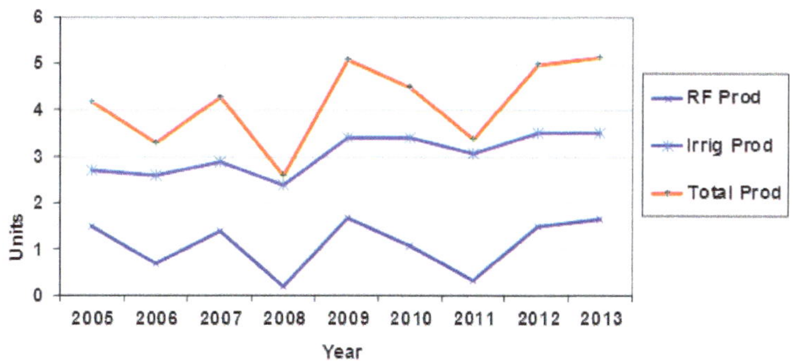

Figure 2. Rainfed & irrigated wheat production (m tons) & productivity (t/ha) trends from 2005 to 2012

Over the past ten years, rainfed wheat has contributed from a low of 7.7% (2008) to a high of about 35.7% (2005) of total wheat production in the country (FAO, 2014). This equals to an average contribution of 26% during past ten years. The contribution by rainfed wheat, thus, has been very erratic (Figure 2) and off course unpredictable with a coefficient of variation of about 53%. The average contribution of 26% looks impressive, however, a share of this or higher magnitude has come only in five out of ten years. Therefore, for assigning a credible role to rainfed wheat in food security assessment of Afghanistan, 15% contribution looks more realistic as rainfed wheat has contributed 15% or more than 15% of wheat production in eight of last ten years (FAO, 2013; MAIL, 2012). Ministry of Agriculture, Irrigation & Livestock (MAIL) of Afghanistan has estimated that country would need about seven million metric tons of wheat by 2022 to achieve self-sufficiency (Waziri et al., 2013). At a 15 per cent share, rainfed wheat could therefore potentially contribute 1.05 million metric ton and the irrigated wheat would have to contribute 5.95 million metric tons in 2022 to achieve self-sufficiency target.

The most viable option to increase wheat productivity in Afghanistan is to bring more farmland under irrigation using the best available options. Afghanistan brought 28,000 hectares of land under irrigation in the year 2011 (APR, 2012). At this rate, country can increase irrigated wheat acreage by about 7 percent every three years. Concerted efforts need to be made to bring more area under irrigation. Proven available technologies like sprinkler irrigation should be promoted to provide supplementary irrigation wherever possible.

5. Afghanistan Climate

Although Afghanistan is a landlocked mountainous country geographically located in central Asia, it is a member of the South Asian Association for Regional Cooperation (SAARC). The country has three distinct geographical regions divided by the Hindu Kush mountains that run on a diagonal from the northeast to the southwest forming central highlands that make up about two thirds of the country's terrain. The northern plains bordering Turkmenistan and southwestern plateau bordering Iran are the other two major geographical regions of the country

Afghanistan's climate can be typically characterized as semi-arid or arid steppe with cold winters and dry summer. The mountainous northeastern region is subarctic climate and experiences cold winters. The country receives most of its precipitation mainly in winter – at higher altitudes it falls as snow. Some eastern areas bordering Pakistan benefit from southeastern monsoon rains. However, the heavy rains do not extend to the southwest, and mountains on the Pakistan side shelter the Northeast. Temperature regimes are greatly modified by altitude. Lower elevations remain almost frost-free with very hot summers, whereas higher areas experience arctic temperatures in winter. The minimum temperature can drop lower than minus 20 degrees Celsius (minus 4 degrees Fahrenheit) in the central highlands while the maximum can be as high as 30 degrees Celsius in the same region during the summer. The maximum temperature can rise to above 40 degrees Celsius in the east and Southwest during the summer months.

6. Climate Trends

Wheat is a cool season crop and Afghanistan's climate offers ample scope for its successful cultivation. Wheat crops can be found in fields year round in different parts of the country. Winter wheat sowings can start in October in the central highlands whereas a spring sown wheat can be harvested after August in the same region. Temperature regimes and precipitation from October to June affect wheat crops in Afghanistan. Depending on how low winter temperatures tread, crop would grow slow or would be under snow for different periods. Similarly, rainfall distribution can also cause million ton variations (Sharma & Habibi, 2012) in country's wheat harvest. A study of average monthly rainfall and minimum and maximum temperatures from 1951 to 2009 (CCAFS, 2014) reveals (Figure 3) that the average monthly rainfall between October and June has reduced by up to 15 millimeters (0.039 inches) per month in the eastern Nangarhar province compared to just 1 millimeter per month in the southern provinces of Kandhar and Helmund. The average reduction in precipitation during these nine months has been about 100 millimeters for eastern Paktiya and Kabul provinces. Increase in mean minimum and mean maximum temperature has been about 2 degrees Celsius for Badakhshan, Ghor, Helmund, Herat and Kandahar provinces. Temperatures have remained relatively unaffected in Kabul and Paktya.

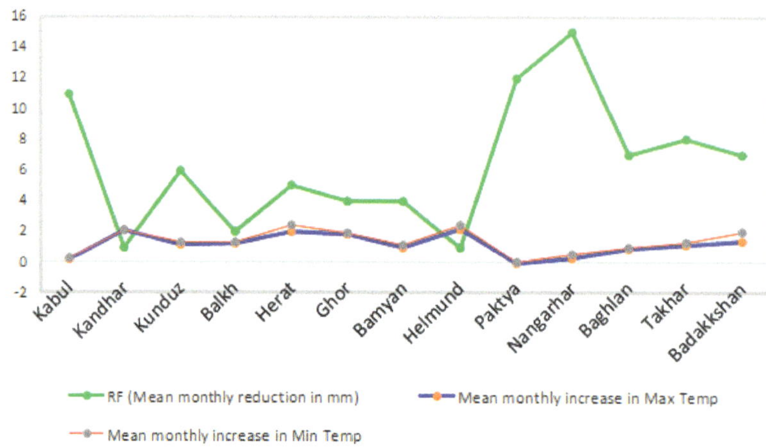

Figure 3. Change in rainfall (RF) and temperature (°C) averages of 2001-2009 compared to 1951-1960 averages at different research stations of Afghanistan

A perusal of wheat phenology reveals that precipitation between March and May is most crucial as grain development and filling takes place during this period. Figure 2 shows drastic dips in wheat production in the country owing mainly to poor precipitation. This is to be noted that although 45 percent of wheat acreage is irrigated, all wheat growing throughout the country benefits from rainfall during these three months.

Figure 4 shows the important relationship between the country's wheat production and the average monthly rainfall between October and June from 2002 to 2009, indicated by a very strong correlation of 0.89 proving the important role played by precipitation during these months. Similarly, national production figures for rainfed wheat are also strongly correlated with rainfall during three months as indicated by correlation coefficient of 0.89 (Figure 5). This finding shows that although the precipitation during the entire crop growth from October to June is important, the amount received in the early growth phase enable crop establishment and the formation of a larger yield potential, but the rains received during March to May allow the realization of established potential.

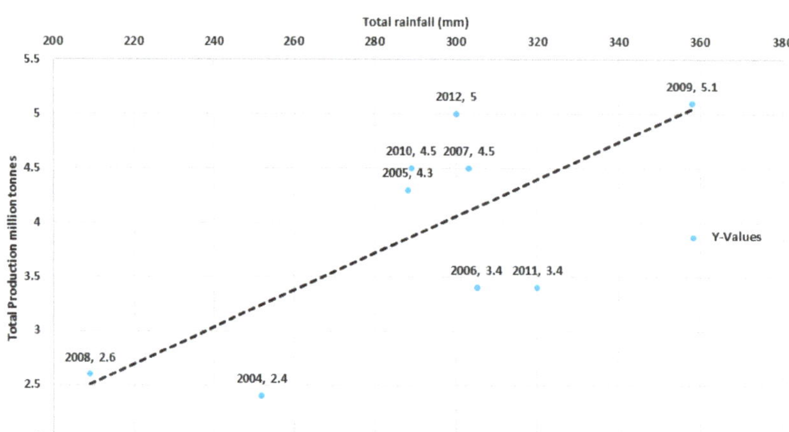

Figure 4. How does total rainfall relate to total production

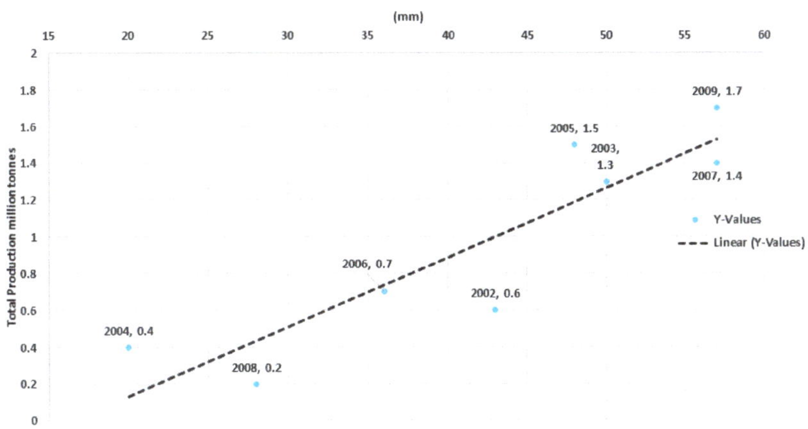

Figure 5. How does March to May rainfall relate to rainfed production

7. Potential Impact of Climate Change on Wheat

The world is experiencing climate change characterized in part by reduced precipitation and warmer temperatures (IPCC, 2014), the two parameters that greatly influence crop growth. Indeed, findings by the IPCC support claims that crop yields have been hit by negative impacts of climate change, expressing high confidence that it is not yet clear whether the balance of impacts has been negative or positive in high-latitude regions. In a recent report, the IPCC expresses medium confidence that climate change has negatively affected wheat and maize yields in many regions and in the global aggregate.

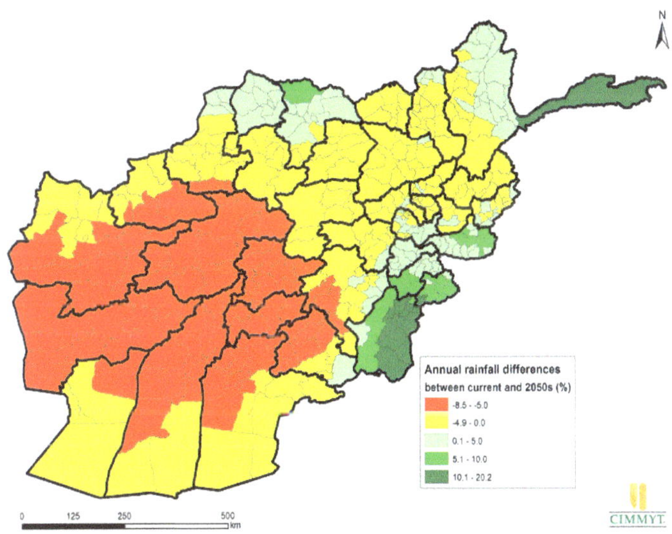

Figure 6. Projected % changes in rainfall in Afghanistan from now and 2050

Over the past 60 years, data from 13 research sites, on many of which CIMMYT currently conducts its trials, reveal reduced rainfall whereas temperatures have increased. Based on these data, Figures 6 to 8 present projected changes in annual mean rainfall, annual mean maximum temperature and annual mean minimum temperatures in Afghanistan from now until 2050 (CCAFS, 2014). Predictions indicate a potential reduction of average mean rainfall in some parts of Afghanistan by up to 8.5 percent, whereas about 20 percent of the country will in fact see an increase in annual mean precipitation by up to 20 percent, thereby benefiting wheat crop in some regions. Similarly, both mean minimum and mean maximum temperatures are expected to increase by up to 4.5 and 5 degrees Celsius, respectively.

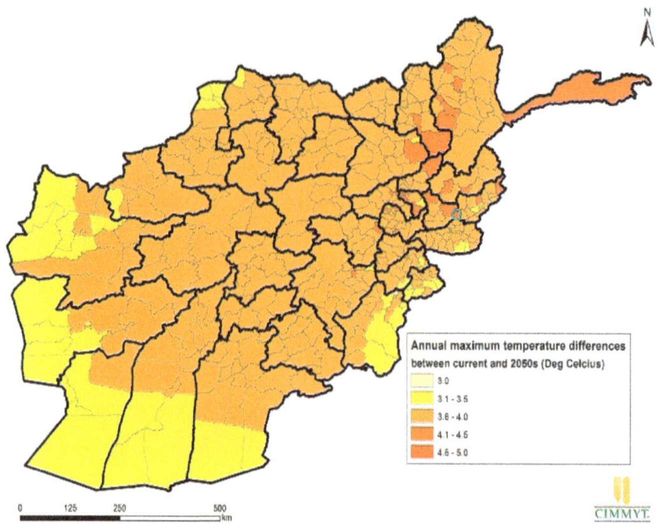

Figure 7. Projected change in annual mean maximum temperature in Afghanistan from now and 2050

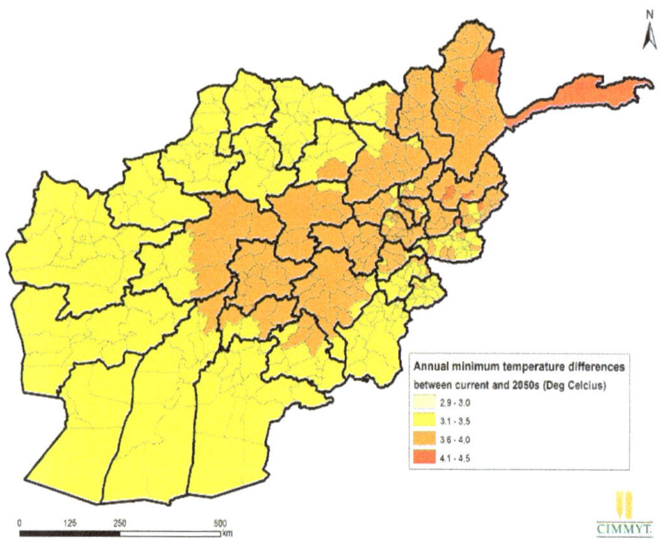

Figure 8. Annual mean minimum temperature projections in Afghanistan from now to 2050

8. Conclusion & Recommendations

Afghanistan has a diverse topography, and cultivates both rainfed and irrigated wheat. These two production domains respond differently to climate change and can thus be impacted differentially. Findings indicate that people in general assume that climate change always leads to a negative impact on crop plants, but that is not the case always. For example, warmer mean temperatures have resulted in considerable yield increases in highland maize when there is a sufficient supply of water.

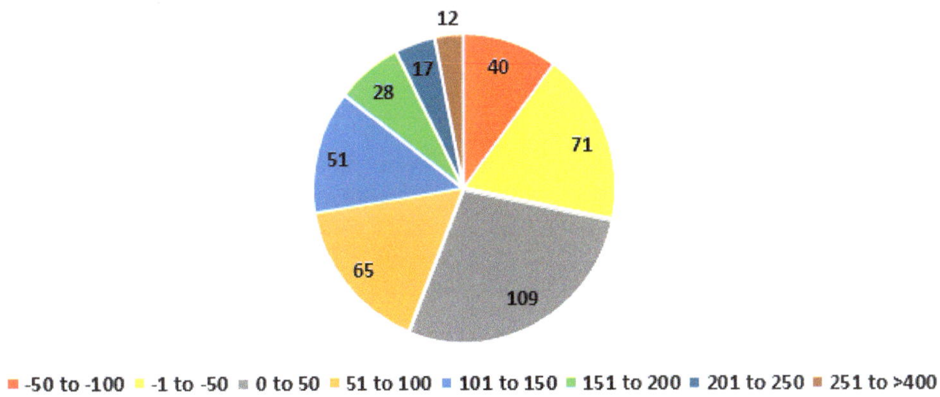

Figure 9. No. of districts (Figure in Pie dia) suffering % change in yield of rainfed wheat in Afghanistan by the year 2050

In Afghanistan, rainfed wheat in areas subjected to warmer mean temperatures and a decline in precipitation, crops might experience severe negative consequences. On the contrary, in cooler areas with enough water when temperatures become warmer yields may not decline or even go up. Assuming sufficient availability of irrigation, higher temperatures are often compensated and could even accelerate growth to higher yields.

On the other hand, the IPCC reports with high confidence that glaciers are shrinking almost worldwide due to climate change and has medium confidence that the phenomenon is affecting runoff and water resources downstream. The IPCC reports with high confidence that climate change is causing permafrost warming and thawing in high-latitude regions and in high-elevation regions. Melting glaciers may lead to problems with the volume of irrigation water feeding crops. However, the general assumption is that at least temporarily river flow will increase due to the thawing leading to greater availability of water. The climate change assessments for 2050 (CCAFS, 2014) indicate that rainfed wheat yields in Afghanistan would decline only in 114 districts to the tune of up to 100 percent, whereas climate change is expected to increase rainfed wheat yields in about 280 districts by up to 400 percent (Figure 9).

Wheat being a C3 crop will be able to use higher future CO_2 concentrations to increase photosynthesis. As a result, in many cases many of the negative effects of climate change may be balanced out. Ultimately, yields may even increase beyond current levels if general suitability doesn't decline. Some researchers discuss this effect but from what is available in literature, it is logical to include this. An analysis CIMMYT undertook on wheat and climate change in Ethiopia showed little decline of yield or increase of yield for the 2050s partly due to the CO_2 effect and the combination of higher temperatures with constant or increases in rainfall.

Acknowledgements

This work was supported in part by funds provided through a grant from the Australian Centre for International Agricultural Research (ACIAR) to CIMMYT. We acknowledge the help of Julie Mollins, CIMMYT-Mexico for editing the article.

References

CCAFS. (2014). Retrieved from http://www.ccafs-climate.org

FAO. (2013). Retrieved from http:/faostat.fao.org

IPC. (2013). *Integrated food security phase classification*. Retrieved from http://www.ipcinfo.org

IPCC. (2014). Retrieved from http://ipcc-wg2.gov/AR5/images/uploads/WG2AR5_SPM_FINAL.pdf

IPCC. (2014). Summary for policymakers. In C. B. Field, V. R. Barros, D. J. Dokken, K. J. Mach, M. D. Mastrandrea,T. E. Bilir, ... L. L. White (Eds.), *Climate Change 2014: Impacts,Adaptation, and Vulnerability. Part A: Global and Sectoral Aspects. Contribution of Working Group II to the Fifth Assessment Report of the Intergovernmental Panel on Climate Change* (pp. 1-32). Cambridge, United Kingdom and New York, NY, USA, Cambridge University Press.

MAIL. (2009). *Agricultural Prospects Report*. Ministry of Agriculture, Irrigation & Livestock, Kabul, Government of Afghanistan.

MAIL. (2012). *Agricultural Prospects Report*. Ministry of Agriculture, Irrigation & Livestock, Kabul, Government of Afghanistan.

Persaud, S. (2012). *Long-Term Growth Prospects for Wheat Production in Afghanistan*. Retrieved from http://www.ers.usda.gov/media/193523/whs11l01_1_.pdf

Sharma, R., & Habibi, H. K. (2012). Wheat secure Afghanistan: assessing priorities. *eWIS, 115*, 11-15.

Waziri, A., Habibi, A., Manan, A. R., Rabbani, H., Kamalzai, H., Alawi, K. S., … Sharma, R. (2013). Making Afghanistan wheat secure by 2022. *eWIS, 116*, 12-14.

WFP. (2014). Retrieved from https://www.wfp.org/countries/afghanistan/overview

World Bank. (2013). Retrieved from http://data.worldbank.org/country/afghanistan

Investigating the Purification of Contaminated Water Supplies by Heavy Metals Such as Cupper and Cadmium Using Diatom Algae

Sara Saadatmand[1] & Atefeh Niazi[1]

[1] Department of Biology, Science and Research Branch, Islamic Azad University, Tehran, Iran

Correspondence: Sara Saadatmand, Department of Biology, Science and Research Branch, Islamic Azad University, Tehran, Iran. E-mail: s_saadatmand@srbiau.ac.ir

Abstract

Using copper and cadmium decontaminating plants has been one of the most important ways in purification of water supplies in recent years. The present study was conducted to investigate the possibility of using monocellular diatom alga (*Nitzchia*) to decontaminate water from copper and cadmium heavy metals. So far, the researchers used four different copper and cadmium heavy metal consistencies of 0.5, 2, 8 and 16 ppm to treat alga. Together with investigating the concentration of the metals absorbed by alga after 14 days of incubation, its growth, chlorophyll a, carotenoids, soluble sugars, and superoxide dismutase and catalase enzymes were also studied. The results then proved a high potential for algae to decontaminate water from Cu and Cd, while the top decontamination rate was found at the highest primary concentration (16 ppm). And also at all treatments, except 0.5ppm and 2 ppm, a descending order in copper growth was apparent, while for chlorophyll a, although it increased in all copper treatments, it had a descending order in all cadmium treatments. The concentration of the carotenoids was highly irregular, although the highest amount was at 8 ppm. A growth was also apparent in reductant glucose measures and the activity of catalase and super oxide dismutase enzymes.

Keywords: diatom, *Nitzchia*, heavy metals, copper, cadmium

1. Introduction

More than 70% of the world surface is covered by water sources such as oceans and seas that are different from lakes, rivers and streams in some ways or another, although, life starts in all of those ecosystems (brine and freshwater) from their producer plants and continues to other creatures. Algae, not different from the other members, are the most simplistic chlorophyll containing members of those ecosystems. Although unlike the other plants, algae have no roots, leaves or pedicels. Such a primary construction is called thallus with the monocellular types not larger than some microns as the smallest member.

Heavy metals are natural substances having a density of more than 5 g cm^{-3}. Accordingly, 53 out of 90 known chemical elements are considered as heavy metals (Weast, 1984; Schytzendubel & Polle, 2002), although not all of them are biologically important. Based on their solvency rate in different physiologic conditions, living cells might have access to 17 heavy metals. According to Greger (1999), because these metals do not exist in the environment, they can't be eliminated from nature. Although there are such natural reasons as volcanic activities that might increase these heavy metals in the environment, there are also some other unnatural reasons caused by human activities such as mining, burning fossil fuels, metal industries, phosphate fertilizers, pesticides, sewage and waste materials that contaminate water, earth and air (Porida et al., 2003; Nalimova et al., 2005; Sebastiani et al., 2004). As a result, filtering these contaminants by different methods, finding new organisms having the ability to save bigger and bigger consistencies of these contaminators inside, and keeping them in safe places is highly important.

Microalgae are aquatic organisms with molecular mechanisms, the thing that let them separate necessary heavy metals from those with less importance. Different decontamination processes conducted by algae, with a more concentration on those containing metallothioneins or phytochelatins are investigated here. As a result, microalgae are known as an acceptable technologic organism to purify water contaminated by heavy metals (Perales-Vela et al., 2006).

Diatoms are monocellular, scarcely thalloid, materials having brown, golden or golden brown colors. They are members of an important group known as epipelic and phytoplankton that live at both freshwater and brine. Diatoms that have siliceous walls, a high density and consequently sink in water, use such methods and characteristics as spiral movements, very small cell dimensions, or having oil and fat in their cell construction to help them stay on top of water, and cell frills such as chitin complements (such as *Thalassiosi rafluvitalis*) to increase friction and decrease sinking in water (McCormick, 1994). Nowadays, diatoms are known as biological indicators of water quality and as an acceptable organism at ecosystems by which we can identify environmental disorders (Lenat et al., 1994).

Heavy metals, naturally or unnaturally, have the potential to make different forms of reactive oxygen species (ROS) (Dipierro et al., 2005; Laspina et al., 2005; Rucinska & Gwozdz, 2005; Demirevska-Kepova et al., 2004; Kopyra & Gwozdz, 2003; Milone et al., 2003; Foyer et al., 1997). In cases that an herbaceous cell can not, by increasing its antioxidant activities, stop increasing ROS, the imbalance between the creation and ROS oxidation metabolism can not be controlled and the chain reactions lead to oxidative stress (Mahalingam & Fedroff, 2003; Bolwell et al., 2002). As a result, the damage to the cells under heavy metals' stress depends on free radical, ROS, and the plants' decontamination mechanisms (Dietz et al., 1999).

Some research findings are related to oxidative damages caused by heavy metals, lipid peroxidation, or oxidative damages to nucleic acids, proteins, chlorophyll and also controlling photosynthesis. One of the most important protective mechanisms of diatoms against heavy metals' stress is chelated the metals inside cytosols using highly active ligands including phytochelatins, metallothioneins, amino acids and organic acids (Rauser, 1999; Clemens, 2001). Phytochelatins (PCs) are a group of peptides complexible to heavy metals with a general structure of (γ-GluCys)n-Gly, in which $n = 2 - 11$. In the presence of some heavy metals, especially cadmium, the PC of the synthase activates and makes PCs out of glutathione (Cobbet, 2000). Then, the metal complex (PC) is transferred into the vacuole. This PC-Cd transmission to the vacuole is provided by Cd^{+2}/H^{+} antiporter and an ABC transporter related to ATP which exists in tonoplast. Attaching sulfide ions into PC-Cd complexes also increases its stability (Ruley et al., 2004). Cadmium is one of the most poisonous metals and causes the highest consistencies of polypeptide PCs attached to metals. It is seen in *Thalassiosira weissflogii* sea diatoms, in which exists a cadmium transpiration of *T.weissflogii* high mineral cadmium consistencies. This transpiration is so much that more than half of the absorbed cadmium is returned to the cultivation medium. In high cadmium consistencies, PC transpiration out of the cells also appears and continues until the exterior cadmium concentration decreases, preventing cadmium and PC transpiration. It is believed that *T. weissflogii* PC complex, releases cadmium as an antitoxin mechanism. Although this complex doesn't seem to be stable in sea water and out of the cell body, PC-cadmium transpiration can be an important survival strategy that helps phytoplankton's survival in contaminated waters by heavy metals (Lee, 1996).

In addition to the effect of heavy metals on enzymatic antioxidants, heavy elements influence on a big number of compounds involving in an oxidative defensive process, while have a light molecular weight. Two different reactions are seen while open to heavy metals. Not only the presence of metallic ions decreases the storage of antioxidants having light molecular weights, an oxidative stress may also activate enzymes that take part in those antioxidants' biosynthesis (Dietz et al., 1999). Non-enzymatic antioxidant systems contain two system redox, ascorbate (ASA) and glutathione (GSH) together with tocopherol, flavonoids, carotenoids, some phenyl compositions, polyamines, and etc.

2. Methods

2.1 Diatom Samples Collection and Identification

Water samples were taken from Tehran, Iran. The samples were investigated looking for diatoms, and those containing such materials were chosen to cultivate and use in the study. To identify the live samples of diatom, they were studied by a microscope having 100 magnifying power (Figure 1).

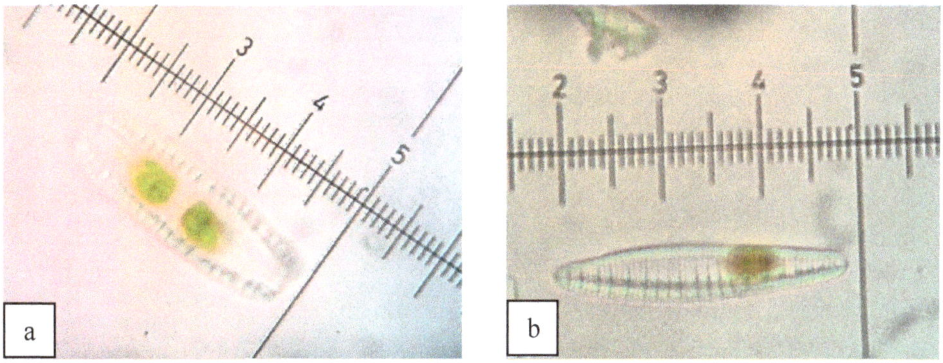

Figure 1. Diatom's microscopic image (100 times magnified), (a): valvic side image, (b): belt side image

Then, a fixed preparation was prepared using Werff's (1995) method. Here, the samples were heated for one hour at 80 °C with 37% H_2O_2. After that, the saturated solution of $KMnO_4$ was added to form precipitation. By adding a little HCl rarified in H_2O_2, the precipitates were separated and investigated using a light microscope (Werff, 1955) (Figure 2).

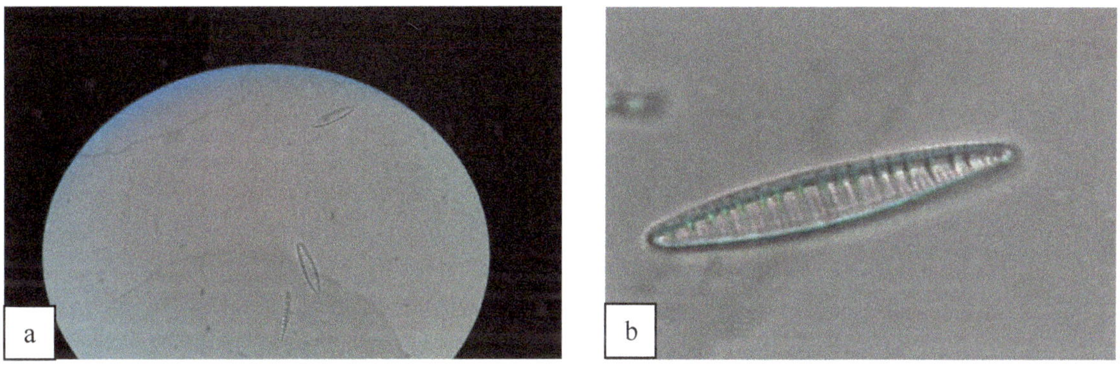

Figure 2. Diatom microscopic image, (a): microscopic image (100 times magnified), (b): details of the image presented at slide a

SEM electron microscope was used to identify diatom samples and to see the interior design of the algae wall (Figure 3).The required diatom material also was prepared for the rest of the investigation using the visual identification key of diatoms at (http://westerndiatoms.colorado.edu) *Nitzschia.*

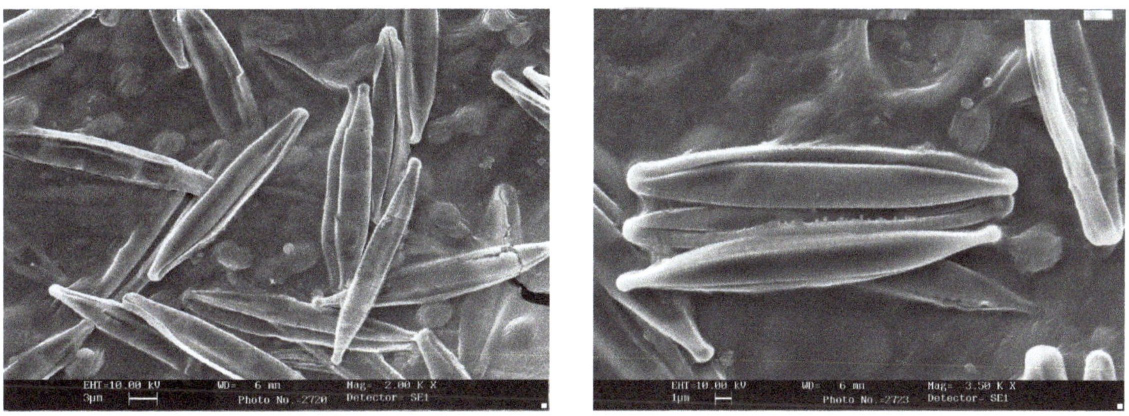

Figure 3. SEM electronic microscope image of diatoms concentration

2.2 Preparing Cultivation Setting and Samples

The Chu's cultivation medium was used to cultivate diatom samples (Bold & Wynne, 1978). The samples, then were distributed under a Laminar flow hood and among 500 ml flasks, and supplied with a 250 ml Chu's Medium sterilized liquid cultivation. After that, the flasks input were covered by cotton, while they were placed in suitable locations under 12-12 photoperiod hour and at 25 °C to grow the samples. After 14 days, a subculture was prepared and relocated at their immediate place. This process continued till the samples became perfectly pure, ready to grow and form the major cultivations and treatments.

2.3 Copper and Cadmium Treatment

Completely pure and standard flasks were used at the major cultivation process. Here, in the presence of a control, the algae were open to four different cadmium and four copper consistencies of 0.5, 2, 8, and 16 ppm. It should also be mentioned that there were 9 repetitions for each treatment in 14 days. Then, the samples growth were compared.

2.4 Measuring the Algae Growth Rate by Spectrophotometer

There are different methods to measure the growth rate of an alga. For example, spectrophotometery method which is a completely new and more accurate one while comparing to the other methods. During the growth rate measurement at this method, some cubic centimeters (CC) of cultivation mediums are taken and their light absorption rate is investigated by spectrophotometer at 530 nm light wave. It should be mentioned that to measure the real absorption rate, the researchers needed to prepare standard solutions for spectrophotometer. Therefore, some CC of the cultivation medium was centrifuged at 2000 g spins for 10 minutes. The resulting liquid would be the standard solution. It is also clear that the samples with the higher absorption level are those having a higher growth rate. Researchers at the present study used this method after 14 days of copper and cadmium insemination to the samples.

After measuring the samples growth and investigating them by a light microscope, they were centrifuged at 1500 g spin for 5 minutes, then separated completely from the cultivation medium and kept at -20 °C for use at comparisons.

2.5 Pigments Investigation (Examining Chlorophyll a and Carotenoids)

Arnon (1949) method was used to investigate the ingredients of chlorophylls, and Davis (1976) method was used to study carotenoids. First, 0.05 gr diatom alga was ground and homogenized with 80% acetone (adding some calcium carbonate while grinding prevents magnesium to exit from chlorophylls ingredients). Then, the resulting homogeneous combination was filtered using number 2 Wathman. To investigate chlorophyll a and carotenoids, the liquid absorption at 645, 663 and 480 nanometer wavelengths was measured in the presence of 80% acetone treatment. And then chlorophyll a and carotenoids were measured.

2.6 Measuring Reducing Sugars by Nelson and Somogyi Method Reductant

As much as 0.025 gr alga was taken, 5ml of water was added and the combination was grinded and filtered by filter paper. Then, 2 ml of the resulting liquid was moved into a test tube containing 2 ml of copper sulfate (40 gr sodium carbonate in 400 C H_2O_2 + 7.5 gr tartaric acid + 4.5 gr hydrous copper sulfate at the final volume of ml) and heated for 8 min at 100 °C. Cu^{2+} (reduction) is the basis for glucose measurement here. Two ml of phosphomolybdic acid was added to the cool test tubes (70 gr phosphomolybdic acid and 10 gr sodium in 70 ml of 5% soda for 40 min at 40 °C + 250 ml orthophosphoric acid) while putting them on a mixer and increasing the volume to one liter. Then, the test tubes were shook very fast till their colors turned into blue, and their absorption rate was measured at 600 nanometer using spectrophotometer. The glucose concentration was also measured using the standard curve of glucose based on micrograms on liter. Here, to draw a 500 mg l^{-1} glucose curve, different consistencies of 0.2, 0.4, 0.6, 0.8, 1, 3, 5 and 7 mg l^{-1} were prepared and 2 ml per each of those solutions were taken, while all were involved in the same process.

2.7 Catalase Enzyme Activity Measurement

Catalase enzyme activity rate measurement was performedwith the method proposed by Pereira (2002). Here, the catalase enzyme activity rate together with H_2O_2 reduction level was investigated by studying light wave changes at 240 nanometers for 1 minute. First, 0.15 gr alga was prepared and ground at 1.5 ml of 50 mill molar potassium phosphate buffer (7.5 pH) containing 1% of 1 millimolar polyvinyl pyrollidone (PVP). All of the extraction phases were done in the presence of ice and then, 3.5 ml H_2O_2 was combined with 50 ml aquapura before taking 70 microliter and adding 2.83 ml potassium phosphate buffer (without PVP and EDTA). And finally, 100 microliter of the resulting substance was added and centrifuged at 4 °C, 2000 spin for 20 minutes.

Then, the clear solution formed was used to measure the catalase enzyme activity. The absorption rate was recorded immediately after 1 minute at 240 nanometers. The control solution was prepared here by adding 2.83 mm potassium phosphate buffer (without PVP and EDTA) to 70 microliter of 2% H2O2, and then adding 100 microliter potassium phosphate buffer containing PVP and EDTA (Pereira et al., 2002).

2.8 Dismutase Superoxide Enzyme Activity Measurement

Dismutase superoxide enzyme activity was measured based on a method proposed by Giannopolitis et al (1997). Here, 3 ml of reaction solution was containing 50 mm potassium phosphate buffer (pH 7.8), 13 mm of 75 μM nitroblute trazolium, 2 μM riboflavin, 0.1 EDTA and 100 microliter enzyme extraction. The reaction started by removing the aluminum foil and exposing the samples to 5000 LUX light for 15 minutes. Then, the samples absorption rate was recorded at 560 nanometers immediately. Two control samples were used here containing no enzyme extraction, one of which was shed by light for 15 minutes, while the other one received no light.An enzyme activity unit is composed of some enzyme which prevents 50% of NBT (reduced) at 560 nanometers (Giannopolitisetal., 1977).

2.9 Heavy Metals Measurement (Copper and Cadmium) Absorbed by Alga

Researchers at the present study measured the metals substances by solving the samples in acid kept in closed tubes, using microwave and atomic spectrophotometer.

To this end, 0.2 gr samples, completely dried, were taken, specific TFM tubes required were located on a flat area and the 0.2 gram samples were added to the tubes together with 6 ml of 65% HNO_3 and 2 ml of 30% H_2O_2. Then, closed the tubes and located them in microwaves adjusted to the following heat control program:

Step	Time	Temperature	Microwave power
1	10 minutes	200 °C	Up to 1000 Watt
2	10 minutes	200 °C	Up to 1000 Watt

After passing the required time, the samples were examined by an atomic spectrophotometry device to measure copper and cadmium levels.

2.10 Statistical Analysis

The present study was carried out as a completely accidental study with 4 repetitions. Treatments used at this study contained different levels of copper and cadmium. The data analysis variance and mean comparisons of the study treatments was conducted based on Duncan comparative test at 5% chance, data distribution was normal and the data analysis was performed using SPSS program. And finally, excel program was applied to design charts and graphs.

3. Results

There was a significant difference in the average growth of the samples after 14 days of incubation at different doses. In all treatments except 0.5 ppm cu and 2 ppm cu, a descending rate in growth was seen, while the highest growth rate was at 0.5 ppm cu treatment. As a result, diatoms growth was under the influence of copper and cadmium heavy metals in a way that the first one increased and the second one decreased their growth rate (Figure 4).

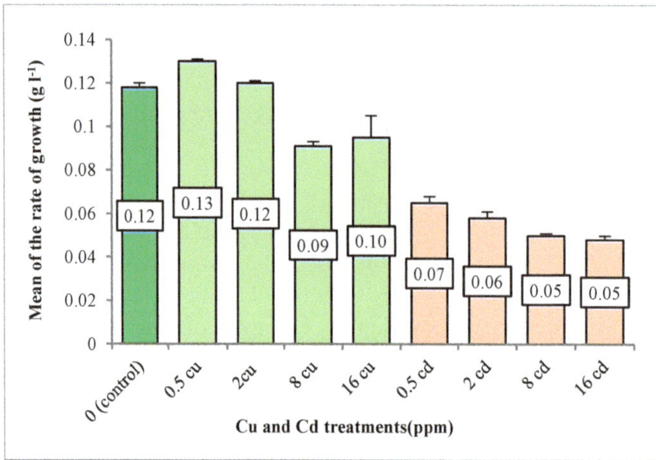

Figure 4. Alga growth after 14 days of different treatments by cadmium and copper

There was a significant difference in chlorophyll a consistencies after 14 days of incubation at different doses. Chlorophyll a measures compared with control, in all copper treatments had an ascending and in all cadmium treatments had a descending growth rate with the highest chlorophyll a concentration at 8 ppm Cu treatment (Figure 5).

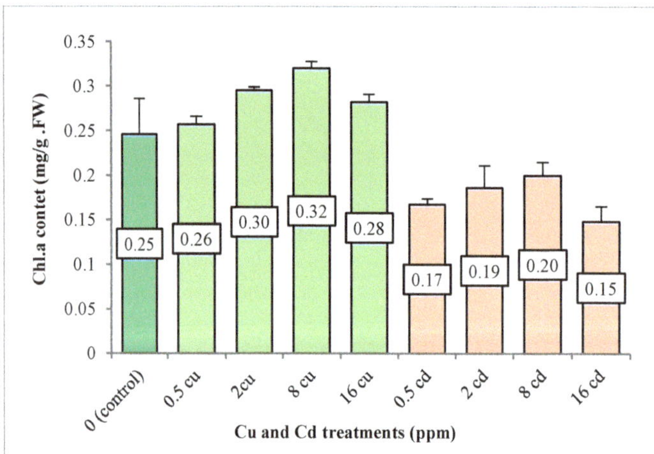

Figure 5. Chlorophyll a concentration in cadmium and copper treatments

There was a significant difference in carotenoid concentration measures after 14 days of incubation at different doses. The highest carotenoid concentration was at 8 ppm cu treatment (Figure 6).

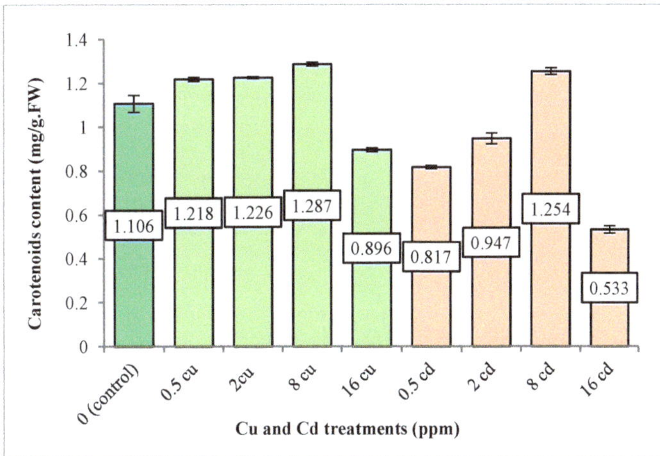

Figure 6. Carotenoid concentration in cadmium and copper treatments

There was a significant difference in reductant glucose concentration measures after 14 days of incubation at different doses. There was also an ascending growth rate in all copper treatments while compared to the control, and the highest soluble sugars concentration average was at 16 ppm cd treatment (Figure 7).

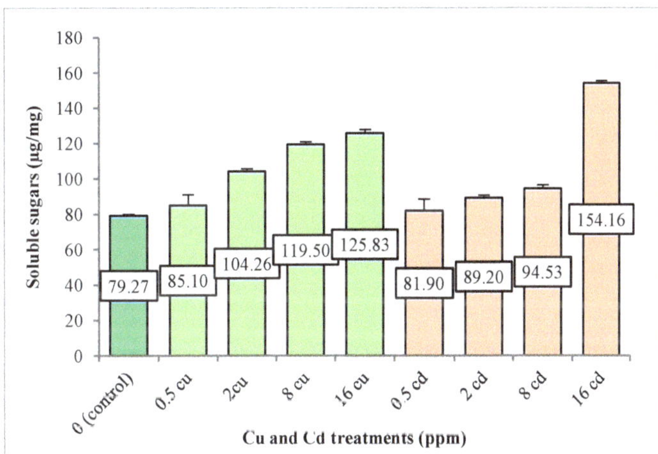

Figure 7. Soluble sugars in cadmium and copper treatments

There was a significant difference in catalase enzyme activity measurement results after 14 days of incubation at different doses. There was also an ascending growth rate in enzyme activities at all treatments while compared to the control, and the highest catalase enzyme activity average after 14 incubation days was at 16 ppm Cd treatment (Figure 8).

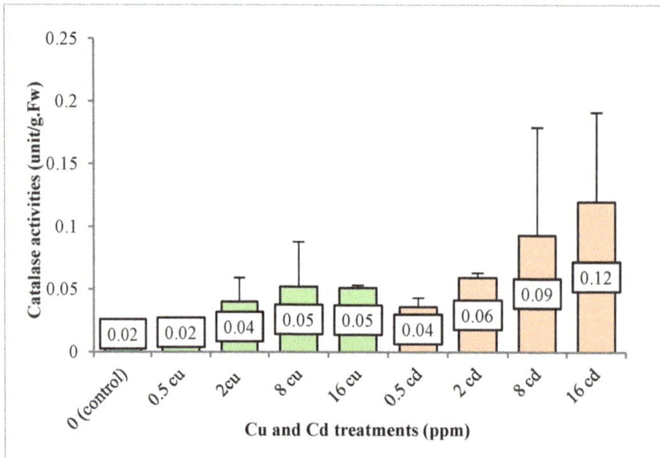

Figure 8. Catalase enzyme activity in different cadmium and copper treatments

There was a significant difference in superoxide dismutase enzyme activity measures after 14 days of incubation at different doses. There was also an ascending growth rate in enzyme activities at all treatments, and the highest superoxide dismutase enzyme activity average was at 16 ppm cd treatment (Figure 9).

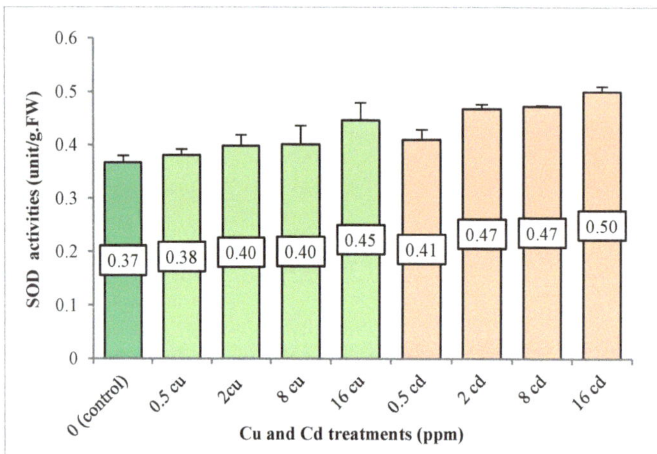

Figure 9. SOD enzyme activity in different cadmium and copper treatments

There was a significant difference in Cu concentration measures (mg l^{-1}) after 14 days of incubation at different doses. The highest Cu concentration measure was at 16 ppm treatment (Figure 10).

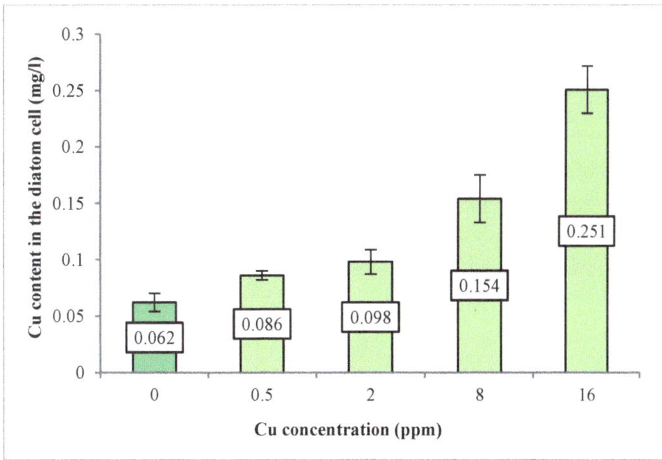

Figure 10. Copper absorption rate by diatom in different treatments

There was a significant difference in Cd concentration measures (mgl^{-1}) after 14 incubation days at different doses. The highest Cd concentration measure average after 14 incubation days was 16 ppm at treatment 8 (Figure 11).

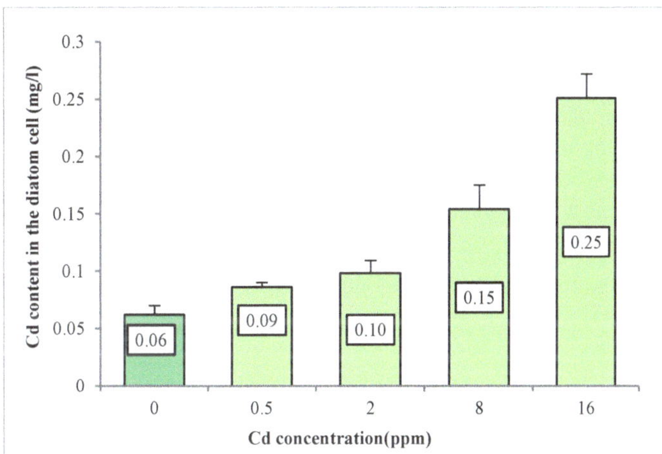

Figure 11. Cadmium absorption rate by diatom in different treatments

4. Discussion

As stated before, Chun's cultivation medium is suitable for algae growth and helps the diatoms to have an acceptable growth rate after 14 days. Copper is a very nutritious substance and an inseparable member of the enzymes being responsible for electron movements speeding up chloroplast reactions in plants (Geatke & Chow, 2003). It should be mentioned that a little zinc, copper, iron, etc. was used at Chun's medium to investigate their effects. As the results indicated, the highest growth rate was recorded at the copper's 0.5 ppm treatment, indicating that copper can improve the algae growth to some extent (2 ppm) (Figure 4). Although, higher than normal consistencies of copper may cause damage on plants tissues (Hall, 2002), like what happened at the present study. High concentration of copper at the plants leaves also may cause changes at processes such as photosynthesis, breathing, enzyme activities, integrity of DNA, etc. that finally lead to some growth issues (Schutzendubel & Polle, 2002; Posmyk et al., 2009).

Regarding cadmium that is not a necessary element for plants growth, a descending growth rate was recorded at all treatments, especially at the 16 ppm one. In the same line with many other studies conducted such as Larbi et al. (1997), Sandalio (2001), or Costa and Spitz (1997) who reported growth failure as one of the most immediate effects of cadmium on plants, the present study also brought about the same results once again.

Heavy metals increase the possibility of the tissues death by increasing lipid peroxidation and producing other active oxygen species. Photosynthesis also is being affected by cadmium, knowing that effective enzymes on CO_2 stabilization are highly under the influence of cadmium (Sandalio, 2001; Larsson et al., 1998). Extra cadmium not only prevents rubisco enzyme activity that plays a key role in Calvin Cycle (Chaffei et al., 2003), but also causes disorder at breathing, providing and absorption of nourishing elements, nitrogen and sulfate metabolism (Balestrasse et al., 2001; Gussarsson et al., 1996; Haag-kever et al., 1999; Lee & Leustek, 1999). Together with appearing disorders at physiological processes, an increase in cadmium levels at different parts of plants may have negative effects on their growth. While a cell receives excessive amounts of heavy metals, different kinds of reactions are provoked against the stress caused by cadmium. To name a few, we can refer to an increase in proteins with small molecular mass or synthesis of peptides (Nakagamy & Hirt, 2004; Wang & Peverly, 1999).

The cells disorders affected on the plant growth may be caused by such different reasons as losing water, elasticity of the cells' walls (Costa & Spitz, 1997), losing necessary ingredients such as K, Ca, Mg and Fe (Gogorcena et al., 2002; Gussarsson et al., 1996), problems brought about by disorders at photosynthesis processes, breathing and nitrogen metabolism as a result of poisonous cadmium consistencies (Balestrasse et al., 2001; Haag-Kever et al.,1999; Larsson et al., 1998).

Most researchers believe that losing chlorophyll while experiencing a heavy metal stress such as the stress caused by cadmium is the main reason of disorders appeared in chlorophyll synthesis in plants, the thing that was proved by Mike et al. (1992) for wheat. The reason is that the plants start absorbing cadmium instead of iron, place it instead of Mg in chlorophyll molecules that prevent chlorophyll synthesis (Polle, 2001). In stressful conditions also the chlorophyll molecule death is certain (Wo'jcik et al., 2006).

Cadmium stress also decreases carotenoids numbers. It is caused by non photochemical suppression of carotenoid, chlorophylls, and finally led to disorders in their structure. Carotenoids may also play an antitoxin role for chlorophylls and decrease the poisonous effects of free radicals, as an example, reacting with chlorophylls to prevent forming free oxygen radicals. It finally leads to their death when acting as a defensive system to protect against oxidative stress imposed on the process (Sanita di Toppi & Gabbrielli, 1999). Losing carotenoid is actually a strategy applied by plants to resist against the oxidative stress imposed by cadmium. At the present study also the samples inseminated by cadmium 16 ppm had which the highest rate of carotenoid loses while compared with the other consistencies of copper and cadmium.

Carotenoid and green pigments decrease during the process as well as antioxidant enzymes increase caused by adding heavy metals consistencies proves the relationship between free radicals production and heavy metals increase. As a result, while a non-enzymatic defensive system becomes weak, the enzymatic system is being activated and helps the plants by opposing free radicals. It also worth mentioning that as proposed by Nyitrai et al. (2003), feeding lighter consistencies, and a combination, of cadmium, massicot, nickel and DCMU to plants or spraying the mixture on their leaves incorporates to improve chlorophyll synthesis process.

A descending rate for green pigments and chlorophyll a production in cadmium treatment, and an ascending rate of that in the copper treatment were seen at the present study, proving an increase in oxidative stress.

To protect themselves against oxidative damages, plants are equipped with a wiper system designed to operate against free radicals. The system contains antioxidant enzymes such as catalase, peroxide, dismutase peroxide and nonenzymic defensive systems like ascorbate, glutathione (Mittler et al., 2004). As presented in Figures 8 and 9, catalase and dismutase superoxide antioxidant enzymic activity increases while being influenced by cadmium and copper stress. Although it should be mentioned that an ascending rate was more apparent in cadmium than copper stress, while the highest point in both enzymes recorded measures was at cadmium 16 ppm treatment.

Increased consistencies of copper in cells leads to H_2O_2 radical hydroxyl production by stepped up reaction with a metal named Haber-Weiss reaction (Palma et al., 1987). Protective mechanisms formed to react against active oxygen types also might, because of high consistencies of heavy metals, increase activities of antioxidant enzymes such as dismutase superoxide and peroxidase (Bueno & Piqueras, 2002), although effective elimination of superoxide and hydrogen peroxide needs an effective activity of a lot of antioxidant containing enzymes. Here, superoxide immediately turns into H_2O_2 under the effect of SOD (Bowler et al., 1992), and then H_2O_2 is being broken into H_2O and H by catalase (Noctor & Foyer, 1998).

Plants have a limited protective power while heavy metals stress increases. A lot of research has proved that high consistencies of heavy metals commonly decreases antioxidant enzyme activities (Polle & Schutzendubel, 2002). Autoxidation and Fenton also seriously weaken enzymatic defensive systems. For example, catalase enzyme activity is being stopped immediately by O_2 superoxide radicals, or hydroxide radicals stop dismutase superoxide enzyme -2n-Cu activity (Casano et al., 1997). As a result, enzyme activity at lower concentrations of heavy metals increases, while it starts to decrease slowly at higher concentrations (based on the plant type) (Cao et al., 2004). It should also be mentioned that a lengthy influence period of heavy metals at first increases enzyme activities, especially peroxides, while after sometimes decreases those activities (Qadir et al., 2004). An increase in heavy metals, especially higher concentrations, stop catalase enzyme activity (Posmyk et al., 2009, Choudhary et al., 2007). Descending dismutase superoxide antioxidant enzyme activity rate at higher concentrations of heavy metals such as cadmium is possibly a result of its deactivation by excessive active oxygen production or disorders caused by synthesizing heavy metals instead of dismutase superoxide (dismutase superoxide).

As presented at Figure 4, reductant sugar measures increases while being stressed by cadmium and copper heavy metals, especially at 16 ppm treatment. The reason might be descending breathing rate and ascending insoluble sugars catalyzer enzyme activities, such as anvertase and sucrose synthesis, which decreases glucose use and increases its production (Shah & Dubey, 1998; Verma & Dubey, 2001).

Soluble sugars kinds are increased as a result of the imposed stress by heavy metals. Cadmium and copper concentration increased sugars in this alga.

Different conclusions are arrived about the potential effects of stress on glucose concentration in plants. Some researchers believed that glucose increases under stress (Jones & Turner, 1980), while the others have opposite views, saying that it decreases (Hanson & Hitz, 1982) or remains changeless (Morgan, 1992). Results of the present study, in the same line with the results presented by Alaoui et al. (2004) regarding copper treatment, also proved that reductant glucose increases under the toxicity influence of copper and cadmium. Some studies also conducted investigating sea diatom growth (*Amphora coffeaeformis*) under the influence of cadmium and copper. These heavy metals influence on the mentioned diatom's parameters of biochemical combination and its growth was measured, while the second parameter was measured by the diatom's chlorophyll content and proved its growth. Results then indicated that heavy metals significantly decreased carbohydrate, protein, amino acid and lipid rates (Anantharaj et al., 2011).

As presented in Figures 10 and 11, while cadmium and copper concentrations were increased in diatom cultivation medium, the alga absorbed more proportions of the heavy metals. As a result, diatom has a high capacity in metal absorption and duplication, the characteristics that increases this microscopic biomes capability in decontaminating wastewater. These diatoms should be identified to solve the problems caused regarding metal poisoning in the water ecosystem contaminations. As a result, complementary investigations should be carried out in industrial scale to decontaminate ecosystems from heavy metals.

5. Conclusion

Due to industrialization and addition of industrial waste water in lakes and rivers, the use of biological methods to clean water has been of interest to researchers. This study has been indicated effects of *Nitzchia* diatom on heavy metal extraction from water. *Nitzchia* is being in city's mere be able to stay in metallic stress and storage a lot of cu and cd ingredient. Therefore, it suggested that *Nitzchia* is a prepare candidate to purgation and cleaning up of industrial swages beside other systems.

References

Alaoui-Sosse, B., Genet, P., Vinit-Dunand, F., Toussaint, A. L., & Epron, D. (2004). Effect of copper on growth in cucumber plants (*Cucumis sativus*) and its relationships with carbohydrate accumulation and changes in ion contents. *Plant Sci., 166*, 1213-1218. http://dx.doi.org/10.1016/j.plantsci.2003.12.032

Anantharaj, K., Govindasamy, C., Natanamurugaraj, G., & Jeyachandran, S. (2011). Effect of heavy metals on marine diatom *Amphora coffeaeformis* (Agardh. Kutz). *Global Journal of Environmental Research, 3*(5). 112-117.

Baker, A. J. M., & Brooks, R. R. (1989). Terrestrial higher plants which hyperaccumulate metallic elements. A review of their distribution, ecology and phytochemistry. *Biorecovery, 1*, 81-126.

Balestrasse, K. B., Gardey, L., Gallego, S. M., & Tomaro, M. L. (2001). Response of antioxidant defence system in soybean nodules and roots subjected to cadmium stress. *Australian Journal of Plant Physiology, 28*, 497-504. http://dx.doi.org/10.1071/PP00158

Bold, H. C., & Wynne, M. J. (1978). *Introduction to the Algae. Structure and Reproduction* (pp. xiv+706). Englewood Cliffs. New Jersey, Prentice-Hall.

Bolwell, G. P., Bindschedler, L. V., Blee, K. A., Butt, V. S., Davies, D. R., Gardner, S. L., ... Minibayeva, F. (2002). The apoplastic oxidative burst in response to biotic stress in plants: a three-component system.

Journal of Experimental Botany, 53, 1367-1376.

Bowler, C., Vanmontagu, M., & Inze, D. (1992). Superoxide dismutase and stress tolerance. *Annual Reviews of Plant Biology and Plant Molecular Biology, 43*, 83-116. http://dx.doi.org/10.1146/annurev.pp.43.060192.000503

Bueno, P., & Piqueras, A. (2002). Effect of transition metals on stress, lipid peroxidation and antioxidant enzyme activities in tobacco cell cultures. *Plant Growth Regulation, 36*(2), 161-167.

Cao, X., Ma, L. Q., & Tu, C. (2004). Antioxidative responses to arsenic in the arsenic-hyperaccumulator Chinese brake fern (*Pteris vittata* L.). *Environmental Pollution, 128*, 317-325

Chaffei, C., Gouia, H., & Ghorbel, M. H. (2003). Nitrogen metabolism in tomato plants under cadmium stress. *Journal of Plant Nutrition, 26*, 1617-1634. http://dx.doi.org/10.108/PLN-120022372

Choudhary, M., Jetley, U. K., Khan, M. A., Zutshi, S., & Fatma, T. (2007). Effect of heavy metal stress on proline, malondialdehyde, and superoxide dismutase activity in the cyanobacterium *Spirulina platensis*-S5. *Ecotoxicology and environmental Safety, 66*(2), 204-209.

Clemens, S. (2001). Molecular mechanisms of plant metal tolerance and homeostasis. *Planta, 212*, 475-486. http://dx.doi.org/10.1155/2012/872875

Cobbet, C. S. (2000). Phytochelatins and their roles in heavy metal detoxification. *Plant Physiol., 123*, 825-832. http://dx.doi.org/10.1146/annurev.arplant.53.100301.135154

Costa, G., & Spitz, E. (1997). Influence of cadmium on soluble carbohydrates, free amino acids, protein content of in vitro cultured *Lupinus albus. Plant Sci., 128*, 131-140. http://dx.doi.org/10.1016/S0168-9452(97)00148-9

Davies, B. H. (1976). Carotenoids. In T. W. Goodwin (Ed.), *Chemistry and biochemistry of plant pigments* (Vol. II, pp. 38-165). New York: Academic Press. http://dx.doi.org/10.1002/9780470122662.ch7

Demirevska Kepova, K., Simova-Stoilova, L., Stoyanova, Z., Holzer, R., & Feller, U. (2004). Biochemical change in barely plants after excessive supply of copper and manganese. *Environ.Exp.Bot., 52*, 253-266. http://dx.doi.org/10.5539/ijb.v4n3p148

Dietz, K. J., Bayer, M., & Kramer, U. (1999). Free radicals and reactive oxygen species as mediators of heavy metal toxicity in plants. In M. N. V. Parasad & J. Hagemeyer (Eds.), *Heavy metal stress in plants: From Molecules to Ecosystem* (pp. 73-79). Berlin: Springer-Verlag. http://dx.doi.org/10.1007/978-94-007-4441-7_2

Dipierro, N., Mondelli, D., Paciolla, C., Brunetti, G., & Dipierro, S. (2005). Change in the ascorbate system in the response of pumpkin (*Cucurbita pepol.*) roots to aluminium stress. *J. Plant Physiol.*, In Press. http://dx.doi.org/10.1016/j.jplph.2004.06.008

Foyer, C. H., Lopez-Delgado, H., & Scott, I. M. (1997). Hydrogen peroxide- and glutathione- associated mechanisms of acclamatory stress tolerance and signaling. *Physiol. Plant., 100*, 241-254.

Giannopolitis, C. N., & Ries, S. K. (1977). Superoxide dismutase I. Occurrence in higher plants. *Plant Physiol., 59*, 309-314.

Gogorcena, Y., Lucena, J. J., & Abadia, J. (2002). Effects of Cd and Pb in sugar beets plants grown in nutrient solution: Induced Fe deficiency and growth inhibition. *Funct. Plant Biol., 29*, 1453-1464. http://dx.doi.org/10.1071/FP02090

Greger, M. (1999). Metal availability and bioconcentration in plants. In M. N. V. Parasad & J. Hagemeyer (Eds.), *Heavy Metal Stress in plants from molecules to Ecosystem* (pp. 1-27). Berlin:Springer-Verlag.

Gussarsson, M., ASP, H., Adalsteinsson, S., & Jensen, P. (1996). Enhancement of cadmium effects on growth and nutrient composition of bireh (Betula pendula) by buthionine sulphoximine (BSO). *J. Exp. Bot. 47*, 211-215.

Haag-Kever, A., Schafer, H. J., Heiss, S., Walter, C., & Rausch, T. (1999). Cadmium expose in Brassica juncea cause a decline in transpiration rate and leaf expansion without effect on photosynthesis. *Journal of Experimental Botany, 50*, 1827-1835.

Hall, J. L. (2002). Cellular mechanisms for heavy metal detoxification and tolerance. *Journal of Experimental Botany, 53*, 1-11. http://dx.doi.org/10.1093/jexbot/53.366.1

Hanson, A. D., & Hitz, W. D. (1982). Metabolic responses of plant water deficit. *Annu. Rev. Plant. Physiol., 23*,

163-203. http://dx.doi.org/10.1146/annurev.pp.33.060182.001115

Jones, M. M., & Turner, N. C. (1980). Osmotic adjustment in expanding and fully expanded leaves of sunflower in response to water deficits. *Aust. J. Plant. Physiol., 7*, 181-192. http://dx.doi.org/10.1071/PP9800181

Kinraide, T. B., Ryan, P. R., & Kochian, L. V. (1992). Interactive effects of Al^{3+}, H^+ and other cations on root elongation considered in terms of cell-surface electrical potential. *Plant Physiol., 99*, 1461-1468. http://dx.doi.org/10.1104/pp.99.4.1461

Kopyra, M., & Gwozez, E. A. (2003). Nitricoxide stimulates seed germination and counteracts the inhibitory effect of heavy metals and salinity on root growth of *Lupinus luteus*. *Plant Physiol. Biochem., 41*, 1011-1017. http://dx.doi.org/10.1016/j.plaphy.2003.09.003

Larbi, A., Morales, F., Abadia, A., Costa, G., & Spitz, E. (1997). Influence of cadmium on soluble carbohydrates, free amino acids, protein content of in vitro cultured *Lupinus albus*. *Plant Sci., 128*, 131-140. http://dx.doi.org/10.1016/S0168-9452(97)00148-9

Larsson, E. H., Bornman, J. F., & ASP, H. (1998). Influence of UV-B radiation and cadmium on chlorophyll fluresecence Brassica napus. *Journal of Experimental Botany, 49*, 1031-1039. http://dx.doi.org/10.1093/jxb/49.323.1031

Lee, J. G. (1996). Export of cadmium and phytochelatin by the marine diatom *Thalassiosir aweissflogii*. *Environmental Science & Technology, 30*(6), 1814-1821. http://dx.doi.org/10.1021/es950331p

Lee, S., & Leustek, T. (1999). The effect of cadmium on sulfate assimilation enzymes in *Brassica juncca*. *Plant Science, 141*, 201-207. http://dx.doi.org/10.1186/1471-2229-14-132

Lenat, D. R., & Crawford, J. K. (1994). Effect of land use on water quality and aquatic biota of three North Carolina Piedmont streams. *Hydrobiologia, 294*, 185-199.

Malik, D., Sheoran, I. S., & Singh, R. (1992). Carbon metabolism in leaves of cadmium treated wheat seedlings. *Plant Physiol. Biochem., 30*, 223-229.

McCormick,P. V. & Cairns, J. (1994). Algae as indicators of environmental change. *Journal of Applied Phycology, 6*, 506-526.

Milon, S., Hovius, R., Vogel, H., & Wohland, T. (2003). Factors influencing fluorescence correlation spectroscopy measurements on membranes: simulation and experiments. *Chem. Phys., 288*, 171-186. http://dx.doi.org/10.1016/S0301-0104(03)00018-1

Mittler, R., Vanderauwera, S., Gollery, M., & Van Breusegem, F. (2004). Reactive oxygen gene network of plants. *Trends Plant Sci., 9*(10), 490-498.

Morgan, J. M. (1992). Osmotic components and properties associated with genotypic differences in osmoregulation in wheat. *Aust. J. Plant .Physiol., 19*, 67-76. http://dx.doi.org/10.1071/PP9920067

Nelson, N. (1944). A photometric adaptation of the Somogyi method for the determination of glucose, *Journal of Biological Chemistry, 153*(2), 375-380.

Noctor, G., & Foyer, C. H. (1998). Ascorbate andglutathione: keeping active oxygen under control. *Annual Review of Plant Physiology and PlantMolecular Biology, 49*, 249-279.

Nyitrai, P., Bóka, K., Gáspár, L., Sárvári, E., Lenti, K., & Karesztes, A. (2003). Characterization of the stimulating effect of low-dose stressors in maize and bean seedlings. *Journal of Plant Physiology, 160*, 1175-1183. http://dx.doi.org/10.1078/0176-1617-00770

Oncel, I., Keles, Y., & Ustun, A. S. (2000). Interactive effects of temperature and heavy metal stress on the growth and some biochemical compounds in wheat seedlings. *Environmental Pollution, 107*, 315-320.

Palma, J. M., Gómez, M., Yáñez, J., & del Río, L. A. (1987). Increased levels of peroxisomal active oxygen related enzymes in copper tolerant pea plants. *Plant Physiology, 85*(2), 570-574. http://dx.doi.org/10.1104/pp.85.2.570

Pereira, G. J. G., Molina, S. M. G., Lea, P. J., & Azevedo, R. A. (2002). Activity of antioxidant enzymes in response to cadmium in *Crotalaria juncea*. *Plant Soil, 239*, 123-132. http://dx.doi.org/10.1023/A:1014951524286

Polle, A. (2001). Dissection of the superoxide dismutase-ascorbate-glutathione pathway by metabolic modeling: computer analysis as a step towards flux analysis. *Plant Physiol., 126*, 445-462.

Posmyk, M. M., Kontek, R., & Janas, K. M. (2009). Antioxidant Enzymes activity and phenolic compounds content in red cabbage seedlings exposed to copper stress. *Ecotoxicology and Environmental Safety, 72*(2), 596-602. http://dx.doi.org/10.1016/j.ecoenv.2008.04.024

Prasad, S., Dwivedi, R., Zeeshan, M., & Singh, R. (2004). UV-B and cadmium induced changes in pigments,photosynthetic electron transport activity, antioxidant levels and antioxidative enzyme activities of *Riccia* sp. *Acta Physiology Plant, 26*, 423-430. http://dx.doi.org/10.1007/s11738-004-0033-8

Qadir, S., Qureshi, M. I., Javed, S., & Abdin, M. Z. (2004). Genotypic variation in phytoremediation potential of Brassica juncea cultivars exposed to Cd stress. *Plant Science, 167*(5), 1171-1181. http://dx.doi.org/10.1016/j.plantsci.2004.06.018

Rauser, W. E. (1999). Structure and function of metal chelators produced by plants. The case for organic acids, amino acids, phytin and metallothioneins. *Cell Biochem Biophys, 31*, 19-48. http://dx.doi.org/10.1007/BF02738153

Sandalio, L. M., Dalurzo, H. C., Gomes, M., Remero-Puertas, M. C., & del Rio, L. A. (2001). Cadmium-induced changes in the growth and oxidative metabolism of pea plants. *J. Exp. Bot., 52*, 2115-2126. http://dx.doi.org/10.1093/jexbot/52.364.2115

Sanita di Toppi, L., & Gabbrielli, R. (1999). Response to cadmium in higher plants - Review. *Environmental and Experimental Botany, 41*, 105-130. http://dx.doi.org/10.1016/S0098-8472(98)00058-6

Schutzendubel, A., & Polle, A. (2002). Plant responses to abiotic stresses: heavy metal induced oxidative stress and protection by mycorrhization. *Journal of Experimental Botany, 53*(372), 1351-1365. http://dx.doi.org/10.1007/978-94-007-4441-7_2

Shah, K., & Dubey, R. S. (1998), A 18 KDa cadmium inducible protein complex, its isolation and characterization from rice (*Oryza sativa* L.) seedlings. *Plant Physiol., 152*, 448-454.

Somogyi, M. (1952). Notes on sugar determination. *Journal of Biological Chemistry, 195*(1), 19-23.

Tziveleka, L., Kaldis, A., Hegedus, A., Kissimon, J., Prombonal, A. Horvath, G., & Arjyroidi-Akoyou, J. (1999). The effect of Cd on chlorophyll and light – Harvesting complex II biosynthesis in greening plants. *Natur. Forsch., 54c*, 740-745.

Verma, S., & Dubey, R. S. (2001), Effect of cadmium on soluble sugars and enzymes of hair metabolism in rice. *Biologia Plantarum., 1*, 117-123. http://dx.doi.org/10.1023/A:1017938809311

Wang, T., & Peverly, J. H. (1999). Iron oxidation states on root surfaces of a wetland plant (*Phragmites australis*). *Soil. Sci. Soc. Am. J., 63*, 247-252.

Weast, R. C. (1984). *CRC handbook of chemistry and physics* (64th ed.). CRC Press, Boca Raton, Fla.

Werff, A. van der. (1955). A new method of concentrating and cleaning diatoms and other organisms. Int. Ver. theor. angew. *Limnol. Verh., 12*, 276-277.

Wo'jcik, M., Skórzyńska-Polit, E., & Tukiendorf, A. (2006). Organic acids accumulation and antioxidant enzyme activities in *Thlaspi caerulescens* under Zn and Cd stress. *Plant Growth Regulation, 48*(2), 145-155. http://dx.doi.org/10.1007/s10725-005-5816-4

Zhang, F., Shi, W., Jim, Z., & Shen, Z. (2003). Response of antioxidative enzymes in cucumber chloroplasts to cadmium toxicity. *J. Plant Nutr., 26*, 1779-1788. http://dx.doi.org/10.1081/PLN-120023282

10

Ecotype and Foliar Fertilization with Florovit Affect Herbage Yield and Quality of Greenhouse-Grown Basil (*Ocimum basilicum* L.)

Joanna Majkowska-Gadomska[1], Emilia Mikulewicz[1], Artur Dobrowolski[1] & Anna Dziedzic[1]

[1] Department of Horticulture, Faculty of Environmental Management and Agriculture, University of Warmia and Mazury in Olsztyn, Olsztyn, Poland

Correspondence: Joanna Majkowska-Gadomska, Department of Horticulture, University of Warmia and Mazury in Olsztyn, Prawocheńskiego 21, 10-957 Olsztyn, Poland. E-mail: majkowska-gadomska@uwm.edu.pl

Abstract

Basil is usually grown in 0.5–0.7 dm^3 pots, at 25 plants per pot. However, potted basil plants wilt and die easily due to high plant density, low substrate volume, insufficient moisture and nutrient depletion. The aim of this study was to determine the herbage yield and nutritional value of greenhouse-grown basil. A two-factorial experiment was performed in a randomized block design with three replications, in 2012–2013. A two-factorial experiment was performed in a randomized block design with three replications, in 2012–2013 Six basil (*Ocimum basilicum* L.) ecotypes were analyzed: sweet basil, 'Queen of Siam' basil, purple basil, cinnamon basil, lemon basil and 'Minette' basil. The second experimental factor was foliar fertilization with Florovit at a concentration of 0.5% and 1%. Basil yield was significantly affected by the ecotype and the interaction between the experimental factors. 'Minette' basil fertilized with 1% Florovit solution was characterized by the highest fresh herbage yield. A statistical analysis revealed that Florovit had no significant effect on basil yield. The concentrations of dry matter, total sugars, L-ascorbic acid and nitrates(V) in basil herbage varied across ecotypes. Foliar fertilization had a significant effect on the organic acid content of basil leaves. The accumulation of the analyzed components in basil herbage was significantly affected by the interaction between the experimental factors. Basil yield was significantly affected by the ecotype. 'Minette' and 'Siam Queen' basil was characterized by the highest fresh herbage yield. The fresh herbage of 'Minette' basil contained the lowest concentrations of dry matter, total sugars, L-ascorbic acid, organic acids and nitrates(V).

Keywords: basil, yield, component, quality, nitrates(V)

1. Introduction

Basil is a popular culinary herb that can be grown in pots and containers Fresh potted basil available on the market is characterized by extended post-harvest shelf-life and quality. Greenhouse production ensures a year-round supply of fresh herbs, spices and seasonings (Aharoni et al., 1993; Jadczak et al., 2006; Biesiada & Kus, 2010). Due to regular control and monitoring of the greenhouse environment, plants grown under cover accumulate lower amounts of undesirable compounds such as nitrates(III), nitrates (V) and heavy metals (Wierzchowska-Renke et al., 1995).

Basil is usually grown in 0.5–0.7 dm^3 pots, at 25 plants per pot. However, potted basil plants wilt and die easily due to high plant density, low substrate volume, insufficient moisture and nutrient depletion. In view of the above, research was undertaken to improve greenhouse basil production conditions. Six basil ecotypes were compared in the study. Basil plants were grown in 3.0 dm^3 containers to ensure adequate substrate solution volume. Another important consideration in greenhouse production is the appropriate use of fertilizers to maximize the yield and improve the biological value of potted herb plants. However, fertilizer rates are usually calculated for basil grown in the field (Nurzynska-Wierdak, 2012). Most research studies conducted to date have investigated simple fertilizers, including foliar application of 0.5% urea solution (Nurzynska-Wierdak & Borowski, 2011), fertigation with the use of ammonium nitrate and potassium sulfate (Nurzyńska-Wierdak et al., 2012), and the application of calcium carbonate to neutralize soil acidity (Dzida, 2010a).

The objective of this study was to determine the herbage yield and nutritional value of greenhouse-grown basil fertilized with the compound fertilizer Florovit.

2. Method

A two-factorial experiment was performed in a randomized block design with three replications, in 2012–2013, in a greenhouse in the Experimental Garden of the University of Warmia and Mazury in Olsztyn. The first experimental factor were six basil (*Ocimum basilicum* L.) ecotypes: sweet basil, 'Queen of Siam' basil, purple basil, cinnamon basil, lemon basil and 'Minette' basil. The second experimental factor was foliar fertilization with Florovit at a concentration of 0.5% and 1%. Florovit is a liquid compound fertilizer that can be applied to plant foliage and soil. It has a pH of 3.3–4.3, and contains N : P : K (3.0 : 0.0 : 2.0), copper (70 mg dm^{-3}), iron (400 mg dm^{-3}), manganese (170 mg dm^{-3}), molybdenum (20 mg dm^{-3}), zinc (150 mg dm^{-3}), calcium, magnesium, sulfur and film-forming agents (inco.pl leaflet).

Basil plants were grown in 3.0 dm^3 containers placed on movable flood tables. Experimental unit area was 1 m^2, and each replicate consisted of 25 pots. Each year, in mid-February, basil seeds were sown in containers (four seeds per container) filled with organic substrate with a pH (H_2O) of 6.7, salt concentration of 0.23 g kg^{-1} and the following chemical composition: N-NH_4 – 25 mg dm^{-3}, N-NO_3 – 10 mg dm^{-3}, P – 41 mg dm^{-3}, K – 129 mg dm^{-3}, Ca – 1050 mg dm^{-3}, Mg – 87 mg dm^{-3}. The recommended cultivation practices for basil were applied during the growing season. Fertilization began two weeks after seedling emergence, and it was continued throughout the growing season at 14-day intervals. Basil leaves were harvested at the beginning of flowering. The plants were cut manually, 5 cm above the ground. The chemical composition of basil herbage was determined at harvest. Average samples of basil herbage were collected from the marketable yield in each treatment, according to Polish Standard PN-72/A-75050. The samples were analyzed in the laboratory of the Department of Horticulture, University of Warmia and Mazury in Olsztyn to determine the content of:

- Dry matter - by drying the plant material at 105 °C to constant weight (PN-90/A-75101/03);
- Total sugars - by the Luff-Schoorl method (PN-90/A-75101/07);
- L-ascorbic acid - by the Tillmans method modified by Pijanowski (PN-90/A-75101/11);
- Organic acids expressed as malic acid equivalents – by titration, as described by Pietersburgski;
- ß-carotene – by column chromatography (PN-90/A-75101/12);
- Nitrates(V) – by the colorimetric method with the use of salicylic acid.

The results were processed statistically by analysis of variance (ANOVA). The significance of differences between means was estimated by Tukey's range test at α = 0.01. All calculations were performed using STATISTICA 10 software.

3. Results and Discussion

A significant correlation was noted between the herbage yield and ecotype of basil. 'Minette' basil was characterized by the highest fresh herbage, dried herbage and crumbled herbage yields. 'Siam Queen' basil produced similar herbage yields. High dry herbage and crumbled herbage yields were also attained in sweet basil. Our results are comparable with those reported by Roslon et al. (2011) in whose study the fresh herbage yield of basil reached 1.04 kg m^{-2}. Cinnamon basil and lemon basil were characterized by the lowest fresh herbage, dried herbage and crumbled herbage yields.

Previous research has shown that the herbage yield and quality of basil are significantly affected by fertilization. In a study by Nurzynska-Wierdak et al. (2011a), fresh herbage yield was significantly higher (3.40 kg m^{-2}, two-year average) in basil plants grown in an unheated plastic tunnel and fertilized with nitrogen. In a pot experiment conducted by Biesiada and Kus (2010), the fresh herbage yield of basil ranged from 2.65 to 4.34 kg m^{-2} depending on fertilizer rate. In our study, a statistical analysis of two-year data demonstrated that Florovit had no significant effect on the fresh herbage, dried herbage or crumbled herbage yields of basil (Table 1).

Table 1. The effect of ecotype and foliar fertilization with Florovit on basil yield (kg m^{-2})

Basil ecotype	Fertilizer rate	Fresh herbage yield	Dry herbage yield	Crumbled herbage yield
		(kg m^{-2})		
Sweet basil	I	0.62	0.25	0.04
	II	0.76	0.20	0.08
	III	0.90	0.16	0.10
Mean		0.76	0.21	0.07
'Siam Queen' basil	I	0.78	0.17	0.05
	II	0.94	0.16	0.07
	III	1.09	0.17	0.11
Mean		0.93	0.17	0.08
'Minette' basil	I	0.98	0.19	0.08
	II	1.07	0.20	0.09
	III	1.11	0.22	0.09
Mean		1.05	0.20	0.08
Purple basil	I	0.39	0.06	0.04
	II	0.51	0.09	0.08
	III	0.64	0.15	0.08
Mean		0.51	0.10	0.07
Cinnamon basil	I	0.22	0.05	0.03
	II	0.33	0.04	0.02
	III	0.32	0.04	0.02
Mean		0.29	0.04	0.02
Lemon basil	I	0.26	0.10	0.02
	II	0.22	0.04	0.03
	III	0.20	0.06	0.03
Mean		0.23	0.07	0.03
Mean for fertilizer rate	I	0.54	0.14	0.04
	II	0.64	0.12	0.06
	III	0.71	0.13	0.07
LSDα=0.01				
Basil ecotype		0.16	0.10	0.02
Fertilizer rate		n.s.	n.s.	n.s.
Interaction		0.2	n.s.	0.02

Note. I – control; II – 0.5% fertilizer rate; III – 1.0% fertilizer rate.

An analysis of the interaction between the experimental factors revealed that 'Minette' basil fertilized with 1.0% Florovit solution produced the highest fresh herbage yield (1.11 kg m^{-2}). Sweet basil grown in the control treatment was characterized by the highest dry herbage yield (0.25 kg m^{-2}). Crumbled herbage yield was highest in 'Siam Queen' basil and sweet basil fertilized with 1.0% Florovit solution (0.11 and 0.10 kg m^{-2}, respectively).

Basil is a popular culinary herb widely used in the Mediterranean cuisine. This spice and medicinal plant is rich in biologically active substances that deliver health benefits (Nurzynska-Wierdak, 2012). In the present study, the dry matter content of fresh basil herbage was significantly affected by the ecotype (Table 2). Lemon basil and cinnamon basil had the highest dry matter content (19.30% and 19.14%, respectively). Purple basil and 'Minette' basil were least abundant in dry matter (11.63% and 13.72%, respectively). Foliar fertilization had no significant effect on the dry matter content of basil herbage, which was significantly influenced by the interaction between the experimental factors. The highest dry matter content was noted in lemon basil grown in the control treatment and in cinnamon basil fertilized with 0.5% Florovit solution. The lowest dry matter content was observed in purple basil, regardless of treatment, and in 'Minette' basil from the control treatment. The highest dry matter content of basil herbage, noted in our study, is similar to the values reported by Dzida (2010b) for basil fertilized with calcium carbonate (28.42% and 29.15%). In experiments carried out by Martyniak-Przybyszewska and Wojciechowski (2004), and Jadczak and Grzeszczuk (2008), the dry matter content of basil herbage reached 12.50% and 12.89%, respectively. In the work of Nurzynska-Wierdak et al. (2011b), basil plants fertilized with various potassium and nitrogen rates had a dry matter content of 13.20%.

Table 2. The effect of ecotype and foliar fertilization with Florovit on the content of dry matter and organic compounds in basil herbage

Basil ecotype	Fertilizer rate	Dry matter	Total sugar	Ascorbic acid	Organic acid	ß-carotene	Nitrate (V)
		(%)	(g $100g^{-1}$ f.m.)	(mg $100g^{-1}$ f.m.)	(mg $100g^{-1}$ f.m.)	(μg·$100g^{-1}$ f.m.)	(mg NO_3·kg^{-1} f.m.)
Sweet basil	I	15.39	7.54	11.08	2.21	563	1127
	II	15.93	7.20	12.04	2.14	537	1471
	III	18.22	8.22	10.69	2.48	576	1098
Mean		16.51	7.65	11.27	2.28	559	1232
'Siam Queen' basil	I	14.18	8.16	9.68	1.74	579	1320
	II	15.05	8.94	11.56	1.94	538	1089
	III	15.27	11.56	10.16	2.28	537	1029
Mean		14.83	9.55	10.47	1.99	551	1146
'Minette' basil	I	12.80	6.60	9.90	1.88	768	1127
	II	14.90	7.20	9.20	1.94	568	751
	III	13.47	7.44	9.51	2.08	534	615
Mean		13.72	7.08	9.54	1.97	623	831
Purple basil	I	11.91	8.98	18.65	2.35	584	551
	II	10.99	6.48	17.84	1.68	553	633
	III	11.99	7.98	18.09	1.94	505	695
Mean		11.63	7.81	18.19	1.99	547	627
Cinnamon basil	I	19.36	6.00	10.81	2.41	593	1274
	II	22.50	10.30	11.87	1.88	559	743
	III	15.56	8.14	10.94	2.21	562	1137
Mean		19.14	8.15	11.21	2.17	571	1051
Lemon basil	I	22.98	7.37	9.99	2.28	518	1904
	II	15.87	9.75	17.44	2.08	364	1198
	III	19.05	8.75	15.70	2.08	564	1319
Mean		19.30	8.62	14.38	2.14	482	1474
Mean for fertilizer rate	I	16.10	7.44	11.69	2.14	601	1217
	II	15.87	8.20	13.33	1.94	520	981
	III	15.59	8.37	12.52	2.18	546	982
LSDα=0.01							
Basil ecotype		2.63	1.50	1.86	n.s.	n.s.	288
Fertilizer rate		n.s.	n.s.	n.s.	0.19	n.s.	n.s.
Interaction		2,21	0.22	0.32	0.22	221	219

Note. I – control; II – 0.5% fertilizer rate; III – 1.0% fertilizer rate.

The concentrations of total sugars in basil herbage were affected by the ecotype and the ecotype x fertilizer interaction. Total sugar content was highest in 'Siam Queen' basil and lowest in 'Minette' basil. Florovit applied at a concentration of 1% contributed to the accumulation of total sugars in basil herbage (11.56 g 100 g^{-1} fresh weight). A higher total sugar content of basil herbage was reported by Nurzynska-Wierdak (2012).

L-ascorbic acid levels in basil herbage ranged from 9.68 to 18.65 mg 100g^{-1} fresh weight, and they were significantly affected by the ecotype and the ecotype x fertilizer interaction. Purple basil had the highest L-ascorbic acid content, at 18.19 mg 100 g^{-1} fresh weight (mean of 2012-2013). Insignificantly higher concentrations of L-ascorbic acid (19.41 mg 100 g^{-1} fresh weight) were reported by Dzida (2011) for purple basil grown in pots.

A statistical analysis revealed that the organic acid content of basil herbage was significantly affected by foliar fertilization with Florovit and the fertilizer x ecotype interaction. The highest organic acid concentrations (2.48 g 100 g^{-1} fresh weight on average) were noted in sweet basil fertilized with 1% Florovit solution, and the lowest – in 'Siam Queen' basil (1.68 g 100 g^{-1} fresh weight). The effect of Florovit on the organic acid content of basil herbage was significant, but it varied across ecotypes depending on rate.

Herb plants are a rich source of antioxidant compounds, including carotenoids (Nurzynska-Wierdak, 2012, Kopsell et al., 2005). According to Daly et al. (2010), the average ß-carotene content of basil herbage is 258 $\mu g \cdot 100\ g^{-1}$. In our study, ß-carotene concentrations were highest in 'Minette' basil grown in the control treatment (768 μg 100 g^{-1}), and lowest in lemon basil fertilized with 0.5% Florovit solution (364 μg 100 g^{-1}).

The nitrate(V) content of herbage is an important consideration in the greenhouse production of herbs and spices. Nitrate accumulation in plant material can be prevented with the use of liquid organic fertilizers, which are particularly recommended for certified organic plantations (Wierzchowska-Renke i in. 1995). No maximum permissible nitrate(V) levels have been established for herb and spice plants. The results of chemical analyses indicate that the herbage of control lemon basil plants had the highest nitrate(V) content (1904 mg $NO_3 \cdot kg^{-1}$ fresh weight). Foliar fertilization did not increase nitrate concentrations, which reached an average of 981 and 982 mg $NO_3 \cdot kg^{-1}$ fresh weight in basil ecotypes fertilized with 0.5% and 1% Florovit solutions, respectively. In a study by Seidler-Łozykowska et al. (2007, 2009), nitrate(V) levels in organically-grown basil ranged from 306.2 to 5250.0 mg $NO_3 \cdot kg^{-1}$ fresh weight, depending on the region and cultivation method. According to Telesiński et al. (2013), the nitrate(V) content of basil herbage can be as high as 9950 mg $NO_3 \cdot kg^{-1}$ fresh weight.

4. Conclusions

Basil yield was significantly affected by the ecotype. 'Minette' and 'Siam Queen' basil was characterized by the highest fresh herbage yield.

Foliar fertilization with Florovit solution had no statistically significant effect on the herbage yield of greenhouse-grown basil.

The fresh herbage of 'Minette' basil contained the lowest concentrations of dry matter, total sugars, L-ascorbic acid, organic acids and nitrates(V).

Acknowledgements

The study was financed as part of research project No. 1014.0804.

References

Aharoni, N., Dvir, O., Chalupowicz, D., & Aharon, Z. (1993). Coping with postharvest physiology of fresh culinary herbs. *Acta Hort, 344*, 69-78.

Biesiada, A., & Kus, A. (2010). The effect of nitrogen fertilization and irrigation on yielding and nutritional status of sweet basil (*Ocimum basilicum* L.). *Acta Sci. Pol., Hortorum Cultus, 9*(2), 3-12.

Daly, T., Jiwan, M. A., O'Brien, N. M., & Aherne, S. A. (2010). Carotenoid content of commonly consumed herbs and assessment of their bioaccessibility using an in vitro digestion model. *Plant Foods Human Nutr, 65*, 164-169. http://dx.doi.org/10.1007/s11130-010-0167-3

Dzida, K. (2010a). Nutriens contents in sweet basil (*Ocimum basilicum* L.) herb depending on calcium carbonate dose and cultivar. *Acta Sci. Pol., Hortorum Cultus, 9*(4), 143-151.

Dzida, K. (2010b). Biological value and essential oil content in sweet basil (*Ocimmum basilicum* L.) depending on calcium fertilization and cultivar. *Acta Sci. Pol., Hortorum Cultus, 9*(4), 153-161.

Dzida, K. (2011). Influence of nitrogen nutrition and cultivar on quality of sweet basil herbs. *Ann. UMCS, 24*(3), 125-132.

Kopsell, D. A., Kopsell, D. E., & Curran-Celentano, J. (2005). Carotenoid and chlorophyll pigments in sweet basil grown in the field and greenhouse. *HortSci., 40*(5), 1230-1233.

Jadczak, D., & Grzeszczuk, M. (2008). Zioła przyprawowe – wartość biologiczna wybranych gatunków. *Panacea, 2*(23), 15-17.

Jadczak, D., Błaszczuk, A., & Rekowska, E. (2006). Effect of cocering on the of macroelements in field of Basil (*Ocimum basilicum* L.) cultivated for bunch harvest. *J. Elementol., 11*(2), 135-141.

Martyniak-Przybyszewska, B., & Wojciechowski, T. (2004). Plonowanie wybranych gatunków roślin przyprawowych w rejonie Olsztyna. *Folia Unic. Agric. Stetin., Agricultura, 239*(95), 245.

Nurzynska-Wierdak, R. (2012). *Ocimum basilicum* L. – wartościowa roślina przyprawowa, lecznicza i olejkodajna. *Ann. UMCS, 21*(1), 20-30.

Nurzynska-Wierdak, R., & Borowski, B. (2011). Dynamics of sweet basil (*Ocimum basilicum* L.) growth affected by cultivar and foliar feeding with nitrogen. *Acta Sci. Pol., Hortorum Cultus, 10*(3), 307-317.

Nurzynska-Wierdak, R., Borowski, B., & Dzida, K. (2011a). Yield and chemical composition of basil herb depending on cultivar and foliar feeding with nitrogen. *Acta Sci. Pol., Hortorum Cultus, 10*(1), 207-219.

Nurzynska-Wierdak, R., Rożek, E., & Borowski, B. (2011b). Response of different basil cultivars to nitrogen and potassium fertilization: total and mineral nitrogen content in herb. *Acta Sci. Pol., Hortorum Cultus, 10*(4), 217-232.

Nurzynska-Wierdak, R., Rożek, E., Dzida, K., & Borowski, B. (2012). Growth response to nitrogen and potassium fertilization of common basil (*Ocimum basilicum* L.) plants. *Acta Sci. Pol., Hortorum Cultus, 11*(2), 275-288.

Roslon, W., Osinska, E., Baczek, K., & Weglarz, Z. (2011). The influence of organic-mineral fertilizers on yield and raw materials quality of chosen plants of the *Lamiaceae* family from organic cultivation. *Acta Sci. Pol., Hortorum Cultus, 10*(1), 147-158.

Seidler-Łozykowska, K., Kozik, E., Golcz, A., & Wójcik, J. (2007). Quality of basil herb (*Ocimum basilicum* L.) from organic and conventional cultivation. *Herba Polonica, 53*(3), 41-46.

Seidler-Łozykowska, K., Mordalski, R., Kucharski, W., Golcz, A., Kozik, E., & Wojcik, J. (2009). Economic and qualitative value of the raw material of chosen species of medicinal plants from organic farming part ii. yield and quality of sweet basil herb (*Ocimum basilicum* L.). *Acta Sci. Pol., Agricultura, 8*(3), 29-35.

Telesiński, A., Grzeszczuk, M., Jadczak, D., Wysocka, G., & Onyszko, M. (2013). Ocena zmian zawartości azotanów (V) w wybranych ziołach przyprawowych w zależności od sposobu ich utrwalania i czasu przechowywania. *Żywność. Nauka. Technologia. Jakość., 5*(90), 168-176.

Wierzchowska-Renke, K., Gajewska, R., & Nabrzynski, M. (1995). Zawartość azotynów i azotanów w ziołach. *Wiad. Zielar., 6*, 13-14.

11

Impacts of Climate Variability and Change on Rainfed Sorghum and Maize: Implications for Food Security Policy in Tanzania

Barnabas M. Msongaleli[1], Filbert Rwehumbiza[2], Siza D. Tumbo[3] & Nganga Kihupi[3]

[1] Department of Geography and Environmental Studies, School of Social Sciences, University of Dodoma, Dodoma, Tanzania

[2] Department of Soil Science, Sokoine University of Agriculture, Morogoro, Tanzania

[3] Department of Agricultural Engineering and Land Planning, Sokoine University of Agriculture, Morogoro, Tanzania

Correspondence: Barnabas M. Msongaleli, Department of Geography and Environmental Studies, School of Social Sciences, University of Dodoma, P. O. Box 359 Dodoma, Tanzania.

Abstract

Concern about food security has increased because of a changing climate, which poses a great threat to food crop productivity. Climate change projections from the Coupled Model Inter-comparison Project phase 5 (CMIP5) and crop models were used to investigate the impacts of climate change on rain-fed cereal production. Calibrated and evaluated crop models simulated maize and sorghum yields over time periods and scenarios across central zone Tanzania with and without adaptation. Simulation outputs without adaptation showed predominant decrease and increase in maize and sorghum yields, respectively. The results showed that maize yields were predicted to decline between 1% and 25% across periods, representative concentration pathways (RCPs) and global circulation models (GCMs). However, sorghum yields were on average predicted to increase between 5% and 21%. Overall when adaptation is incorporated toward mid-century, yields are projected to increase for both crops. The yield projections variation between cereal crops highlights the importance of location and crop specific climate change impact assessments. Despite the uncertainties in predicting the impacts of climate change on rainfed crops, especially on cereals (maize and sorghum) which are important staple food crops in semi-arid Tanzania, the findings of this study enable policy makers to develop plans aimed at sustainable food security. In conclusion, the results demonstrate the presumption that sorghum productivity stands a better chance than maize under prospects of negative impacts from climate change in central zone Tanzania.

Keywords: agronomic adaptation, cereals, climate change, policy, simulation modeling, uncertainty

1. Introduction

Although several studies project the net effect of climate change on cereal yields to be negative in sub-Saharan Africa (SSA), the direction of yield change in any given area depends on the physiology of the crop concerned and the current climatic condition under which it is grown because different species have different base and optimum temperatures for development (Porter & Semenov, 2005; Lobell et al., 2008; Challinor et al., 2014). Rainfall projections from an assessment of 12 CMIP3 (AR4) GCMs over eastern Africa suggest an increase in rainfall by the end of the 21st Century (Shongwe, van Oldenborgh, van den Hurk, & van Aalst, 2011; IPCC, 2013), though confounded by extreme precipitation changes (droughts and heavy rainfall) during the last 30-60 years (Williams & Funk, 2011; Lyon & DeWitt, 2012). Rainfall projections for Tanzania appear consistent with those of eastern Africa, which indicate increase in annual rainfall with the ensemble range spanning changes of -4 to +30% by the 2090s (McSweeney, New, & Lozano, 2010). However, rainfall projections data from CMIP5 (Taylor, Stouffer, & Meehl, 2012) used in the Fifth Assessment Report (AR5) escalate the uncertainty in that, while overall rainfall is projected to increase by 9% for central Tanzania as observed by Wambura, Ngongolo, Mlonganile, Sangalugembe, and Tumbo (2014), analyses by Taylor et al. (2013) of both the Fourth Assessment Report (AR4) and AR5 multi-model ensembles attributed the projected increases to extreme monthly rainfall rather than changes to mean rainfall. Previous studies had indicated a decline in rainfall in future for central areas of Tanzania (Paavola, 2003; Matari, Chang'a, Chikojo, & Hyera, 2008).

Because of the uncertainties in processes underpinning the changing climate especially on rainfall projections, more research is needed to understand the influence of the projections on crop production on local conditions. Information is required to enable an understanding of how the projected changes in climate will impact smallholder dryland farmers who need to strategically respond and adapt to the ill-effects of changing climate (Hatfield et al., 2011; Rurinda et al., 2014). It is due to this wide range of projections and the possible impacts on crop production at different scales and over different time periods that detailed studies are needed, because the underlying uncertainties exacerbate concerns about food security and impede decision making on food security policy and climate change issues (Ingram, Gregory, & Izac, 2008; Thornton, Jones, Ericksen, & Challinor, 2011; Yao, Qin, Zhang, Lin, & Boken, 2010).

Overall, although rainfall projections in eastern Africa depict a glimpse of hope regarding possible impacts on crop productivity (Doherty, 2009), other studies indicate contrasting conclusions, such as enhanced crop yields or no change in crop yields (Kurukulasuriya & Mendelsohn, 2008) and yield decreases (Thornton, Jones, Alagarswamy, & Andresen, 2009). Other studies in SSA show some evidence of negative climate change impact on crop yield for major staple cereal food crops like maize, sorghum and millet (Schlenker & Lobell, 2010; Knox, Hess, Daccache, & Wheeler, 2012; Berg, De Noblet-Ducoudre, Sultan, Lengaigne, & Guimberteau, 2013; Zinyengere, Crespo, & Hachigonta, 2013; Waha, Müller, & Rolinski, 2013). These studies give generalized and broad conclusions about the impact of climate change which are not manifest in crop production to address increased food security concerns. In Tanzania, in particular, several studies indicate that maize production is projected to decline in the future (Mwandosya, Nyenzi, & Luhanga, 1998; Arndt, Farmer, Strzepek, & Thurlow, 2011; Rowhani, Lobell, Linderman, & Ramankutty, 2011; Kilembe, Thomas, Waithaka, Kyotalimye, & Tumbo, 2013).

The projected yields of crops under a range of climate scenarios, however, suffer from the limitations associated with the difficulty in obtaining data on local conditions or crop characteristics (Ruane et al., 2013, Watson & Challinor, 2013; Thornton, Jones, Alagarswamy, Andresen, & Herrero, 2010); uncertainties in climate data (e.g. Ramirez-Villegas, Challinor, Thornton, & Jarvis, 2013) and uncertainties in crop models' processes (Ainsworth, Leaky, Ort, & Long, 2008). Currently, however, the weaknesses with regard to data availability and uncertainties are addressed in the methodological procedures of the Agricultural Model inter-comparison and Improvement Project (AgMIP) through aggregation of geographic data regarding the spatial distribution of climate (daily weather), topography, soils, land-use, farm-level management, socioeconomic conditions, and reported yields (Rosenzweig et al., 2013a, 2013b). Moreover, the use of statistical methods to evaluate and understand uncertainty in the outputs of climate change impacts has been recognized as it enhances drawing of robust conclusions regarding model applications (Falloon et al., 2014).

Due to high dependence on rain-fed agriculture in Tanzania, it is clear that smallholder farmers are sensitive to possible adverse changes in climate and they are faced with the question of how to adapt to climate change. Therefore, they need information on the potential impact of climate change for the next few decades. Recent studies have established that an undertaking of climate change impact assessments at local scales is essential as it allows exploration of local agronomic management practices and their incorporation into adaptation strategies formulation (Zinyengere, Crespo, Hachigonta, & Tadross, 2014). In Tanzania, there is paucity of information on impacts of climate change on sorghum and maize though studies elsewhere seem to suggest that impacts on sorghum are predicted to be less significant compared to maize. For example, according to Lobell et al. (2008) maize yield in Southern Africa is projected to decline by about 30% compared with a decrease of only 2% for sorghum by 2030. Other studies show that sorghum will increase by a range of 19 to 72% across eastern and southern Africa (Zinyengere et al., 2014; Turner & Rao, 2013).

Detailed crop simulation studies at various scales are required due to spatial variability of climate especially rainfall, in order to provide relevant knowledge on impacts and for evaluating possible adaptation options under farm and policy levels (Thompson, Berrang-Ford, & Ford, 2010; White, Hoogenboom, Kimball, & Wall, 2011). The use of multiple-models in climate change assessment has shown to enhance the quantification of uncertainties, as different models differ in structure and parameter values (Rötter, Carter, Olesen, & Porter, 2011). A significant proportion of uncertainty in climate impact projections has been attributed to variations among crop models (Asseng et al., 2013). In this study the Agricultural Production Systems sIMulator (APSIM) and Decision Support System for Agrotechnology Transfer (DSSAT) crop models were used to quantify impacts of climate variability and change on rainfed sorghum and maize productivity across five locations having contrasting soil properties and crop management practices. The specific objectives were; first, to link APSIM and DSSAT with GCMs to simulate rainfed sorghum and maize production under the CMIP5;

second, to simulate scenarios representing some agronomic strategies feasible under conditions of dryland farming to provide insight into their potential for adaptation; and third to evaluate uncertainty in climate change impacts on sorghum and maize in order to provide relevant information to policy makers and others.

2. Methods

2.1 Description of the Study Area

Central Tanzania (Singida and Dodoma regions) has been identified as one of livelihood zones based on FAO (Figure 1). The zone is designated as "sorghum-livestock" and is most relevant to sorghum production. The central regions account for three-quarters of Tanzania's 500,000 to 800,000t annual sorghum harvests. The zone is one of the most sensitive to climate variability and change mainly owing to temperature and rainfall variability. Soils in this zone are mainly sandy and loamy of low fertility and seasonally waterlogged or flooded pockets of clays. Weather stations at five (5) locations were identified from which weather data for running crop simulation models were obtained (Table 1). Observed maize and sorghum yields across the zone were obtained from the Tanzania National Panel Survey (TNPS) of 2010-2011 (NBS, 2012) for the 2009/10 season.

Table 1. Geographical locations and rainfall characteristics of five weather stations in study area

Station	Latitude	Longitude	Altitude (m.a.s.l)	Annual rainfall (mm)	JFM (mm)	OND (mm)
Mpwapwa	-6.2	36.30	1007	584	330	183
Dodoma	-6.167	35.67	1118	567	366	107
Hombolo	-5.75	35.95	1062	627	379	180
Singida	-4.48	34.45	1377	797	419	228
Manyoni	-5.44	34.50	1245	695	385	222

Note. m.a.s.l = metres above sea level; JFM = January February March; OND = October November December.

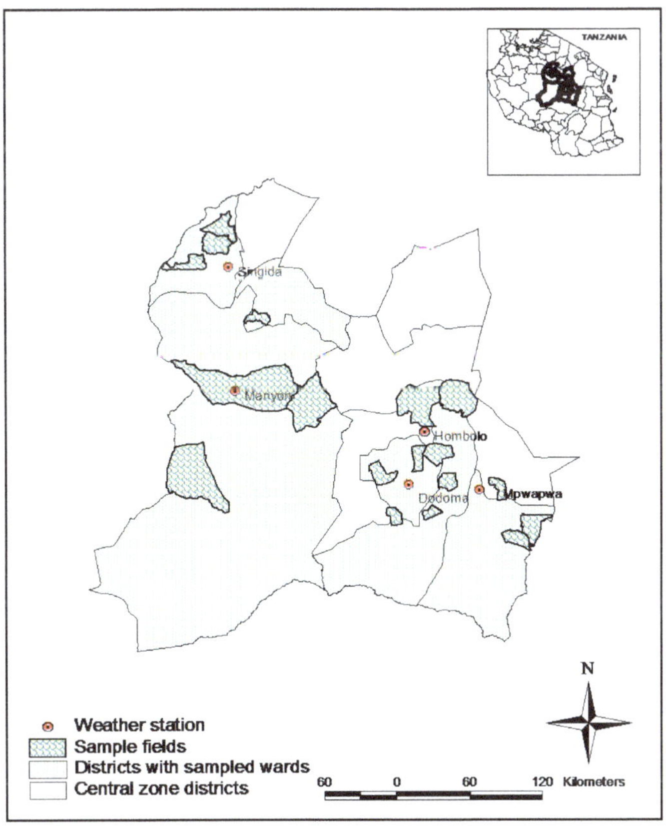

Figure 1. Map showing locations of fields with reported sorghum and maize yields used in model evaluation

2.2 Model Description, Calibration and Evaluation

Calibration and evaluation of APSIM (Keating et al., 2003) model used in this study was done based on experimental data for two growing seasons. APSIM model was evaluated for simulation of days after sowing to flowering and maturity, dry matter accumulation (biological yield) and grain yield. Genetic coefficients used by APSIM and DSSAT for sorghum are shown in Tables 2 and 3, respectively. Data for calibrating and evaluating both models for maize (*Situka* variety) were obtained from literature (Mourice et al., 2014).

Table 2. Crop parameters in APSIM for three sorghum varieties

Parameter		Source	Units	Macia	Tegemeo	Pato
Thermal time accumulation	End of juvenile phase to panicle initiation	C	°C day	230	270	275
	Flag stage to flowering	C	°C day	195	170	175
	Flowering to start of grain filling	C	°C day	80	80	100
	Flowering to maturity	C	°C day	675	760	760
	Maturity to seed ripening	L	°C day	1	1	1
Photoperiod	Day length photoperiod to inhibit flowering	D	H	11.5	11.5	11.5
	Day length photoperiod for insensitivity	D	H	13.5	13.5	13.5
	Photoperiod slope	L	°C/h	0.01	0.01	0.01
	Base temperature	L	°C day	8	8	8
	Optimum temperature	D	°C day	30	30	30
	Plant height (max)	O	mm	1290	1650	1780

Note. C: calibrated; D: Default; L: literature; O: observed.

Table 3. Genetic coefficients in DSSAT for three sorghum varieties

Coefficient	Definition	Macia	Tegemeo	Pato
TBASE	Base temperature below which no development occurs, °C	8.0	8.0	8.0
TOPT	Temperature at which maximum development rate occurs during vegetative stages°C	34.0	34.0	34.0
ROPT	Temperature at which maximum development rate occurs for reproductive stages °C	34.0	34.0	34.0
P1	Thermal time from seedling emergence to the end of the juvenile phase (expressed in degree days above a base temperature of 8°C) during which the plant is not responsive to changes in photoperiod.	300	440	460
P_2O	Critical photoperiod or the longest day length (in hours) at which development occurs at a maximum rate. At values higher than P20, the rate of development is reduced.	12.5	12.5	12.5
P_2R	Extent to which phasic development leading to panicle initiation (expressed in degree days) is delayed for each hour increase in photoperiod above P20.	1	1	1
P5	Thermal time (degree days above a base temperature of 8°C) from beginning of grain filling (3-4 days after flowering) to physiological maturity.	520.0	650.0	650.0
G1	Scaler for relative leaf size.	15	15	15
G2	Scaler for partitioning of assimilates to the panicle (head).	6.5	6.0	6.0
PHINT	Phylochron interval; the interval in thermal time (degree days) between successive leaf tip appearances.	49.0	49.0	49.0

2.3 Crop Management and Soil Input Data

Macia sorghum variety was chosen for subsequent analyses because it is high-yielding with early maturity, and currently highly preferred by smallholder farmers. Similarly, Situka maize variety was chosen based on its early maturity and tolerance to low nitrogen typical of the study area farm characteristics. Crop management information including planting dates, plant population, varieties used, organic and inorganic fertilizer use, intercropping and measured yields, were obtained through key informants structured interview and augmented by the information from TNPS. This information was used in the construction of a farm survey template as per AgMIP protocols (Rosenzweig et al., 2013b). For the current analysis, a sample of 63 fields planted with maize and 48 fields planted with sorghum were extracted from the database (based on available reported yields), to construct farm survey template in central Tanzania. A Quick and Dirty [File Translation] User Interface (QUADUI) tool (http://www.agmip.org/tools) was used to translate the templates into ready model run files for the simulations.

Soil profile data used to parameterize soil modules within APSIM and DSSAT were mainly extracted from available databases e.g. Africa Soil Information Service (AfSIS) (Leenaars, 2012) and WISE (2008), but also data from freshly dug soil profiles were used. A total of 5 soil profiles were identified to represent variable soils available within central Tanzania. Considering the variability of soils across farms, the soil profiles were deliberately subdivided to capture the soil quality based on increasing or decreasing the amount of organic carbon and amount of available water by 20% as described in the AgMIP Handbook (Rosenzweig et al., 2013b). The subdivisions resulted in a total of 15 soil profiles with classes of poor, average and good quality soils. A range of analytical data for soil profiles used in the models is presented in Table 4.

Table 4. Range of soil analytical data for soil profiles used in simulations

Depth of bottom (cm)	Clay (%)	Silt (%)	Organic carbon (%)	pH in water	Cation Exchange Capacity (cmol/kg)	Lower Limit (LL) (cm^3/cm^3)	Drained upper limit (DUL) (cm^3/cm^3)	Saturation (SAT) (cm^3/cm^3)
15-25	13-55	6-20	0.35-1.26	5.4-5.9	12.8-24.2	0.077-0.355	0.128-0.441	0.349-0.499
30-45	9-59	4-16	0.24-0.76	5.1-4.8	9.0-22.1	0.057-0.375	0.108-0.454	0.358-0.517
46-80	12-59	2-18	0.1-0.6	4.5-4.8	7.5-21.8	0.102-0.386	0.216-0.456	0.324-0.508
102-115	9-55	3-6	0.04-0.49	4.0-4.8	6.8-23.0	0.150-0.305	0.362-0.441	0.376-0.529

2.4 Climate and Data Scenarios

Climate change scenarios for near-term (2010-2039), mid-century (2040-2069) and end-century (2071-2099) periods were generated using 20 GCMs from CMIP5 (Table 5) for two Representative Concentration Pathways (RCPs): RCP 4.5 and RCP 8.5 bias-corrected using the method of Hempel et al. (2013). The simulations were performed for the three climate change scenarios using data from all 20 GCMs. Moreover, climate data of five GCMs namely: CCSM4, GFDL-ESM2M, HadGEM2-ES, MIROC5, and MPI-ESM-MR, were separately analysed for mean changes in projected climate compared with baseline (1980-2010). These GCMs were selected due to their long history of development and evaluation, a preference for higher resolution, and established performance in monsoon regions (Rosenzweig et al., 2013b).

RCPs usually refer to the portion of the concentration pathway extending up to 2100, for which Integrated Assessment Models have produced corresponding emission scenarios (IPCC, 2013). The RCP8.5 is a high emissions scenario, corresponding to projections of high human population (12 billion by 2100), high rates of urbanization and limited rates of technological change, all resulting in emissions approaching 30 Gt of carbon by 2100 compared with 8Gt in 2000 (Riahi et al., 2007). The RCP4.5 scenario is an intermediate mitigation scenario characterized by continuously increasing human population but at a rate lower than in the RCP8.5 scenario, intermediate levels of economic development and less rapid and more diverse technological change (Moss et al., 2010).

2.5 Evaluation of Adaptation Options

According to NAPA (URT, 2007) potential adaptation measures include adjustments in management practices such as planting density, fertilizer application and planting date. Planting densities above those currently practised by smallholder dryland farmers were adopted assuming improvement in extension services and farmers' reception and adoption of improved agricultural technologies. Planting density and fertilizer application were combined with two planting dates as agronomic management scenarios for each crop (Table 6). The selected agronomic management scenarios were based on local expert recommendations under conditions of central Tanzania and affordability by the local farmers. These were obtained from agricultural reports in respective districts. Fertilizer amounts are those recommended by experts for the low-input systems predominant in semi-arid areas of Tanzania. An early planting date (EP) corresponds to onset of rains, which have shown early trends and a late planting date (LP) serves to test the possibility of shifts of rainfall pattern in the future.

Table 5. Coupled Model Intercomparison Project phase 5 (CMIP5) general circulation models considered in this study

	Modelling centre	Country	Model	Lat.	Lon.	Res.
i	Commonwealth Scientific and Industrial Research Organisation/Bureau of Meteorology (CSIRO-BOM)	Australia	ACCESS1.0	1.87	1.25	MR
ii	Beijing Climate Centre, China Meteorological Administration	China	BCC-CSM1.1	2.81	2.79	LR
iii	College of Global Change and Earth System Science, Beijing Normal University	China	BNU-ESM	2.81	2.79	LR
iv	Community Climate System Model, Climate and Global Dynamics Division/ National Centre for Atmospheric Research	USA	CCSM4			
v	Community Earth System Model, Climate and Global Dynamics Division/ National Centre for Atmospheric Research	USA	CESM1-BGC			
vi	Commonwealth Scientific and Industrial Research Organisation/Queensland Climate Change Centre of Excellence (QCCCE)	Australia	CSIRO-Mk3.6	1.87	1.87	MR
vii	Canadian Centre for Climate Modelling and Analysis	Canada	CanESM2	2.81	2.79	LR
viii	Geophysical Fluid Dynamics Laboratory	US-NJ	GFDL-ESM2G	2.5	2.0	LR
ix		US-NJ	GFDL-ESM2M	2.5	2.0	LR
x	Met Office Hadley Centre	UK-Exeter	HadGEM2-CC	1.87	1.25	MR
xi		UK-Exeter	HadGEM2-ES	1.75	1.25	MR
xii	*Institut Pierre-Simon Laplace*	France	IPSL-CM5A-LR	3.75	1.89	LR
xiii			IPSL-CM5A-MR	2.50	1.26	LR
xiv	Atmosphere and Ocean Research Institute (University of Tokyo), National Institute for Environmental Studies and Japan Agency for Marine-Earth Science and Technology	Japan	MIROC-ESM	2.81	2.79	LR
xv		Japan	MIROC5	1.40	1.40	HR
xvi	Max Planck Institute for Meteorology (MPI-M)	Germany	MPI-ESM-LR	1.87	1.87	MR
xvii		Germany	MPI-ESM-MR	1.87	1.87	MR
xviii	Meteorological Research Institute	Japan	MRI-CGCM3-	1.12	1.12	HR
xix	Norwegian Climate Centre	Norway	Nor-ESM1-M	2.50	1.89	LR
xx	Institute for Numerical Mathematics	Russia	INM-CM4	2.0	1.5	MR

Table 6. Agronomic management scenarios for maize and sorghum

Crop	Management		
	Planting density (Plants/ha)	Fertilizer Application (kg N/ha)	Dates (period) of planting
Maize	33,000	40, 60	Early: Early-mid December Late: Early-mid January
Sorghum	90,000	20,40	Early: Early-mid December Late: Early-mid January

2.6 Uncertainty and Confidence Assessment

Uncertainty in the projected impact of climate change on crops was assessed through two measures namely, sign of mean yield change and comparison of interannual yield variability using coefficients of variation (CV), similar to methods used by Ruiz-Ramos and Minguez (2010) and Zinyengere et al. (2014). The sign of mean yield change was determined for each crop, crop model, GCM and RCP. Coincidence of GCMs and crop models with the same sign of change across RCP was used to ascertain the degree of confidence in the direction of yield change. Mean CVs were compared for GCMs and RCPs, thereby identifying the sources of large uncertainty through high interannual variability.

3. Results and Discussion

3.1 Climate Change Projections

Climate models consistently projected increased temperatures for selected weather stations in the Central zone of Tanzania. Projected temperature changes showed a mean increase in the range of 1.4–2.8 °C (Table 7). Dodoma station showed both the highest projected mean increase in temperature, recorded under the HADGEM2-ES (2.8 °C) and the lowest mean temperature increase with GFDL-ESM2M (1.4°C). In contrast, the projected change in rainfall varied from one location to the other but consistently showing decline across all GCMs, except MIROC5, which showed an increase +4.5–7.3% (Table 7). Projected mean rainfall changes were small, within a -1.7% average, although varying considerably across GCMs. While projected rainfall changes were variable and uncertain, the projected temperature changes showed strong consistency with an upward trend.

Table 7. Mean change in projected climate between baseline (1980-2010) and future (2040-2069) periods for RCP 8.5

Station	GCM	Temperature (°C)			Rainfall (%)
		Average	Minimum	Maximum	
Dodoma	CCSM4	1.9	1.9	2.0	-2.5
	GFDL-ESM2M	1.4	1.7	1.2	-8.5
	HADGEM2-ES	2.8	2.9	2.8	-1.4
	MIROC5	2.2	2.1	2.4	7.3
	MPI-ESM-MR	2.4	2.4	2.5	-0.4
Manyoni	CCSM4	1.9	1.8	2.0	-8.9
	GFDL-ESM2M	1.8	1.8	1.7	-3.0
	HADGEM2-ES	2.7	2.6	2.8	-5.2
	MIROC5	2.3	2.1	2.4	7.0
	MPI-ESM-MR	2.3	2.1	2.4	-0.2
Singida	CCSM4	1.9	1.8	1.9	-10.3
	GFDL-ESM2M	1.8	1.8	1.7	-1.9
	HADGEM2-ES	2.7	2.6	2.8	-2.7
	MIROC5	2.3	2.1	2.4	4.5
	MPI-ESM-MR	2.6	2.7	2.4	-0.3

3.2 Model Evaluation and Crop Yield Projections

Statistical indicators showing the simulation efficiency of APSIM model in simulating sorghum are shown in Table 8. Root mean square error (RMSE) which is an overall measure of model performance and compares the agreement of simulated versus observed values show a good response because the lower the values of RMSE the better the model in explaining most of the variations in the dataset. Moreover, data indicate that the simulated grain and biomass yield values reasonably matched observed values, owing to the agreement index (d-statistic) ranging from 0.6 to 0.9 across the varieties. The d-statistic values close to 1 are regarded as better simulations and according to these statistical indicators the model performance was deemed satisfactory. Simulated and observed (survey data) yields for the 2009/2010 season are shown in Figure 2. The models APSIM and DSSAT appear appropriate owing to the R^2 values of 0.75 and 0.69 for maize and 0.82 and 0.61 for sorghum, respectively (Figure 2). The calibrated models were used to simulate maize and sorghum grain yields under the three scenarios (near-term, mid-century and end-century).

Table 8. Statistical indicators of model performance

Parameters/Cultivar	Macia		Tegemeo		Pato	
	RMSE (kg/ha)	d-Stat	RMSE (kg/ha)	d-Stat	RMSE (kg/ha)	d-Stat
Grain yield	133	0.73	87	0.62	140	0.60
Biomass	178	0. 93	418	0.66	236	0.83

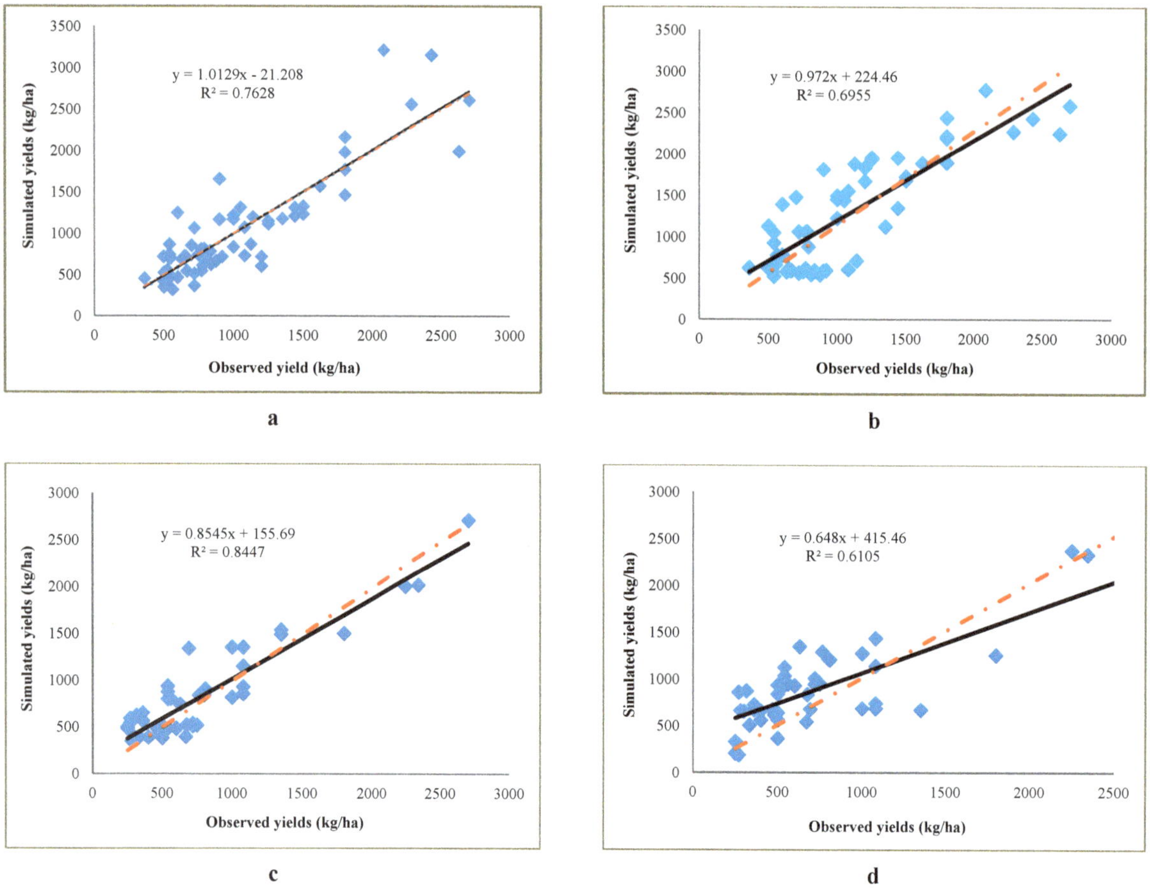

Figure 2. Relationship between simulated yields of maize (a & b) and sorghum (c & d) and reported yields for the season 2009-2010 (dashed line is 1:1)

Projected mean yield change showed a consistent decline for maize yields (Table 9). Average yield decline across all GCMs varied between 1% and 2.4% in the near term, between 3.7% and 7.1% towards mid-century and between 4.6% and 25.3% towards end of the century. In contrast, the projected sorghum yields show an increase varying between 5.4% and 6.9% in the near term, between 7.5% and 14.5% towards mid-century and between 5.7% and 20.7% towards end of the century (Table 10). The magnitude of yield change is higher in DSSAT than in APSIM. This is consistent to results by Rosenzweig et al. (2014) who reported model agreement on direction of yield change but varying in magnitude. Sultan et al. (2014) observed high consistence across climate and crop models in climate and impacts projections on sorghum between the western and eastern parts of the Sahel. In East Africa, a study by Thornton et al. (2009) similarly showed yield decreases for maize over the region ranging between 1% and 15% across emissions scenarios and climate models, largely as a result of temperature increase.

Increase in sorghum yields shown by almost all GCMs under both RCPs, may be attributed to increase in temperatures and the slight changes in projected rainfall which appear to create conducive conditions for sorghum growth, being more tolerant to heat and water stress. The results are in agreement with the observations by Turner and Rao (2013) and Gwimbi, Thomas and Hachigonta (2013), which show sorghum gaining in terms of grain yields from higher temperatures in specific regions with lower baseline temperatures (below 20 °C). However, Tingem, Rivington, and Bellochi (2009) reported that for the future, little or no change or even decreases in maize and sorghum yields are projected in eight agricultural regions of Cameroon. Projections of increased crop yields as a result of climate change have not been explored in Tanzania in particular. Recently, Kilembe et al. (2013) simulated yields for the climates of 2010 (baseline) and 2050 using DSSAT and showed a decrease in sorghum yields ranging from 5 percent to more than 25 percent of the baseline. These results suggest an overall decline in sorghum yields towards midcentury, contrary to results from the present study. This could be attributed to lack of consideration of local environments in terms of weather, soils, varieties and planting dates in the study by Kilembe et al. (2013). On the other hand, the International Model for Policy Analysis of Agricultural Commodities and Trade (IMPACT) showed significant increase in sorghum yields with climate change in Tanzania taking into account technological advancement by 2050. While results from the current study agree and/or disagree with some large scale studies, their strength emanate from the consideration of local crop specific varieties and management practices and the consideration of relevant weather information. In the central zone, the study showed that the magnitude of maize yield decline at least to mid-century is not likely to exceed 25%, results which are in agreement with Moore et al. (2012). In contrast, Mwandosya et al. (1998) projected a decline in maize yields of between 80 and 90% towards the end of the century.

Considering uncertainties introduced by the crop models' processes, it has been apparent that DSSAT being able to exhibit the effects of [CO2] showed a relatively higher magnitude level of impacts compared with APSIM at highest projected [CO2] i.e. 801 ppm for end century RCP8.5 (Tables 9 and 10). However, some studies have suggested that C4 plants (e.g., maize and sorghum) do not respond much to elevated levels of CO_2 (Sultan et al., 2013). Other studies have further shown that variable responses of the crop models to input parameters, is another source of uncertainty. For example Sadras and Calvino (2001) indicate that CERES model is more sensitive to soil water deficit whereas APSIM is relatively sensitive to physical and chemical characteristics of the soil (Wang et al., 2009).

Table 9. Percentage mean maize yield changes between baseline and three future periods for twenty Global Circulation Models (GCMs) and two Representative Concentration Pathways (RCPs): 4.5 and 8.5. (Shaded figures represent yield change for the five selected GCMs)

	DSSAT						**APSIM**					
GCMs	NEAR-TERM		MIDCENTURY		ENDCENTRUY		NEAR-TERM		MIDCENTURY		ENDCENTRUY	
	RCP4.5	RCP8.5	RCP4.5	RCP8.5	RCP4.5	RCP8.5	RCP4.5	RCP8.5	RCP4.5	RCP8.5	RCP4.5	RCP8.5
ACCESS10	-2.9	-2.7	-11.5	-12.8	-13.7	-42.7	-3.0	-4.0	-9.0	-5.8	-4.7	-20.8
BCC-CSM11	0.0	-1.1	-1.8	-6.0	-3.2	-16.0	-0.8	-1.5	-1.4	-1.4	-2.5	-8.7
BNU-ESM	-0.8	-0.8	-1.8	-4.0	-4.0	-14.6	-5.5	-4.6	-5.4	-7.0	-7.3	-12.5
CANES	-0.6	-1.7	-3.3	-5.5	-4.1	-18.0	-1.2	-1.7	-3.7	-3.7	-5.6	-12.3
CCSM4	2.0	-1.4	1.0	-2.9	-0.5	-11.9	-0.5	-1.6	-3.7	-4.0	-3.3	-10.3
CESMI	0.0	-1.6	-1.1	-3.6	-2.4	-11.9	-2.8	-3.1	-4.4	-4.3	-4.9	-8.2
CSIRO	-3.7	-2.9	-10.9	-13.7	-17.3	-36.7	-4.0	-2.8	-5.2	-5.8	-7.3	-19.3
GFDL-ESM2G	-1.7	-3.3	-10.9	0.1	0.1	-10.1	-4.6	-3.6	-3.6	-2.7	1.4	-6.5
GFDL-ESM2M	-4.5	3.7	-0.1	0.3	3.6	-0.5	-0.9	3.6	0.7	0.5	3.0	-0.1
HADGEM2-CC	-1.3	-1.1	-11.1	-21.7	-10.2	-61.9	-1.9	-1.5	-2.7	-5.2	-1.3	-21.5
HADGEM2-ES	0.0	-3.3	-7.7	-15.9	-21.1	-58.1	-0.6	-2.7	-1.7	-10.4	-8.2	-21.8
INMCM4	0.6	0.6	0.1	0.0	-1.1	-6.4	-2.2	-1.8	-3.3	-3.6	-4.2	-6.2
IPSL-LR	-1.6	-2.6	-4.5	-11.8	-10.4	-46.5	-1.7	-3.9	-0.9	-2.7	-4.1	-17.8
IPSL-MR	-1.9	-1.2	-3.6	-9.9	-11.7	-45.1	-3.1	-3.2	-1.3	-6.9	-9.6	-34.4
MIROC5	-0.1	-1.3	-3.2	-5.2	-8.5	-16.0	0.2	-2.3	-1.7	-4.1	-5.9	-6.0
MIROC-ESM	1.7	3.4	0.3	-2.3	-2.2	-19.1	-3.3	-3.4	-6.5	-8.2	-8.6	-17.1
MPI-LR	-0.3	-0.9	-2.4	-8.4	-2.6	-31.1	-2.9	-2.6	-3.9	-3.7	-2.0	-15.7
MPI-MR	0.2	-3.2	-4.1	-11.7	-4.8	-33.8	-4.4	-3.1	-5.1	-5.8	-5.1	-16.8
MRI	-5.8	-2.4	-7.5	-6.4	-5.3	-15.1	-3.6	-1.7	-9.6	-5.2	-7.4	-11.8
NORESMI	1.5	0.3	0.1	-1.1	-0.9	-9.5	-0.4	-3.2	-2.2	-2.9	-4.4	-11.1
Mean	-1.0	-1.2	-4.2	-7.1	-6.0	-25.3	-2.4	-2.4	-3.7	-4.6	-4.6	-13.9

Table 10. Percentage mean sorghum yield changes between baseline and three future periods for twenty Global Circulation Models (GCMs) and two Representative Concentration Pathways (RCPs): 4.5 and 8.5. (Shaded figures represent yield change for the five selected GCMs)

	DSSAT						APSIM					
	NEAR-TERM		MIDCENTURY		ENDCENTURY		NEAR-TERM		MIDCENTURY		ENDCENTURY	
	RCP4.5	RCP8.5	RCP4.5	RCP8.5	RCP4.5	RCP8.5	RCP4.5	RCP8.5	RCP4.5	RCP8.5	RCP4.5	RCP8.5
ACCESS10	-6.2	-6.0	-2.7	6.1	4.7	13.0	7.3	7.6	9.1	10.3	11.1	10.7
BCC-CSM11	6.9	8.0	12.9	15.9	12.7	10.2	7.8	7.4	9.0	9.8	7.8	8.2
BNU-ESM	3.3	3.7	7.1	12.6	10.9	19.7	4.3	4.2	4.4	5.4	4.7	6.2
CANES	4.5	6.5	9.4	13.1	10.1	20.8	4.0	5.8	4.6	3.9	2.4	5.3
CCSM4	7.3	7.1	12.8	14.6	14.3	21.4	5.7	8.3	4.6	7.6	7.9	7.9
CESMI	6.0	6.7	11.6	15.4	13.8	22.1	6.5	7.5	6.7	9.0	8.0	8.7
CSIRO	6.2	8.1	13.0	16.1	16.2	22.9	9.2	9.2	10.8	10.8	11.0	10.0
GFDL-ESM2G	4.7	7.1	11.8	14.9	15.0	20.0	6.7	9.2	8.4	6.8	10.8	7.2
GFDL-ESM2M	3.6	7.6	9.4	13.1	8.9	15.1	8.0	5.7	8.6	7.3	3.6	3.2
HADGEM2-CC	7.1	8.8	14.4	16.0	17.9	25.2	7.1	7.4	9.4	8.7	9.8	11.4
HADGEM2-ES	10.2	7.8	14.9	18.4	16.9	24.8	7.8	7.3	9.6	10.3	9.5	5.6
INMCM4	4.9	5.1	9.3	12.8	2.9	18.6	6.9	6.3	7.2	6.6	7.8	7.3
IPSL-LR	7.8	9.0	14.9	18.1	17.1	25.8	6.3	6.7	8.8	9.7	8.7	11.7
IPSL-MR	8.1	8.0	13.1	18.1	15.3	22.7	7.2	7.0	8.5	7.4	5.5	4.4
MIROC5	8.9	8.8	13.0	15.7	13.4	23.4	8.6	7.8	10.4	9.6	9.7	11.2
MIROC-ESM	4.5	4.5	9.1	9.9	9.5	22.4	3.8	1.5	1.6	1.2	-6.7	7.6
MPI-ESM-LR	7.0	5.6	10.4	15.3	14.1	21.8	6.7	6.5	6.7	8.2	0.6	7.6
MPI-ESM-MR	4.7	7.4	10.7	15.1	14.7	23.0	5.3	8.5	6.1	9.3	0.8	9.5
MRI	2.5	5.1	7.9	12.7	10.5	19.4	8.6	8.1	6.8	7.3	-0.5	6.9
NORESMI	6.6	6.0	11.1	15.2	13.4	20.7	7.4	5.9	7.7	8.4	0.8	7.2
Mean	5.4	6.2	10.7	14.5	12.6	20.7	6.8	6.9	7.5	7.9	5.7	7.9

3.3 Effect of Adaptation Options

To identify adaptation strategies for rainfed sorghum and maize in the study area, four options were evaluated under present and future (mid-century RCP8.5) climate using a subset of five GCMs. Adaptation options combine sowing dates, plant density and inorganic fertilizer applications (Table 6). Mean grain yields of both sorghum and maize from the two crop models under the evaluated adaptation options are shown in Figures 3 and 4. The influence of the improved agro-systems on crop sensitivity to climate change is still a matter of debate (Turner & Rao, 2013; Sultan et al., 2014). While on the one hand, Turner and Rao (2013) show minimum stress from warming temperatures under the current low-input production systems (no-N fertilizer added) compared to improved systems (with adequate N fertilization), Sultan et al. (2014) show that increasing fertilizer inputs in the Sahel agricultural system could make it more responsive to climatic stresses and produce more negative impacts (in a relative sense, %) in crop yields under climate change. Results from both studies suggest that sorghum yields under current smallholders' low-input systems would be resilient or even increase under increasing temperatures. Moreover, the results seem to suggest that micro-dosing with Nitrogen could significantly increase yields even in the hottest and driest locations.

Figure 3 shows that sorghum yields from which 20 kg N ha^{-1} of inorganic fertiliser is applied, are approximately twice as high as the yields obtained without using fertilizer and are three times higher when 40 kg N ha^{-1} of fertilizer is used. Considering variability among GCMs, GFDL-ESM and HADGEM2-ES though having contrasting characteristics (Table 7), produced highest mean sorghum yields. At the higher level of fertilizer application (40 kg N ha^{-1}) yield increases projected for sorghum were consistently higher under early planting than with late planting across all GCMs. Simulated sorghum yields, however, indicate that even when planting is delayed by up to one month there is no significant reduction in yields. Traore et al. (2014) similarly reported that a one month delay in planting sorghum and maize did not significantly affect the final yields. Results of simulated maize yields under evaluated adaptation options are shown in Figure 4. Even though rain-fed maize production is determined by the adequacy, reliability and timeliness of rainfall, simulated grain yields were increased with inorganic fertilizer application amount. However, farmers are averse to taking risks and therefore, not ready to invest in inputs and improvements if they are not sure of securing good yields in a particular season, as a result low levels of productivity persist (Bezabih & Di Falco, 2012). Application of fertilizers will become more critical if farmers are to reduce their vulnerability to the impacts of climate change. Besides application of fertilizers, adjustments in planting densities and sowing dates will also be of major importance.

Unlike sorghum, there is an appreciable difference in maize yields among the GCMs with GFDL-ESM giving the highest and HADGEM2-ES giving the lowest yield. This could be due to the effect of projected increase in temperature between the two GCMs (Table 7) where GFDL-ESM projects the lowest increase in average temperature by mid-century whereas the converse is true for HADGEM2-ES. Studies have shown that increased temperatures and change in rainfall patterns will affect major staple cereal food crops such as maize, sorghum and millets because of possible yield decline in future (Zinyengere et al., 2013; Lobell, Banziger, Magorokosho, & Vivek, 2011). For example, analyses by Lobell et al. (2011) show that each degree day spent above 30 °C reduces maize grain yield by 1% under optimal rain-fed conditions, and by 1.7% under drought conditions in Africa.

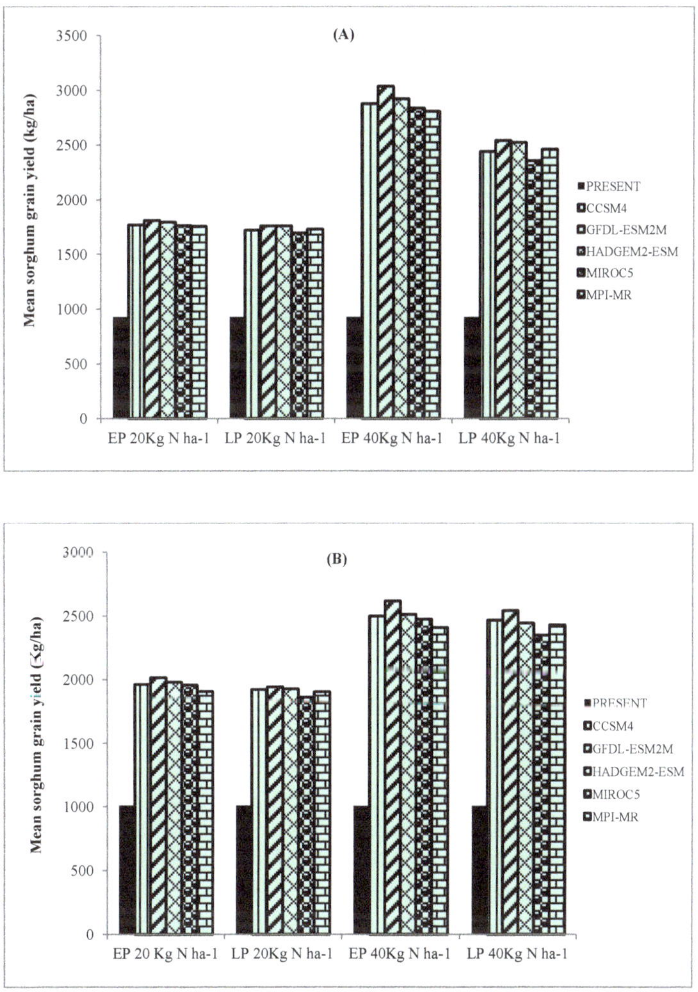

Figure 3. Mean simulated sorghum grain yields under different adaptation options for present (baseline) and midcentury (RCP8.5) using GCMs with APSIM (A) and DSSAT (B)

Note. EP = Early planting; LP = Late planting.

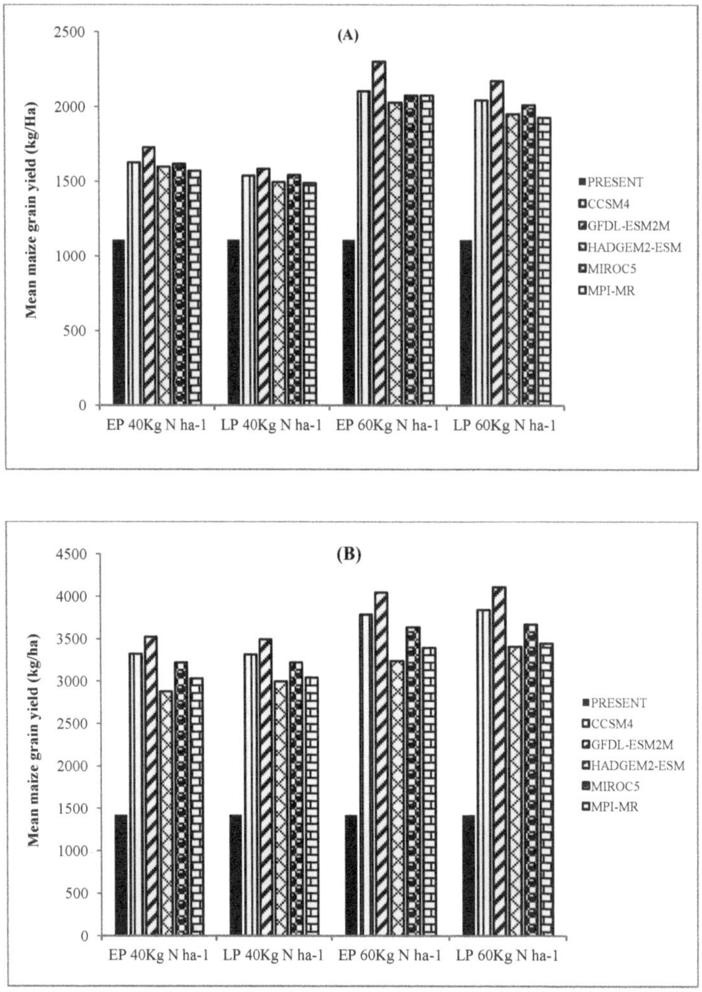

Figure 4. Mean simulated maize grain yields under different adaptation options for present (baseline) and midcentury (RCP8.5) using GCMs with APSIM (A) and DSSAT (B)

Note. EP = Early planting; LP = Late planting.

3.4 Uncertainty on Climate Change Impact and Adaptation Options

According to the simulation results from the five GCMs, a clear trend that is consistent across crop models and RCPs has been established in as far as yield change is concerned (Tables 9 and 10). Interannual variability of yield under both crop models ranged from a mean CV of 37% to 55% for sorghum and 56% to 70% for maize (Figure 5). Mean CVs were higher for APSIM than for DSSAT and also higher for maize than for sorghum. The results confirm the uncertainty brought about by the crop models due to differences in parameters (Asseng et al., 2013). The higher range of predictions for maize across GCMs, reflects the uncertainty of climate prediction impacts using GCMs. Similar results were obtained by Moore et al. (2012) who reported between 20% and 30 % decrease in maize yields towards mid-century in Morogoro. They attributed the uncertainty to combined effects of greenhouse gas emissions and land cover land use change (LCLUC). Similar results were obtained in Zimbabwe where a GCM and CERES-maize showed that maize yields would decrease by approximately 11–17%, under irrigated and non-irrigated conditions (Stige et al., 2006).

The coefficient of variation (CV) from simulations incorporating agronomic adaptation options is illustrated in Figure 6. While uncertainty remains a factor, a clear trend was established in that yield variability was least influenced by GCMs as shown by the limited differences in CVs, except for HADGEM-ES under DSSAT and early planting scenario which showed increased variability over and above other GCMs (Figure 6D). To a large extent, yield variation appears to have been driven by agronomic adaptation options (Figure 6). Agronomic adaptation strategies influenced uncertainty considerably (low CVs of less than 18%) as shown in Figure 6D) compared with high CVs of up to 59% when simulations were run under current agronomic management practices i.e. mainly without fertilizer application. The results are in agreement with Walker and Schulze (2006) who reported reduced variability for treatments using inorganic fertiliser under all the future climate scenarios modelled compared with the CV of maize yields under previous climate conditions.

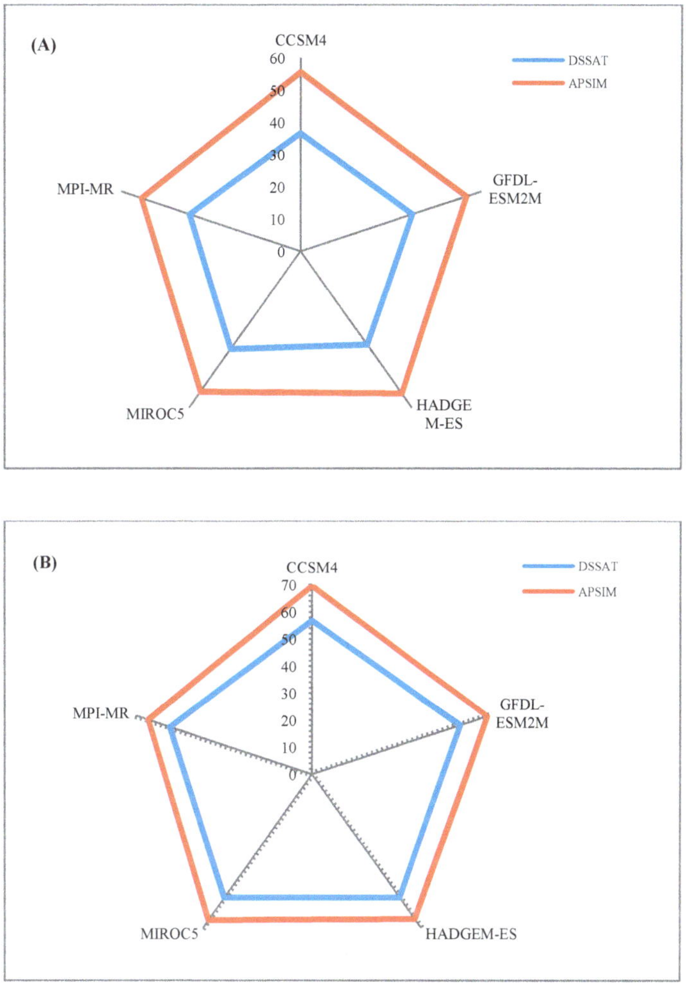

Figure 5. Mean percentage values of coefficient of variation for Sorghum (A) and maize (B) clustered by GCMs

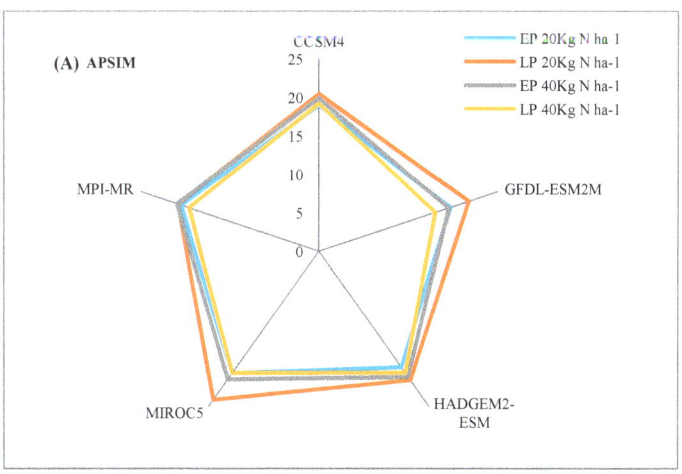

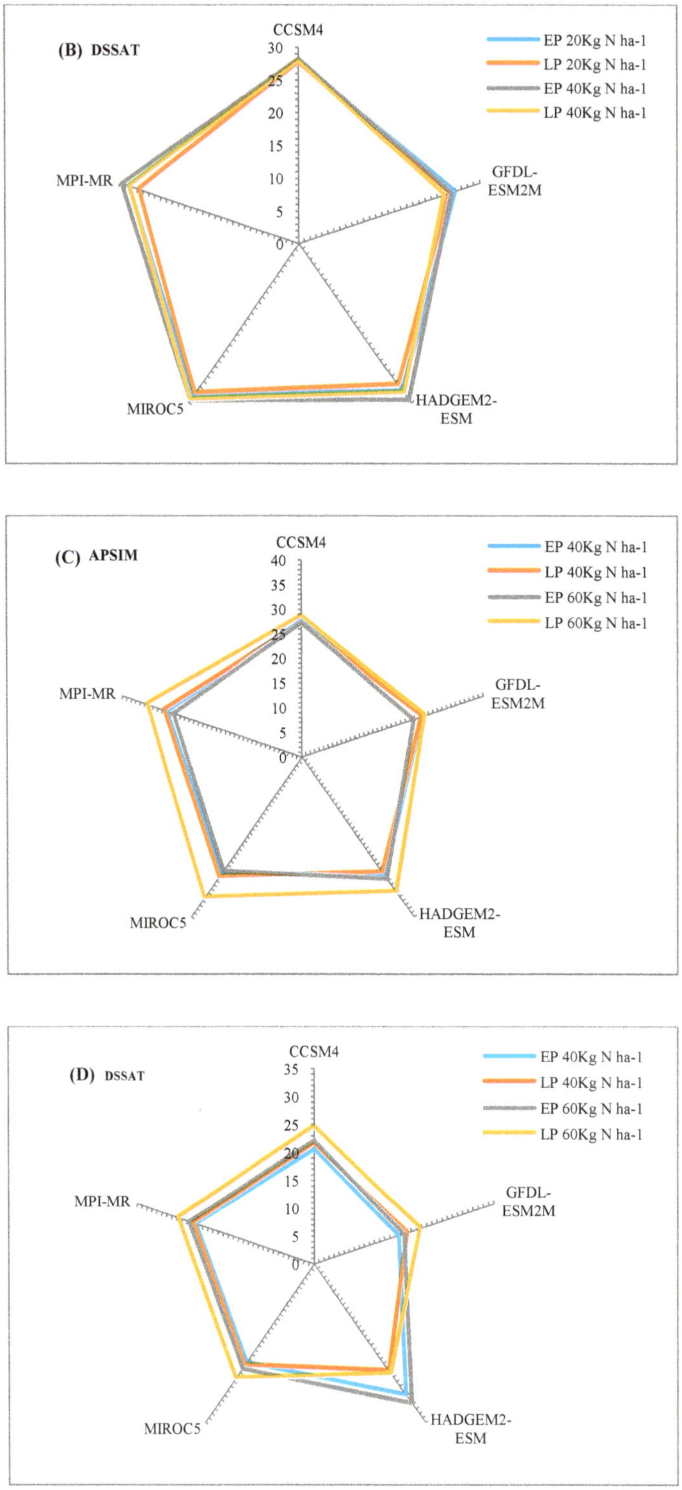

Figure 6. Mean percentage values of coefficient of variation for Sorghum (A & B) and maize (C& D) yields with respect to adaptation options clustered by GCMs

4. Conclusions

Results from this study demonstrate how crop simulation models coupled with GCMs could play a role in policy decisions with respect to climate change considerations. It has been shown that sorghum yields will consistently increase over different time periods with up to 25% increase towards the end of the century. On the other hand, overall maize yields, by contrast, have been projected to decline. These findings are consistent with those of other studies that have shown similar negative effects on maize, under current varieties and crop management practices, without adaptation strategies. Basic agronomic adaptation options such as fertilizer applications, appropriate planting density and planting dates appear to be ideal for future climate uncertainties. This has a bearing on agricultural plans and policies which may need to be reoriented for enhanced crop productivity.

Despite the uncertainty in crop models and GCMs, the results enhance people's understanding of current climate variability as well as the anticipated climate change which is appropriate for informed agricultural management decisions. Furthermore, the results accentuate the uncertainty that comes from using different models in climate change assessments. All in all the study has contributed to a better understanding of large-area modelling because existing large-area crop models do not currently simulate the non-climatic (e.g. local varieties and crop management practices) determinants of crop yield; factors which also need considerations if useful insights are to be provided for future decision making in a rapidly changing climate.

Acknowledgements

The Regional Universities Forum for Capacity Building in Agriculture (RUFORUM) is greatly appreciated for funding this PhD study from which this paper is a part. Enhancing Climate Change Adaptation in Agriculture and Water Resources in the Greater Horn of Africa (ECAW) project supported the crop model training with DSSAT while the Agricultural Model Intercomparison and Improvement Project (AgMIP) enabled further training with APSIM and DSSAT.

References

Ainsworth, E. A., Leaky, A. D. B, Ort, D. R., & Long, S. P. (2008). FACE-ing the facts: inconsistencies and interdependence among field, chamber and modelling studies of elevated [CO_2] impacts on crop yield and food supply. *New Phytology, 179*, 5-9. http://dx.doi.org/10.1111/j.1469-8137.2008.02500.x

Arndt, C., Farmer, W., Strzepek, K., & Thurlow, J. (2011). *Climate change, agriculture and food security in Tanzania* (p. 26). Working Paper No. 2011/52 United Nations University-World Institute for Economic Development Research (UNU-WIDER).

Asseng, S., Ewert, F., Rosenzweig, C., Jones, J. W., Hatfield, J. L., Ruane, A. C., ... Wolf, J. (2013). Uncertainty in simulating wheat yields under climate change. *Nature Climate Change, 3*, 827-832. http://dx.doi.org/10.1038/nclimate1916

Berg, A., de Noblet-Ducoudr'e, N., Sultan, B., Lengaigne, N., & Guimberteau, M. (2013). Projections of climate change impacts on potential C4 crop productivity over tropical regions. *Agricultural and Forest Meteorology, 170*, 89-102. http://dx.doi.org/10.1016/j.agrformet.2011.12.003

Bezabih, M., & Di Falco, S. (2012). Rainfall variability and food crop portfolio choice: Evidence from Ethiopia. *Food Security, 4*(4), 557-567. http://dx.doi.org/10.1007/s12571-012-0219-7

Challinor, A. J., Watson, J., Lobell, D. B., Howden, S. M., Smith, D. R., & Chhetri, N. (2014). A meta-analysis of crop yield under climate change and adaptation. *Nature Climate Change, 4*, 287-291. http://dx.doi.org/10.1038/nclimate2153

Falloon, P., Challinor, A., Dessai, S., Hoang, L., Johnson, J., & Koehler, A. (2014). Ensembles and uncertainty in climate change impacts. *Frontiers in Environmental Science, 2*(Article 33), 1-7. http://dx.doi.org/10.3389/fenvs.2014.00033

Gwimbi, P., Thomas, T. S., & Hachigonta, S. (2013). Assessing the vulnerability of agriculture to climate change in Lesotho. In S. Hachigonta, G. C. Nelson, T. S. Thomas, & L. M. Sibanda (Eds.), *Southern African Agriculture and Climate Change: A Comprehensive Analysis*. International Food Policy Research Institute, Washington, D.C. http://dx.doi.org/10.2499/9780896292086

Hatfield, L., Boote, K. J., Kimball, B. A., Ziska, L. H., Izaurralde, R. C., Ort, D., ... Wolfe, D. (2011). Climate Impacts on Agriculture: Implications for Crop Production. *Agronomy Journal, 103*, 351-370. http://dx.doi.org/10.2134/agronj2010.0303

Hempel, S., Frieler, K., Warszawski, L., Shewe, J., & Piontek, F. (2013). A trend- reserving bias correction – the ISI-MIP approach. *Earth Systems Dynamics, 4*, 219-236. http://dx.doi.org/10.5194/esd-4-219-2013

Ingram, J. S. I., Gregory, P. J., & Izac, A. M. (2008). The role of agronomic research in climate change and food security policy. *Agriculture Ecosystems and Environment, 126*, 1-12.

http://dx.doi.org/10.1016/j.agee.2008.01.009

IPCC. (2013). Summary for Policymakers. In T. F. Stocker, D. Qin, G.-K. Plattner, M. Tignor, S. K. Allen, J. Boschung, A. Nauels, Y. Xia, V. Bex & P. M. Midgley (Eds.), *Climate Change 2013: The Physical Science Basis. Contribution of Working Group I to the Fifth Assessment Report of the Intergovernmental Panel on Climate Change*. Cambridge University Press, Cambridge, United Kingdom and New York, NY, USA.

Jones, J. W., Hoogenboom, G., Porter, C. H., Boote, K. J., Batchelor, W. D., Hunt, L. A., ... Ritchie, J. T. (2003). The DSSAT cropping system model. *European Journal of Agronomy, 18*, 235-265. http://dx.doi.org/10.1016/S1161-0301(02)00107-7

Keating, B. A., Carberry, P. S., Hammer, G. L., Probert, M. E., Robertson, M. J., Holzworth, D., ... Smith, C. J. (2003). An overview of APSIM, a model designed for farming systems simulation. *European Journal of Agronomy, 18*, 267-288. http://dx.doi.org/10.1016/S1161-0301(02)00108-9

Kilembe, C., Thomas, T. S., Waithaka, M., Kyotalimye, M., & Tumbo, S. (2013). *Tanzania*. In M. Waithaka, G. C. Nelson, T. S. Thomas & M. Kyotalimye (Eds.), *East African Agriculture and Climate Change: A Comprehensive Analysis* (pp. 313-343). IFPRI, Washington, DC.

Knox, J., Hess, T., Daccache, A., & Wheeler, T. (2012). Climate change impacts on crop productivity in Africa and South Asia. *Environmental Research Letters, 7*(034032), 8.

Leenaars, J. G. B. (2012). Africa Soil Profiles Database Version 1.0. A compilation of georeferenced and standardized legacy soil profile data for Sub Saharan Africa (with dataset). *ISRIC report 2012/03*. Africa Soil Information Service (AfSIS) project and ISRIC-World Soil Information, Wageningen, the Netherlands

Lobell, D. B., Burke, M. B., Tebaldi, C., Mastrandrea, M. D., Falcon, W. P., & Naylor, R. L. (2008). Prioritizing climate change adaptation needs for food security in 2030. *Science, 319*, 607-610. http://dx.doi.org/10.1126/science.1152339

Lobell, D. B., Bänziger, M., Magorokosho, C., & Vivek, B. (2011). Nonlinear heat effects on African maize as evidenced by historical yield trials. *Nature Climate Change, 1*, 42-45. http://dx.doi.org/10.1038/nclimate1043

Lyon, B., & DeWitt, D. G. (2012). A recent and abrupt decline in the East African long rains. *Geophysical Research Letters, 39*(2). http://dx.doi.org/10.1029/2011GL050337

Matari, E. E., Chang'a, L. B., Chikojo, G. E., & Hyera, T. (2008). Climate change scenarios development for second national communication, Tanzania. *TMA Research Journal, 1*, 8-18.

McSweeney, C., New, M., & Lozano, G. (2010). *The UNDP Climate change country profiles, Tanzania. Improving the accessibility of observed and projected climate information for studies of climate change in developing countries*. Retrieved from http://country-profiles.geog.ox.ac.uk

Moore, N., Alagarswamy, G., Pijanowski, B., Thornton, P., Lofgren, B., Olson, J., ... Qi, J. (2012). East African food security as influenced by future climate change and land use change at local to regional scales. *Climatic Change, 110*, 823-844. http://dx.doi.org/10.1007/s10584-011-0116-7

Moss, R. H., Edmonds, J. A., Hibbard, K. A., Manning, M. R., Rose, S. K., van Vuuren, D. P., ... Wilbanks, T. J. (2010). The next generation of scenarios for climate change research and assessment. *Nature, 463*, 747-756. http://dx.doi.org/10.1038/nature08823

Mourice, S. K., Rweyemamu, C. L., Tumbo, S. D., & Amuri, N. (2014). Maize Cultivar Specific Parameters for Decision Support System for Agrotechnology Transfer (DSSAT) Application in Tanzania. *American Journal of Plant Sciences, 5*, 821-833. http://dx.doi.org/10.4236/ajps.2014.56096

Mwandosya, M. J., Nyenzi, B. S., & Luhanga, M. L. (1998). *The Assessment of Vulnerability and Adaptation to Climate Change Impacts in Tanzania*. Dar-es-Salaam, Tanzania: Centre for Energy, Environ. Sci. Technol. (CEEST).

NBS. (2012). *Tanzania National Panel Survey Report Wave 2, 2010 -2011, Living Standards Measurements Survey*. National Bureau of Statistics, Dar-es-Salaam, Tanzania.

Paavola, J. (2003). *Vulnerability to Climate Change in Tanzania: Sources, Substance and Solutions*. A paper presented at the inaugural workshop of Southern Africa Vulnerability Initiative (SAVI) in Maputo, Mozambique June 19-21.

Porter, J. R., & Semenov, M. A. (2005). Crop responses to climatic variation. *Philosophical Transactions of the Royal Society, B*(360), 2021-5. http://dx.doi.org/10.1098/rstb.2005.1752

Ramirez-Villegas, J., Challinor, A, Thornton, P., & Jarvis, A. (2013). Implications of regional improvement in global climate models for agricultural impact research. *Environmental Research Letters, 8*, 12. http://dx.doi.org/10.1088/1748-9326/8/2/024018.

Riahi, K., Grübler, A., & Nakicenovic, N. (2007). Scenarios of long-term socio-economic and environmental development under climate stabilization. *Technological Forecasting and Social Change, 74*, 887-935. http://dx.doi.org/10.1016/j.techfore.2006.05.026

Rosenzweig, C., Elliott, J., Deryng, D., Ruane, A. C., Müllere, C., Arneth, A., ... Jones, J. W. (2014). Assessing agricultural risks of climate change in the 21st century in a global gridded crop model intercomparison. *Proceeding of National Academy of Sciences*. http://dx.doi.org/10.1073/pnas.1222463110

Rosenzweig, C., Jones, J., Hatfield, J., Antle, J., Ruane, A., Boote, K., ... Mutter, C. (2013b). *AgMIP Guide for Regional Integrated Assessments: Handbook of Methods and Procedures Version 5.0.*

Rosenzweig, C., Jones, J. W., Hatfield, J. L., Ruane, A. C., Boote, K. J., Thorburn, P., ... Winter, J. M. (2013a). The agricultural model intercomparison and improvement project (AgMIP). Protocols and pilot studies. *Agriculture and Forest Meteorology, 170*, 166-182. http://dx.doi.org/10.1016/j.agrformet.2012.09.011

Rötter, R. P., Carter, T. R., Olesen, J. E., & Porter, J. R. (2011). Crop-climate models need an overhaul. *Nature Climate Change, 1*, 175-177. http://dx.doi.org/10.1038/nclimate1152

Rowhani, P., Lobell, D. B., Linderman, M., & Ramankutty, N. (2011). Climate variability and crop production in Tanzania. *Agricultural and Forest Meteorology, 151*, 449-460. http://dx.doi.org/10.1016/j.agrformet.2010.12.002

Ruane, A. C., Cecil, L. D., Horton, R. M., Gordon, R., McCollum, R., Brown, D., ... Rosenzweig, C. (2013). Climate change impact uncertainties for maize in Panama: Farm information, climate projections, and yield sensitivities. *Agriculture and Forest Meteorology, 170*, 132-145. http://dx.doi.org/10.1016/j.agrformet.2011.10.015

Ruiz-Ramos, M., & Minguez, M. I. (2010). Evaluating uncertainty in climate change impacts on crop productivity in the Iberian Peninsula. *Climate Research, 44*, 69-82. http://dx.doi.org/10.3354/cr00933

Rurinda, J., Mapfumo, P., van Wijk, M. T., Mtambanengwe, F., Rufino, M. C., Chikowo, R., & Giller, K. E. (2014). Comparative assessment of maize, finger millet and sorghum for household food security in the face of increasing climatic risk. *European Journal of Agronomy, 55*, 29-41. http://dx.doi.org/10.1016/j.eja.2013.12.009

Sadras, V. O., & Calviño, P. A. (2001). Quantification of grain yield response to soil depth in soybean, maize, sunflower, and wheat. *Agronomy Journal, 82*, 577-583. http://dx.doi.org/10.2134/agronj2001.933577x

Schlenker, W., & Lobell, D. B. (2010). Robust negative impacts of climate change on African agriculture. *Environmental Research Letters, 1*, 14010. http://dx.doi.org/10.1088/1748-9326/5/1/014010

Shongwe, M. E., van Oldenborgh, G. J., van den Hurk, B., & van Aalst, M. (2011). Projected changes in mean and extreme precipitation in Africa under global warming. Part II: East Africa. *Journal of Climate, 24*(14), 3718-3733. http://dx.doi.org/10.1175/2010JCLI2883.1

Stige, L. C., Stave, J., Chan, K. S., Ciannelli, L., Pretorelli, N., Glantz, M., ... Stenseth, N. C. (2006). The effect of climate variation on agro-pastoral production in Africa. *Proceedings of the National Academy of Science, 103*, 3049-3053.

Sultan, B., Guan, K., Kouressy, M., Biasutti, M., Piani, C., Hammer, G. L., ... Lobell, D. B. (2014). Robust features of future climate change impacts on sorghum yields in West Africa. *Environmental Research Letters, 9*(104006), 13. http://dx.doi.org/10.1088/1748-9326/9/10/104006

Sultan, B., Roudier, P., Quirion, P., Alhassane, A., Muller, B., Dingkuhn, M., ... Baron, C. (2013). Assessing climate change impacts on sorghum and millet yields in the Sudanian and Sahelian savannas of West Africa. *Environmental Research Letters, 8*, 14040. http://dx.doi.org/10.1088/1748-9326/8/2/014040.

Taylor, K. E., Stouffer, R. J., & Meehl, G. A. (2012). An overview of CMIP5 and the experiment design. *Bulletin of American Meteorological Society, 93*, 485-98. http://dx.doi.org/10.1175/BAMS-D-11-00094.1

Taylor, R. G., Todd, M. C., Kongola, L., Maurice, L., Nahozya, E., Sanga, H., & MacDonald, A. M. (2013). Evidence of the dependence of groundwater resources on extreme rainfall in East Africa. *Nature Climate Change, 3*, 374-378. http://dx.doi.org/10.1038/nclimate1731

Thompson, H. E., Berrang-Ford, L., & Ford, J. D. (2010). Climate Change and Food Security in Sub- Saharan Africa: A Systematic Literature Review. *Sustainability, 2*, 2719-2733. http://dx.doi.org/10.3390/su2082719

Thornton, P. K., Jones, P. G., Alagarswamy, A., & Andresen, J. (2009). Spatial variation of crop yield responses to climate change in East Africa. *Global Environmental Change, 19*, 54-65. http://dx.doi.org/10.1016/j.gloenvcha.2008.08.005

Thornton, P. K., Jones, P. G., Alagarswamy, G., Andresen, J., & Herrero, M. (2010). Adapting to climate change: Agricultural system and household impacts in East Africa. *Agricultural Systems, 103*, 73-82. http://dx.doi.org/10.1016/j.agsy.2009.09.003

Thornton, P. K., Jones, P. G., Ericksen, P. J., & Challinor, A. J. (2011). Agriculture and food systems in sub-Saharan Africa in a 4 °C + world. *Philosophical Transactions of the Royal Society, 369*, 117-136.

Tingem, M., Rivington, M., & Bellocchi, G. (2009). Adaptation assessment for crop production in response to climate change in Cameroon. *Agronomy Sustainable Development, 29*, 247-256. http://dx.doi.org/10.1051/agro:2008053

Traore, B., van Wijk, M. T., Descheemaeker, K., Corbeels, M., Rufino, M. C., & Giller, K. E. (2014). Evaluation of climate adaptation options for Sudano-Sahelian cropping systems. *Field Crops Research, 156*, 63-75. http://dx.doi.org/10.1016/j.fcr.2013.10.014

Turner, N. C., & Rao, K. P. C. (2013). Simulation analysis of factors affecting sorghum yield at selected sites in eastern and southern Africa, with emphasis on increasing temperatures. *Agricultural Systems, 121*, 53-62. http://dx.doi.org/10.1016/j.agsy.2013.06.002

URT. (2007). *United Republic of Tanzania, National Adaptation Programme of Action (NAPA)*. Division of Environment, Vice President's Office, Dar es Salaam.

Waha, K., Müller, C., & Rolinski, C. (2013). Separate and combined effects of temperature and precipitation change on maize yields in sub-Saharan Africa for mid- to late-21st century. *Global Planetary Change, 106*, 1-12. http://dx.doi.org/10.1016/j.gloplacha.2013.02.009

Wambura, F., Ngongolo, H., Mlonganile, P., Sangalugembe, C., & Tumbo, S. (2014). *CMIP5 climate change projections for Tanzania*. A paper presented at the International Conference on Reducing Climate Change Challenges through Forestry and other Land Use practices 1st – 3rd April 2014, Nashera Hotel, Morogoro, Tanzania.

Wang, J., Wang, E. L., Luo, Q. Y., & Kirby, M. (2009). Modelling the sensitivity of wheat growth and water balance to climate change in Southeast Australia. *Climatic Change, 96*, 79-96. http://dx.doi.org/10.1007/s10584-009-9599-x

Watson, J., & Challinor, A. (2013). The relative importance of rainfall, temperature and yield data for a regional-scale crop model. *Agriculture and Forest Meteorology, 170*, 47-57. http://dx.doi.org/10.1016/j.agrformet.2012.08.001

White, J. W., Hoogenboom, G., Kimball, B. A., & Wall, G. W. (2011). Methodologies for simulating impacts of climate change on crop production. *Field Crops Research, 124*, 357-368. http://dx.doi.org/10.1016/j.fcr.2011.07.001

Yao, F. M., Qin, P. C., Zhang, J. H., Lin, E., & Boken, V. (2011). Uncertainties in assessing the effect of climate change on agriculture using model simulation and uncertainty processing methods. *Chinese Science Bulletin, 56*, 729-737. http://dx.doi.org/10.1007/s11434-011-4374-6.

Zinyengere, N., Crespo, O., & Hachigonta, S. (2013). Crop response to climate change in southern Africa: A comprehensive review. *Global and Planetary Change, 111*, 118-126. http://dx.doi.org/10.1016/j.gloplacha.2013.08.010

Zinyengere, N., Crespo, O., Hachigonta, S., & Tadross, M. (2014). Local impacts of climate change and agronomic practices on dry land crops in Southern Africa. *Agriculture, Ecosystems and Environment, 197*, 1-10. http://dx.doi.org/10.1016/j.agee.2014.07.002

12

Risk and Maximum Residue Limits: A Study of Hops Production

Ruojin Zhang[1], Thomas L. Marsh[2] & David H. Gent[3]

[1] School of Insurance, Southwestern University of Finance and Economics, China

[2] School of Economic Sciences, Washington State University, Pullman, WA, USA

[3] Department of Botany and Plant Pathology, Oregon State University, Corvallis, OR, USA

Correspondence: Ruojin Zhang, School of Insurance, Southwestern University of Finance and Economics, Chengdu, Sichuan Province, China.

Abstract

This paper examines how maximum residue limits (MRLs) affect the optimal choice by growers of chemical applications to control pests and diseases. In practice, growers who export balance both yield risk and pesticide residue uncertainty when making chemical application decisions. To address these issues we specify an expected utility model and calibrate it to data collected from a 2012 survey of hop growers in the Pacific Northwest. Then we simulate hop grower/exporter decisions subject to MRLs across a myriad of scenarios. As anticipated, risk preferences contribute to explaining higher chemical use. All else equal, more stringent MRLs tend to induce risk averse growers to apply fewer chemicals than do risk neutral growers because of the increasing likelihood of crop rejection due to exceeding an MRL. Under specific circumstances risk preferences coupled with underlying MRLs have the potential to tip the decision towards less chemical uses with potential for more growers implementing integrated pest management strategies or, alternatively, greater use of other pesticides not subject to MRL restrictions.

Keywords: expected utility, hop production, maximum residue limits, risk preferences

1. Introduction

Maximum residue limits (MRLs) are a maximum concentration of chemical residue to be legally permitted on food and agricultural products. MRLs are heterogeneous across countries and regions and can be applied in both domestic and international trade of food and agricultural products. Extensive literature exists on agricultural production from continuous stochastic income due to price or output uncertainty (e.g., Just & Zilberman, 1983; Babcock & Hennessy, 1996). In contrast this article develops a model of a grower's input decisions under yield risk and uncertain pesticide residue with MRL constraints imposed on exports. We specify a theoretical model that (i) incorporates the grower's risk preferences; (ii) emphasizes the role of MRLs under forward contracts (Note 1); and (iii) proposes a decision rule that balances both yield risk and pesticide residue. The simulation model illustrates hop production and exports in the Pacific Northwest. A wide range of scenarios are simulated to assess hop grower response to key production, trade, MRL, and policy parameters.

On one hand, hop growers face a substantial degree of production risk from pest/disease infestation. For example, during the 1998 season some growers experienced a 60% reduction in yield due to the two-spotted spider mite (TSSM) injury. Overall, Washington production was down an average of 10% in 1998 due to TSSM attack (Crop Profile for Hops in Washington, 2001). The disease powdery mildew can similarly reduce yield by up to 90% if left untreated (Mahaffee, Engelhard, Gent, & Grove, 2009); yield damage of 20% from omitting a fungicide application late in the season also been documented (Gent et al., 2014). On the other hand, hop growers/exporters increasingly face more stringent MRLs determined exogenously by importing countries.

The paper makes several primary contributions not present in the literature. First, it demonstrates tradeoffs that growers balance between yield risk and pesticide residues when using chemicals to control pests and diseases. Second, it addresses decisions for both risk neutral and risk averse growers. Third, it specifically examines decisions for hop production providing guidance to grower/exporters previously not available.

The remainder of the article is structured as follows. In the next section, we provide a theoretical model. This is followed by specification of the empirical model and simulation procedures. These simulations provide experimental evidence of hypothetical economic decisions and consequences. Results and implications are then provided. We end with concluding remarks and discuss future directions for research.

2. Theoretical Model

2.1 Hop Production and Marketing Process

We assume a stylized hop production and marketing timeline: contract stage, production stage and marketing stage. In the contract stage, the hop merchant (buyer) offers a multi-year forward contract (typically 3 to 5 years) that specifies the hop price per pound ($\overline{p}$), size of contract (i.e. purchase quantity) ($\overline{y}$), and the pesticide tolerance level ($\overline{q}$). Given a hop contract, the representative hop grower makes a decision on the choice of chemical levels at the beginning of the production stage. Here define $x = (x_1, \ldots, x_k)$ as a vector of chemical inputs such as fungicides, miticides, insecticides, etc.

After the growing season the grower realizes the outcomes of hop yield y and pesticide residue q at the end of this stage. Here q is interpreted as the level of pesticide residue (inversely related to quality). Terms related to hop quality are often specified in a contract (Note 2).

In the marketing stage, the hop buyer/importer decides whether to accept or to reject the contracted hops. We assume the buyer/importer will accept the hops only when the pesticide residue is below a specified tolerance level $q < \overline{q}$. Otherwise the buyer/importer rejects the hops delivery.

2.2 Pesticide Residue and the MRL

For the purpose of theoretical analysis, we assume hop yield, $y \in [0, a]$, is stochastic. Let $q \in [0, b]$ be the pesticide residue which is also stochastic. Define a conditional joint probability density function as:

$$f(y, q \mid x) \tag{1}$$

where,

x is the vector of chemical inputs. This distribution is taken to be realized at the end of the production stage. In a general sense this function structures the relationship among chemical use, crop yield, and residue. More specific functional relationships are defined in the simulation section.

A pesticide tolerance level in the form of a maximum residue limit, or MRL, $\overline{q}$, is exogenously determined by a third party such as the government or an importing agent. From a purely statistical perspective, identifying a MRL is equivalent to estimating an extreme percentile of an unknown distribution (MacLachlan & Hamilton, 2010). In this manner one can have a pre-specified confidence that an estimate exceeds the true percentile.

2.3 Input Decisions under Risk

We assume a hop grower makes input decision at the beginning of production stage. If hops are accepted and marketed, i.e., $q < \overline{q}$, growers will receive a higher contracted price $\overline{p}$ and, hence, a higher profits π_1. In this case the grower will receive deterministic revenue of $\overline{p} \cdot \overline{y}$ by fulfilling the contract. If hops are not accepted by the buyer nor marketed, i.e., $q \geq \overline{q}$, the grower will receive a lower price $\hat{p} < \overline{p}$ and, hence, lower profit $\pi_2 < \pi_1$ (Note 3). Combining revenue with costs defines the growers profit function:

$$\begin{aligned} \pi_1 &= \overline{p} \cdot \overline{y} - w \cdot x - \gamma, \quad q < \overline{q} \\ \pi_2 &= \hat{p} \cdot \overline{y} - w \cdot x - \gamma, \quad q \geq \overline{q}. \end{aligned} \tag{2}$$

Here,

$w \cdot x$ is the growers marginal cost of pesticide use and γ represents the remaining production costs.

The grower chooses the chemical input under the joint distribution in Equation (1) and solves the following optimality problem:

$$\max_x V(\pi) = \max_x \int_0^a \int_0^{\overline{q}} u(\pi_1) f(y, q \mid x)\, dq dy + \int_0^a \int_{\overline{q}}^b u(\pi_2) f(y, q \mid x)\, dq dy \tag{3}$$

where the hop grower's risk preferences are represented by the von Neumann-Morgenstern utility function $u(\cdot)$. Risk aversion is assessed using the Arrow-Pratt absolute risk aversion coefficient $r \equiv -(\partial^2 U / \partial^2 \pi)/(\partial U / \partial \pi)$, with $r > 0$ representing risk aversion (Pratt, 1964). For a constant r it is known as constant absolute risk aversion.

The optimal level of chemical input solves the first order condition (FOC):

$$\int_0^a \int_0^{\bar{q}} u(\pi_1) \frac{\partial f(y,q|x)}{\partial x} dqdy + \int_0^a \int_{\bar{q}}^b u(\pi_2) \frac{\partial f(y,q|x)}{\partial x} dqdy$$
$$-w \cdot \left(\int_0^a \int_0^{\bar{q}} u'(\pi_1) f(y,q|x) dqdy + \int_0^a \int_{\bar{q}}^b u'(\pi_2) f(y,q|x) dqdy \right) = 0. \tag{4}$$

The first line in Equation (4) is the marginal effect of chemical input on the distribution of yield and pesticide residue. The second line is the marginal effect of chemical input on the two different utility levels. As shown in Babcock and Hennessy (1996), Falco and Chavas (2009) and Antle (2010), we would expect an increase in input level to change the shape of the distribution. In general, we expect the mean of pesticide residue distribution increases with chemical input. However, the variance effect of an addition chemical input is ambiguous. Thus we cannot rule out that an increase in the chemical input may increase the probability of acceptance and decrease the probability of rejection, or an increase in the chemical input increases the probability of both acceptance and rejection.

If we assume y and q are independent with marginal distributions of $g(y|x)$ and $h(q|x)$. Then the effect of x on the joint distribution

$$\frac{\partial f(y,q|x)}{\partial x} = \left(\frac{\partial g(y|x)}{\partial x} h(q|x) + \frac{\partial h(q|x)}{\partial x} g(y|x) \right),$$

can be decomposed into the effects on the marginal distributions. The FOC becomes

$$\int_0^a \int_0^{\bar{q}} u(\pi_1) \left(\frac{\partial g(y|x)}{\partial x} h(q|x) + \frac{\partial h(q|x)}{\partial x} g(y|x) \right) dqdy$$
$$+ \int_0^a \int_{\bar{q}}^b u(\pi_2) \left(\frac{\partial g(y|x)}{\partial x} h(q|x) + \frac{\partial h(q|x)}{\partial x} g(y|x) \right) dqdy$$
$$-w \cdot \left(\int_0^a \int_0^{\bar{q}} u'(\pi_1) g(y|x) h(q|x) dqdy + \int_0^a \int_{\bar{q}}^b u'(\pi_2) g(y|x) h(q|x) dqdy \right) = 0. \tag{5}$$

Consider the case when $\bar{q} = 0$, i.e., the MRL stipulates absolutely no tolerance for pesticide residue. In this case selected terms in the integration of the first line and the third line become zero in Equation (5) and the FOC reduces to

$$\int_0^a \int_0^b u(\pi_2) \cdot \left(\frac{\partial g(y|x)}{\partial x} \cdot h(q|x) + \frac{\partial h(q|x)}{\partial x} \cdot g(y|x) \right) dqdy$$
$$-w \cdot \int_0^a \int_0^b u'(\pi_2) \cdot g(y|x) \cdot h(q|x) dqdy = 0. \tag{6}$$

Because the MRL is sufficiently stringent, the optimal rate of chemical inputs is the one that maximizes expected utility from π_2. In addition when hop yield and residue are independent, the effect of the pesticide breaks down into the effect on each marginal distribution.

Now consider the comparative static effect when $\bar{q} > 0$. Applying Leibniz's rule yields

$$\frac{\partial x}{\partial \bar{q}} = -\Delta^{-1} \cdot \Bigg[u(\pi_1) \left(\int_0^a h(\bar{q}) \frac{\partial g(y)}{\partial x} dy + \int_0^a \frac{\partial h(\bar{q})}{\partial x} g(y) dy \right)$$
$$- \left(\int_0^a u(\hat{\pi}_2) h(\bar{q}) \frac{\partial g(y)}{\partial x} dy + \int_0^a u(\hat{\pi}_2) \frac{\partial h(\bar{q})}{\partial x} g(y) dy \right)$$
$$-w \cdot \int_0^a \left(u'(\pi_1) - u'(\hat{\pi}_2) \right) \cdot h(\bar{q}) \cdot g(y) dy \Bigg]. \tag{7}$$

In Equation (7) $\pi_1 = \bar{p} \cdot \bar{y} - w \cdot x$, $\hat{\pi}_2 = \hat{p}(\bar{q}) \cdot y - w \cdot x$ is the profit when $q = \bar{q}$. The second order condition $\Delta = \partial^2 V / \partial x^2$ is assumed to be negative. A positive sign in Equation (7) would mean a more stringent government MRL provides incentives for chemical reduction. However, the sign on Equation (7) is ambiguous.

3. Materials and Methods

3.1 Hops Survey and Data

A survey was carried out of hop growers' production and management practices. The survey which was conducted in 2012 targeted the population of hop growers in Pacific Northwest (Note 4). The response rate is 8.75% and it covers about 10% of the hop hectares in the Pacific Northwest.

Table 1 provides descriptive statistics for selected variables from the survey data. The average yield per hectare was 2401 kg with average revenue of \$18061/ha. The average contract price was \$7.99/kg with a minimum of \$4/kg and maximum of \$15/kg. The average contract length was 3 years with a minimum less than 1 year and maximum of 5 years.

Table 1. Descriptive statistics for selected variables from the 2012 grower survey

Variable	Mean	Standard Deviation	Minimum	Maximum	N
Yield (kg/ha)	2401.94	656.07	927.72	3801.44	46
Revenue ($/ha)	$18061.09	$6334.08	$9622.25	$37721.05	46
Contract (years)	2.74	1.48	0	5	46
Parcel (hectare)	32.20	33.17	0.20	122.22	46
Price ($/kg)	7.99	3.30	4.41	15.44	46

Note. The value of 0 for contract length was one organic field. A parcel is in hectares reported across all varieties and across all farms.

Other descriptive statistics were reported but not presented in the Table 1. Twenty different hop varieties were grown with Columbus/Tomahawk (20%), Cascade (11%), and Zeus (10%) leading the varieties reported. Growers contracted crops with different agents, including hop merchants (67%), large brewers (31%), and others (2%). None of the growers described themselves as an "organic" grower but one did grow a three hectare parcel of organic hops under a very short term contract (< 1 year). Average yield loss estimated by growers to powdery mildew (2.71%), downy mildew (2.43%) and spider mites (1.14%) which totaled 6.3% with a maximum of 18% and minimum of 0% across growers. On average chemicals were sprayed 6-7 times per season to control powdery mildew, downy mildew and spider mites, in line with other pesticide survey data (Sherman & Gent, 2014). Revenue, costs, and other outcomes were consistent with Galinato et al. (2011).

3.2 Simulation Procedures and Assumptions

To complement our theoretical model, we conduct an illustrative simulation study of hops production and chemical use calibrated to grower survey data. To establish a baseline we simulate expected utility under yield risk with no MRL restrictions ($q < \bar{q}$):

$$EU(X=x) = \frac{1}{N}\sum_{i=1}^{N} U(\pi_{1i}) \tag{8}$$

Next we focus, in particular, on simulating the effects of yield risk, risk attitudes and uncertainties of MRLs on input choices. The expectation functional form of Equation (3) is specified as,

$$EU(X=x) = \frac{N_1}{N}\sum_{i=1}^{N_1} U(\pi_{1i}) + \frac{N_2}{N}\sum_{i=1}^{N_2} U(\pi_{2i}) \tag{9}$$

where the first summation is taken over the observations when the MRL is met, $q < \bar{q}$, and the second summation is taken over the observations when the MRL is exceeded, $q < \bar{q}$. The rate of inputs that result in the highest value of Equation (9) is taken to be the expected utility maximizing rate. Similar to Babcock and Hennessey (1996) we select a constant absolute risk aversion (CARA) utility function (Pratt, 1964) as $\mu(\pi;r) = 1 - \exp(-r\pi)$ with r as the coefficient of absolute risk aversion.

A hypothetical yield function is specified dependent upon the number of chemical applications is defined by $y_i = \alpha + \beta x_i - \phi x_i^2$ $(\alpha = 1000, \beta = 2000, \phi = 250)$ where the parameters are calibrated to an illustrative yield profile. When the MRL is not binding the optimal level of chemical is 6 applications, which is close to the mean number of applications reported in the grower survey. A damage function (Damage Profile 1) that reduces yield is initially defined to be symmetric across choices of chemical inputs contingent on responses from the grower survey (scenarios 1-15). Damage Profile 1 considers 4, 5, 6, 7, and 8 applications of chemicals that reduce yield by 10%, 5%, 0%, 5%, and 10% respectively. Variations of this damage profile are examined below in further model scenarios.

Pesticide residue is assumed to be positively correlated with chemical use. MacLachlan and Hamilton (2010) report distributions of pesticide residues are often unknown, and studies have used normal, log normal, Weibull, exponential, or power distributions. In the simulations below the residue is assumed to be distributed in the family of extreme value distributions. Given the outcomes, the hypothetical binding levels of the MRL used in the analysis are 0.4 and 0.2. The exact level of MRL will depend upon the chemical used, time since application, and agents determining its level. Sensitivity analysis is used to examine model outcomes to deviations from the parameter assumptions.

A Monte Carlo simulation is completed to generate data and to realize outcomes from the structure above. Fifty iterations are completed for each simulation scenario with 1000 observations drawn each iteration. The optimal level of chemical use is defined when the expected utility achieves a maximum over a range of choices.

3.3 Scenarios

Scenarios are setup beginning with a baseline model (Scenario 1) and variations of it to provide sensitivity analysis to key model parameters and policy variables. The baseline model is simulated when hop yield has mean of 2241 kg/ha with standard deviation σ_y = 567 (see Table 2). The MRL is not binding and the hop grower exhibits risk neutrality ($r = 0.00001$). Given responses from the grower survey the contract price, $\overline{p}$, is chosen to be \$11/kg. The contract size, $\overline{y}$, is 2241 kg/ha. The hop side agreement price, $\hat{p}$, is \$6/kg.

Table 2. Primitive parameters used to calibrate the baseline model

$\overline{p}$	Hop base price (\$/kg)	11.03	Chosen based on the contract
$\overline{y}$	Contract size (kg/ha)	2241.19	Chosen based on the contract
$\hat{p}$	Hop spot market price (\$/kg)	6.62	A average price based on USDA estimates
y_{mean}	The mean of hop yield (kg/ha)	2241.19	Chosen based on the survey data
σ_y	Standard deviation	567	Chosen based on the survey data
r	Coefficient of risk aversion	0.00001	Chosen to represent risk neutrality

4. Results

Model results for scenarios 1-15 are presented in Table 3, which are deviations from the baseline model. For the baseline model the optimal level of chemical applications is 6 with expected profit of \$4843.30. Scenarios 2 and 3 introduce a binding MRL of 0.4 and 0.2, respectively. Profit reduces to \$3573.87 for Scenario 2 and \$2806.34 for Scenario 3 with the number of chemical applications remaining at 6 for Scenario 2 but decreasing to 5 for Scenario 3. Scenarios 4 and 5 introduce a risk aversion coefficient of 0.02 but the results are nearly identical to Scenarios 2 and 3.

Table 3. Simulation results for risk averse and risk neutral growers producing hops

Scenario	Absolute Risk Aversion (r)	Optimal Chemical Usage (x^c) Expected Utility	Optimal Chemical Usage (x^c) Expected Profit	Expected Profit ($/ha)	Standard Deviation	MRL
1-Baseline	0.00001	6.000	6.000	4843.30	σ_y, σ_{res}	Not Binding
2	0.00001	5.960	6.000	3573.87	σ_y, σ_{res}	0.4
3	0.00001	5.000	5.100	2806.34	σ_y, σ_{res}	0.2
4	0.02	5.960	6.000	3349.12	σ_y, σ_{res}	0.4
5	0.02	5.000	5.080	2672.47	σ_y, σ_{res}	0.2
6	2.0	5.020	6.000	3390.72	σ_y, σ_{res}	0.4
7	2.0	4.740	5.180	2700.95	σ_y, σ_{res}	0.2
8	0.02	6.000	6.000	2901.68	$4\sigma_y, \sigma_{res}$	0.4
9	0.02	5.840	5.120	2483.50	$4\sigma_y, \sigma_{res}$	0.2
10	2.0	4.200	6.000	3592.08	$.5\sigma_y, \sigma_{res}$	0.4
11	2.0	4.000	5.140	3165.65	$.5\sigma_y, \sigma_{res}$	0.2
12	2.0	4.440	5.160	2901.52	$\sigma_y, 2\sigma_{res}$	0.4
13	2.0	4.340	4.860	1576.88	$\sigma_y, 2\sigma_{res}$	0.2
14	2.0	4.000	5.1600	2464.09	$.5\sigma_y, 2\sigma_{res}$	0.4
15	2.0	4.000	4.760	1586.15	$.5\sigma_y, 2\sigma_{res}$	0.2

Scenarios 6 and 7 introduce a larger risk aversion coefficient of 0.20. For Scenario 6, under the MRL = 0.40, the number of chemical applications is reduced to nearly 5 for the risk averse grower and remain at 6 for the risk neutral grower. Under the MRL = 0.20, or Scenario 7, the number of chemical applications is reduced to 4.74 for the risk averse grower and 5.14 for the risk neutral grower. This demonstrates that under certain circumstances the risk averse grower will reduce the number of chemical applications and use fewer chemical applications than the risk neutral grower.

Additional simulations are examined to explore sensitivity of the above results to specifications of the yield and pesticide residue distributions. For scenarios 8 and 9 we inflate the standard deviation of yield by a factor of 4. Under the MRL = 0.40, or Scenario 8, the number of chemical applications is the same as the baseline. However, when the decreases to MRL = 0.2 the number of applications is 5.84 for the risk averse grower and 5.12 for the risk neutral grower. The important observation is that increased riskiness in yield can overshadow restrictions from the MRL, and the number of chemical applications is greater under risk aversion than risk neutrality. For scenarios 10 and 11 we deflate the standard deviation of yield by a factor of 0.5. The opposite is observed in that chemical applications are again now less for the risk averse grower.

Under Scenarios 12 and 13 the standard deviation of the pesticide residue is multiplied by a factor of 2.0, increasing its uncertainty. It is anticipated that increasing the uncertainty of the residue would further decrease the number of chemical applications. For Scenario 12, the number of applications go from about 5 (Scenario 6) to 4.44 under risk aversion. For Scenario 13, the number of applications go from about 4.74 (Scenario 6) to 4.34 under risk aversion. In effect, all else equal, the more growers are uncertain about the pesticide residue the fewer chemical applications will be applied. Finally, Scenarios 14 and 15 further demonstrates as yield risk decreases and variance of the pesticide residue increases then the number of chemical applications decrease.

These results provide several important observations worth discussing. First, all else equal, a binding MRL decreases chemical applications. Second, all else equal, increasing the variance associated with the pesticide residue further decreases the number of chemical applications. However, there is an important balancing act for the grower. As demonstrated above, the riskiness of yield and variance of the residue tradeoff with one another provided an interesting but not clear cut outcomes to guide chemical use. Finally, in circumstances where reductions in chemical applications are optimal, there is an opportunity for integrated pest management strategies or, conversely, greater use of other pesticides not subject to the MRL.

4.1 Alternative Damage Profiles

Table 4 presents outcomes from different damage profiles. Initially we considered 4, 5, 6, 7, and 8 applications of chemicals with damages of 10%, 5%, 0%, 5%, and 10%, respectively (Damage Profile 1). Damage Profile 2 for 4, 5, 6, 7, and 8 applications of chemicals are damages of 50%, 25%, 0%, 25%, and 50%, respectively (Damage Profile 2). There is no difference between outcomes for the Baseline and Scenario 16. Indeed, because of the larger magnitude of damage, even with a MRL = 0.20 (Scenario 17), there is no discernible difference in chemical applications relative to the baseline.

Table 4. Further simulation results under alternative damage profiles

Scenario	Damage Profile	Absolute Risk Aversion (r)	Optimal Chemical Usage (x^c) Expected Utility	Optimal Chemical Usage (x^c) Expected Profit	Expected Profit ($/ha)	MRL
1-Baseline	1	0.00001	6.000	6.000	4843.30	Not Binding
16	2	0.00001	6.000	6.000	5225.90	Not Binding
17	2	2.0	5.920	6.000	2978.31	0.2
18	3	0.00001	7.000	7.000	4824.44	Not Binding
19	3	2.0	4.560	5.000	3340.17	0.2

Now consider an asymmetric profile with damages of 10%, 5%, 2.5%, 0%, and 5%, respectively (Damage Profile 3). Scenario 18 with no binding MRL simply increases the number of applications from 6 to 7. This is not surprising with the shift in yield from the asymmetric damage profile. Imposing a MRL = 0.20 in Scenario 19 reduces the number of applications to 4.56 for the risk averse grower and 5.00 for the risk neutral grower. This is consistent with Scenario 7 but exhibits a larger decrease in the number of applications. Under this profile more applications of chemicals control better yield damage but also increase chemical residue. It is clear from the above scenarios that the likelihood and magnitude of change will depend on various factors including the shape of the yield and damage functions.

5. Conclusion

Many forces affect a grower's chemical input usage during the hop production season and marketing process. This paper focuses on how pesticide residue and yield risk affect a hop grower's chemical input decision. Intuitively, it is likely that a grower/exporter uses less chemical inputs when he or she is facing more stringent tolerance limits on pesticide residue (i.e., MRLs). But other tradeoffs exist. Consequently, an expected utility model with risk preferences is specified and simulated to identify grower tradeoffs between yield risk and uncertainty in pesticide residue with MRLs. The empirical model illustrates hop production and exports in the Pacific Northwest.

Model results provide several important observations worth discussing. First, all else equal, a binding MRL tend to decrease chemical applications. Second, all else equal, increasing the variance associated with the pesticide residue (more uncertainty) further decreases the number of chemical applications. However, there is an important balancing act for the grower. As demonstrated above, the riskiness of yield and variance of the pesticide residue tradeoff with one another providing and interesting but not clear cut outcomes to guide chemical use. Finally, in circumstances where reductions in chemical applications are optimal, there is an opportunity for integrated pest management strategies or, alternatively, greater use of other pesticides not subject to MRL restrictions.

The model and empirical results are hypothetical and not a definitive study of MRLs in agricultural production. Nevertheless they provide an initial assessment of the situation in hop production and interesting insights to growers' decisions and the consequences of these decisions. Because of limited data we simulate scenarios with average chemical costs and stylized damages. We focused only on yield damage in the present analysis, but reductions in crop quality also impact decisions on pesticide use. Quality defects due to pests may be equally important to yield depression in some instances. However, damage functions for quality defects are difficult to assess and may interact with customer demand (Zadoks, 1985). As a result our findings are general observations, as opposed to specific recommendations. Future research should include gathering more extensive data on yield and quality damage, and the relationship between pesticide use and resultant residue levels. Moreover, it should examine the issue of ambiguity loss as well as risk aversion.

Acknowledgements

This research was funded by USDA-NIFA and the WSU IMPACT Center.

Disclaimer

The use of trade, firm, or corporation names in this publication is for the information and convenience of the reader. Such use does not constitute an official endorsement or approval by the United States Department of Agriculture or the Agricultural Research Service of any product or service to the exclusion of others that may be suitable.

References

Antle, J. M. (2010). Do Economic Variables Follow Scale or Location-Scale Distributions? *American Journal of Agricultural Economics, 92*(1), 196-204. http://dx.doi.org/10.1093/ajae/aap010

Babcock, B. A., & Hennessy, D. A. (1996). Input Demand under Yield and Revenue Insurance. *American Journal of Agricultural Economics, 78*, 416-427. http://dx.doi.org/10.2307/1243713

Falco, S. D., & Chavas, J. (2009). On Crop Biodiversity, Risk Exposure, and Food Security in the Highlands of Ethiopia. *American Journal of Agricultural Economics, 91*, 599-611. http://dx.doi.org/10.1111/j.1467-8276.2009.01265.x

Galinato, S. P., George, A., & Hinman, H. (2011). *2010 Estimated Cost of Producing Hops in the Yakima Valley, Washington State*. Washington State University Extension Publication FS028E.

Gent, D. H., Grove, G. G., Nelson, M. E., Wolfenbarger, S. N., & Woods, J. L. (2014). Crop Damage Caused by Powdery Mildew on Hop and its Relationship to Late Season Management. *Plant Pathology, 63*(3), 625-639. http://dx.doi.org/10.1111/ppa.12123

Just, R. E., & Zilberman, D. (1983). Stochastic Structure, Farm Size and Technology Adoption in Developing Agriculture. *Oxford Economic Papers, 35*(2), 307-328.

Lichtenberg, E., & Zilberman, D. (1986a). The Econometrics of Damage Control: Why Specification Matters. *American Journal of Agricultural Economics, 68*, 261-273. http://dx.doi.org/10.2307/1241427

MacLachlan, D. J., & Hamilton, D. (2010). Estimation methods for Maximum Residue Limits for pesticides. *Regulatory Toxicology and Pharmacology, 58*(2), 208-18. http://dx.doi.org/ 10.1016/j.yrtph.2010.05.012

Mahaffee, W. F., Engelhard, B., Gent, D. H., & Grove, G. G. (2009). Powdery mildew. In W. M. Mahaffee, S. J. Pethybridge & D. H. Gent (Eds.), *Compendium of Hop Diseases and Pests* (pp. 25-31). APS Press, St. Paul, MN.

Pratt, J. W. (1964). Risk Aversion in the Small and in the Large. *Econometrica: Journal of the Econometric Society*, 122-136. http://dx.doi.org/10.2307/1913738

Sexton, S. E., Lei, Z., & Zilberman, D. (2007). The Economics of Pesticides and Pest Control. *International Review of Environmental and Resource Economics,* 271-326. http://dx.doi.org/10.1561/101.00000007

Sherman, J., & Gent, D. H. (2014). Concepts of Sustainability, Motivations for Pest Management Approaches, and Implications for Communicating Change. *Plant Disease, 98*(8), 1024-1035. http://dx.doi.org/10.1094/PDIS-03-14-0313-FE

Zadoks, J. C. (1985). On the Conceptual Basis of Crop Loss Assessment: The Threshold Theory. *Annual Review of Phytopathology, 23*, 455-473. http://dx.doi.org/10.1146/annurev.py.23.090185.002323

Notes

Note 1. For example, hops have historically been purchased using multi-year forward production contracts. Only a small share of the hops produced targets the spot market. On average, over 90% of the crop has been contracted in advance of harvest where hop's price and purchase quantity are "locked in" at the time the contract is issued.

Note 2. In general, hop's descriptive quality attributes include hop cone's color, size, moisture etc. In the current paper we are interested in the quality issues related to pesticide residue, which we assume can be tested and determined by the hop merchant.

Note 3. Personal communications with hop growers in the Pacific Northwest indicate that growers often have side agreements with other growers by which they purchase hops to fulfill contracts in deficit events. Hence, we assume for convenience that the contact size is met with a lower net price.

Note 4. The majority of hop farms are located in Washington's Yakima Valley. In 2008 for example, Washington State produced 12381 hectares of hops, which made up about 75% of the US commercial hop's production. Behind Washington was Oregon with 2578 hectares and Idaho with 1592 hectares which make up around 15.5% and 9.5% of the US commercial hop production respectively.

Efficacy of Dairy Cattle Slurry in Preventing Zinc Deficiency of a Silage Corn (*Zea mays* L.) Grown on a Sandy Soil

Saad Drissi[1], Abdelhadi Aït Houssa[2], Ahmed Bamouh[1], El Madani Daoudi[3] & Mohamed Benbella[2]

[1] Plant Production, Protection and Biotechnology Department, Hassan II Agronomy and Veterinary Institute, Rabat, Morocco

[2] Department of Agronomy and Plant Breeding, National Agricultural School of Meknes, Meknes, Morocco

[3] Moroccan Laboratory of Agriculture (LABOMAG), Casablanca, Morocco

Correspondence: Saad Drissi, Plant Production, Protection and Biotechnology Department, Hassan II Agronomy and Veterinary Institute, Rabat, Morocco. E-mail: drissi_agro@yahoo.com

Abstract

Corn silage (*Zea mays* L.), grown on sandy soil, was severely affected by Zinc (Zn) deficiency stress. Mineral and organic Zn sources were required to prevent this deficiency. The objective of this study is to evaluate the effectiveness of dairy cattle (*Bos taurus*) slurry as an organic source of Zn in preventing Zn deficiency in corn silage grown on sandy soil. A field experiment was conducted and six rates of dairy cattle slurry were spread just before sowing: 0 or no slurry spread; 50; 100; 150; 200 and 300 t ha^{-1}. These slurry rates were compared to an adequate mineral Zn supply of 5 mg kg^{-1} applied to soil as ($ZnSO_4 \cdot 7H_2O$). Regression analysis has shown that slurry rates ranging between 50 and 60 t ha^{-1} had almost similar efficacy to the adequate Zn mineral supply. Both of them corrected Zn deficiency symptoms, enhanced plant growth and increased silage yield by 37.5%, compared to no Zn or slurry applications. A high slurry rate of 280 t ha^{-1} maximized silage yield but may pose a threat to the environment due to its high content of nitrogen (N). At the next cropping season, a sufficient DTPA residual Zn soil content of 0.9 mg kg^{-1} was recorded only with a high slurry rate (300 t ha^{-1}).

Keywords: corn silage, zinc deficiency, dairy cattle slurry, Zn mineral, mineral content, growth, yield

1. Introduction

Zn deficiency is a worldwide nutritional constraint that reduces the yield of many crops, especially cereals (Cakmak, 2008). Maize (*Zea mays* L.) is known to be very sensitive to Zn deficiency (Lindsay & Norvel, 1977). Under Zn deficiency stress, maize has shown white bands between the midrib and the margin of leaves (Singh, Natesan, Singh, & Usha, 2005), accompanied by an obvious decline in plant height (De Vasconcelos, Clístenes, & Fernando, 2011) and in dry matter production (Van Biljon, Wright, Fouche, & Botha, 2010). These symptoms are explained by the important physiological functions performed by Zn, which include protein synthesis and energy production (Hansch & Mendel, 2009). Sandy soils are generally deficient in Zn (Bhupinder, Senthil, Singh, & Usha, 2005). In this case, an adequate soil Zn supply as mineral or organic forms can prevent this deficiency (Murphy & Walsh, 1972). In the context of many dairy farmlands from Loukkos perimeter (North-West of Morocco), corn silage grown on sandy soil has shown clear Zn deficiency symptoms. Consequently, an important silage yield decline was recorded compared to adequate fertilized plots receiving either Zn mineral supply or farmyard manures. In this regard, Wallingford, Murphy, Powers and Manges (1975) reported that farmyard manures can enhance soil Zn availability. Among farmyard manures applied, dairy cattle slurry is an important effluent of dairy farms. Many studies reported its advantageous effects in improving soil content in macronutrients, micronutrients, organic matter (Nikoli & Matsi, 2011; Matsi, Lithourgidis, & Gagianas, 2003), and thereby corn silage yield (Schröder, Ten Holte, & Brourwer, 1997). In a recent study, Nikoli and Matsi (2011) reported that liquid cattle manure as a source of micronutrients could also indirectly improve the availability of soil native micronutrients through the action of its organic matter. Also, Sinclair and Edwards (2008) reported that cattle slurry contains an amount of Zn ranging between 11 and 18 g t^{-1} (wet weight). On the other hand, Nikoli and Matsi (2011) stated that farmyard manures increased soil content in micronutrients chelated forms, and many authors like Ortega-Blu and Molina-Roco (2007), Chatterjee and Mandal (1985) and Chand, Randhawa, and Bhumbla (1981) reported that Zn chelated forms were more efficient than inorganic ones. Still, an over application of farm manures may threaten the environment because of nitrate leaching (Beckwith, Cooper, Smit, & Shepherd, 1998) and heavy metals accumulation (Berenguer, Cela, Santiveri, Boixadera, & Lloveras, 2008).

Despite the numerous researches discussing the advantages and disadvantages of farm manures application on soil nutrients enrichment, we find little information about the possibility of substituting Zn mineral supply either by dairy cattle slurry or by other animal manures. For this reason, the following work aims to compare the effectiveness of dairy cattle slurry rates to an adequate mineral Zn supply, in terms of preventing Zn deficiency in corn silage grown on sandy soil.

2. Method

2.1 Site Description

A field experiment was carried out on an agricultural farm (Bassita II) located in Loukkos perimeter (34°96′N lat., 6°21′W long., 60 m above the sea level, North West of Morocco). The climate is maritime. During the growing season (May to August), the average maximum and minimum temperatures were 18.5 °C and 27.3 °C, respectively. Also, a low precipitation rate (11 mm) was recorded during this period. A climatic comparison between the studied growing season and the long term climatic data of the last 40 years is provided in Table1.

Table 1. Climatological data of the experimental site during the growing season (May to August) of 2012 and the last 40 years average (1971-2010) *

	Min temperature (°C)		Max temperature (°C)		Evapotranspiration (mm day^{-1})		Rainfall (mm)	
	2012	Long term	2012	Long term	2012	Long term	2012	Long term
May	15.54	13.76	26.32	22.24	3.64	2.75	7	34.71
June	18.23	16.67	27.66	25.04	4.07	3.56	0	9.17
July	20.58	18.69	29.58	27.57	4.56	4.06	0	0.93
August	19.93	18.94	25.70	27.57	4.02	4.09	4	1.82

*Source: Taken from National Directorate of Meteorology (DMN), Morocco.

2.2 Soil and Dairy Cattle Slurry Characteristics

The soil is sandy (88.8% of sand, 7.5% of clay and 5.3% of silt), deficient in DTPA extractable Zn with an amount of 0.13 mg kg^{-1}, which was below the critical level of 0.8 mg kg^{-1} required for corn production (Lindsay & Norvell, 1977). It's not calcareous (0.1% of total $CaCO_3$), with a pH of 6.1 (soil/H_2O as 1/5) and a low amount of organic matter (0.4%) (Walkley & Black, 1934). The other chemical properties of soil are presented in Table 2.

Table 2. Soil characteristics

Soil property	
pH H_2O (1/5)	6.10
Cation exchange capacity (meq 100 g^{-1}) (Cobaltihexamine Chloride method)	4.40
Organic matter (%) (Walkley & Black method)	0.40
Extractable P_2O_5 (mg kg^{-1}) (Olsen method)	49
Extractable K_2O (mg kg^{-1})(Ammonium Acetate method)	81
Extractable MgO (mg kg^{-1}) (Ammonium Acetate method)	101
Extractable CaO (mg Kg^{-1}) (Sodium Acetate method)	868
DTPA extractable Zn (mg kg^{-1})	0.13
DTPA extractable Fe (mg kg^{-1})	17.45
DTPA extractable Mn (mg kg^{-1})	17.40
DTPA extractable Cu (mg kg^{-1})	0.06

The dairy cattle (*Bos taurus*) slurry used in this experiment was a mixture of feces, urine, feed residue and washing water. It had 79% of moisture. Its dry matter contained 106.79 mg kg^{-1} of total Zn and 1.91% of total N. The other chemical properties are listed in Table 3.

Table 3. Dairy cattle slurry characteristics

Total mineral content on slurry dry matter			
(%)		(mg kg^{-1})	
N	1.91	Zn	106.79
P	0.52	Cu	19.75
K	1.80	B	26.49
Mg	0.58	Mn	179.95
Ca	3.19	Fe	5534.28
Cl	0.81	Na	8466.76
Organic matter (%)		65.60	
C/N		19.84	
Moisture content (%)		79	

2.3 Crop Management and Experimental Design

Corn silage cv. Panama was seeded on May 13, 2012 in 45 cm double rows spacing, 90 cm between double rows spacing and 12 cm seed distance to approximate 120,000 plants ha^{-1}. The experimental design was a randomized complete block with five replications. The plot size was 4 by 8 m and contained three double rows. There were 4 m between adjacent plots in the same block and 1.4 m between adjacent blocks. Six rates of dairy cattle slurry were tested: 0; 50; 100; 150; 200 and 300 t ha^{-1}. The slurry rate of 50 t ha^{-1} produces a Zn amount of 1.1 kg ha^{-1} and N amount of 201 kg ha^{-1}. These slurry treatments were compared to an adequate mineral Zn supply of

5 mg kg^{-1} applied as ($ZnSO_4{\cdot}7H_2O$; 22.5% of Zn). The mineral Zn treatment was recognized in a previous study as the optimal supply for corn silage grown on a sandy soil (Drissi, unpublished MA thesis). One day before sowing, the total rates of slurry were spread using a spreading machine. Few hours after, the slurry was incorporated into the soil using a rotavator. Mineral Zn treatment was applied as a solution of Zn sulfate in the middle of each double cropping row. It was split at three different times during the growing season: i) 50% just after sowing, ii) 25% at 4-5 leaf stage, and iii) 25% at 8-9 leaf stage.

In order to make sure that Zn was the only nutrient limiting corn silage growth and yield, all treatments were adequately fertilized with 395 Kg ha^{-1} of N as ammonium nitrate and diammonium phosphate (DAP), 334 Kg ha^{-1} of P_2O_5 as DAP, 33 Kg ha^{-1} of K_2O as potassium sulfate, 3.3 Kg ha^{-1} of Mn as manganese sulfate, 6.2 Kg ha^{-1} of Cu as copper sulfate and 4.6 Kg ha^{-1} of B as boron sulfate, these amounts were split adequately during the growing season.

Each line of plants was equipped with a drip line irrigation system with 1.2 L h^{-1} emitters and 0.4 m as emitters spacing. Watering was done whenever required throughout the experiment. In order to control weeds, pre-emergence herbicides (Pendimethaline, Mesotrione, Terbuthylazine and S-metolachlor) were sprayed. Insects (Heliothis, spodoptera and sesamia) were controlled with Indoxocarb and fungal disease (Helmintosporium) was controlled with Epoxiconazole.

2.4 Measurements

According to the description of Zn deficiency symptoms in corn reported by many authors (Alloway, 2008; Singh et al., 2005), the evolution of these symptoms was described during the growing season on ten randomly chosen plants for each treatment's replication. Stem height and stem diameter were measured at harvest on ten randomly chosen plants for each treatment's replication. Also, the total leaf area was measured on three randomly chosen plants per each treatment's replication using Formula (1) (Dwyer & stewart, 1986 in Mokhtarpour et al., 2010):

$$\textit{Total leaf area per plant} = \sum_{n=1}^{n=j}(L \times W \times 0.75) \qquad (1)$$

Where,

L, W, and j are leaf length, leaf maximum width, and last leaf of a plant, respectively.

The harvest was done on August 25, 2012, approximately at a shoot moisture content of 54%. Ten randomly chosen plants per each treatment's replication were cut close to the soil surface, and the fresh weights of the harvest were measured.

For shoot dry matter partitioning, a subsample of three plants per each replication's treatment was taken and partitioned into leaves, stem, kernels, cob and husks. Then, the plants parts were dried to constant weight. Three randomly chosen ears per each replication's treatment were taken to determine kernels yield components: total number of ovules per ear using Formula (2), pollination rate using Formula (3) and thousand kernels dry weight.

$$\textit{Total number of ovules per ear} = \textit{Number of kernels} + \textit{Number of sterilized ovules} \quad (2)$$

$$\textit{Pollination rate} = (\textit{Number of kernels per ear} \,/\, \textit{Total number of ovules per ear}) \times 100 \quad (3)$$

In order to determine corn mineral concentration at harvest, three whole aerial tissues from 0, 100, 200 and 300 t ha^{-1} of slurry treatments and Zn mineral supply were dried in an oven at 70 °C and ground to powder. Thereafter, three subsamples of 2 g from the three chopped plants, cited above, were digested with a mixture of nitric, perchloric and sulfuric acids to analyze Zn concentration (Chintala et al., 2012a; Chintala et al., 2012b). Three other subsamples of 0.6 g were digested with a mixture of salicylic and sulfuric acids to analyze N, P, K, Ca and Mg. K, Ca, Mg and Zn were determined using an atomic absorption spectrophotometer (Varian AA 240 FS; air acetylene flame). While a continuous flow analyzer (Skalar San++) was used to determine N and P (Chintala et al., 2013; Chintala et al., 2014). Furthermore, three soil samples (0-20 cm depth) from each treatment, except 50 and 150 t ha^{-1} slurry rates, were sampled at the next growing season. Thereafter, they were ground to pass a 2 mm sieve then analyzed for available Zn by DTPA extractant method using the same atomic absorption spectrophotometer.

2.5 Statistical Analysis

All data were subjected to the analysis of variance (ANOVA) at $P \leq 0.05$ level. Regression analysis was applied to evaluate slurry rates, and the best regression lines were fitted using the highest r^2. The difference between Zn mineral supply and dairy cattle manure rates was analyzed using the Dunett test at $P \leq 0.05$ level. These statistical analyses have been done using the program SPSS (Version 17.0).

3. Results

3.1 Evolution of Visual Zn Deficiency Symptoms

Zn deficiency symptoms appeared at an earlier stage (4-5 leaf stage) for all treatments, but the severity of these symptoms was less manifested either with increasing slurry rates or with mineral Zn supply. A clear linear white area between the midrib and the margin of leaves appeared on plants grown without Zn or slurry applications while some pale green linear stripes were observed with high slurry rates (200-300 t ha^{-1}). Approximately at 8-9 leaf stage, all treatments showed a clear recovery from Zn deficiency stress. This recovery capacity manifested itself more and earlier with increasing slurry rates. From anthesis to harvest, all treatments, except for no Zn or slurry applications, didn't show any Zn deficiency symptoms.

3.2 Plant Growth

At harvest, stem height, stem diameter and leaf area showed significant quadratic responses towards slurry rates. Regression analysis showed that slurry rates around 50 t ha^{-1} induced similar growth responses to Zn fertilization. Both of them improved stem height, stem diameter and leaf area approximately by 9%, 10% and 25%, respectively, compared to no Zn or slurry applications (Figures 1A, 1B and 1C). Maximum growth was recorded with high slurry spread between 240 and 300 t ha^{-1}. It provided an increase that reaches almost 19%, 29% and 92 % on stem height, stem diameter, and leaf area, respectively.

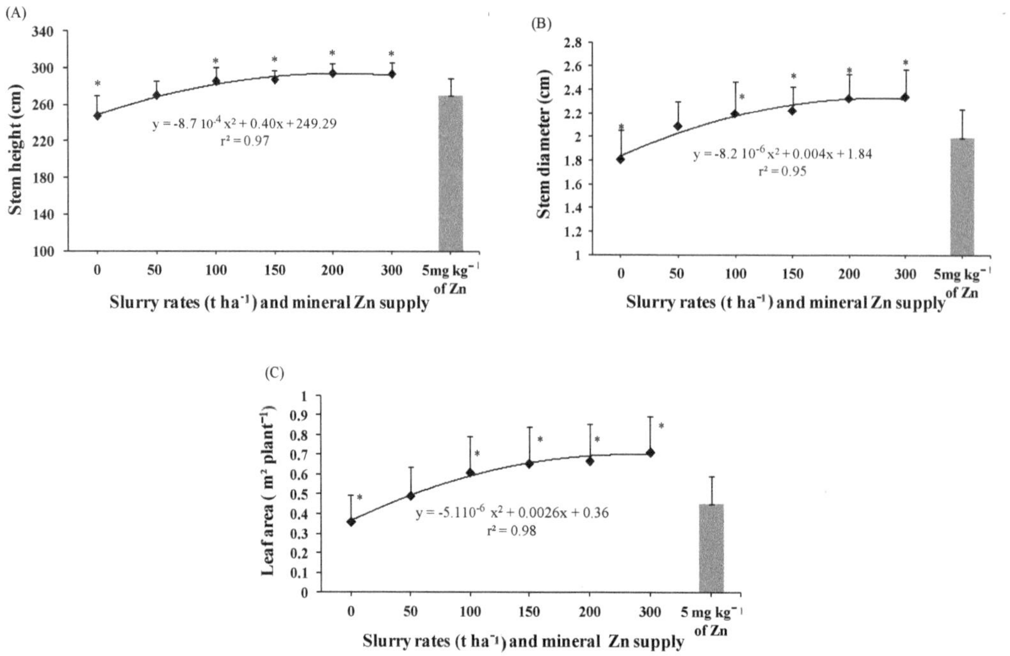

Figure 1. Stem height (A) stem diameter (B) and leaf area (C) of corn silage at harvest following application of dairy cattle slurry rates or mineral Zn. Vertical bars denote standard deviation. * The value differed from Zn treatment using Dunett test ($p \leq 0.05$)

3.3 Silage Yield and Shoot Dry Matter Partitioning

Dairy slurry or mineral Zn applications provided a significant rise in silage yield. This latter had quadratic response towards slurry spread. A rate of 63 t ha^{-1} had similar response to mineral Zn supply. Both of them induced a silage yield rise of 37.5% compared to no Zn or slurry applications (Figure 2). The maximum yield was recorded with high slurry rate of 292 t ha^{-1}, which induced an increase of 94.78% compared to no Zn or slurry applications.

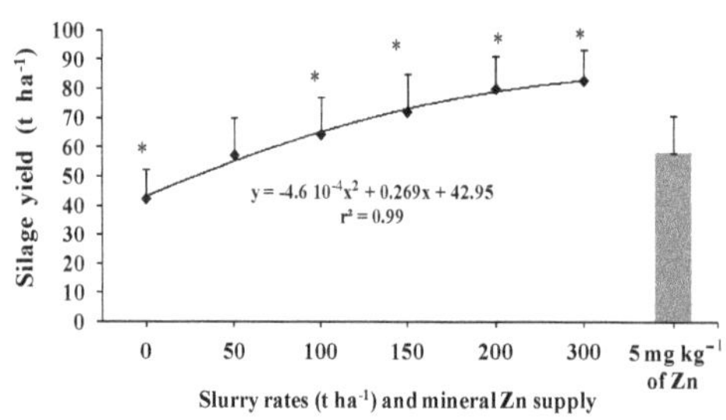

Figure 2. Corn silage yield following application of dairy cattle slurry rates or mineral Zn. Vertical bars denote standard deviation. * The value differed from Zn treatment using Dunett test ($p \leq 0.05$)

Through the analysis of shoot dry matter partitioning, we noted that leaves, stem and ears dry weights increased significantly with the addition of slurry or Zn (Figure 3A). 50 t ha^{-1} of slurry or Zn supply provided increases of approximately 28.4%, 37% and 10.5% on ear, stem and leaves dry weights, respectively, compared to no Zn or slurry applications. These increases were approximately of 43.9%; 101% and 82.5%, respectively, from slurry rate of 150 t ha^{-1}. The ear biomass partitioning analysis showed that kernels, husks and cob dry weights responded significantly to slurry and Zn treatments (Figure 3B). Compared to no Zn or slurry applications, Zn supply resulted in kernels dry weight rise of 31.2%, which was similar to the increase achieved with slurry spread of 50 t ha^{-1} while slurry application of 50 t ha^{-1} provided an increase of approximately 47.5%. Husks and cob dry weights showed significant increases of 35.1% and 36.3%, respectively, either with slurry or Zn applications.

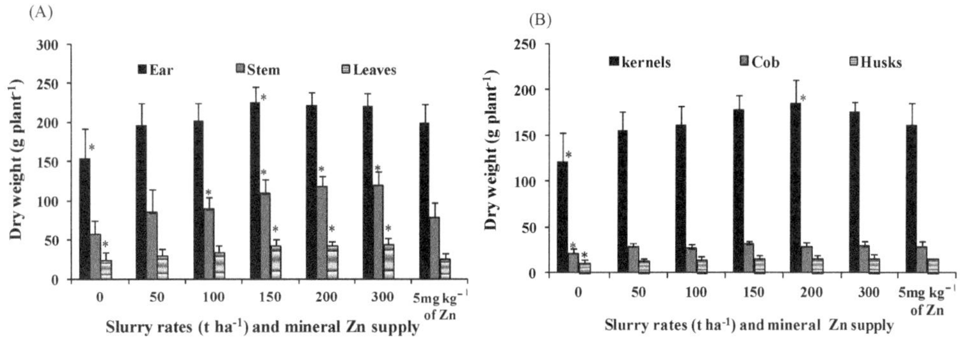

Figure 3. Shoot dry weight partitioning on different plant parts of corn silage (A and B) following application of dairy cattle slurry rates or mineral Zn. Vertical bars denote standard deviation. * The value differed from Zn treatment using Dunett test ($p \leq 0.05$)

In order to explain the kernels dry weight response, different kernels yield components were measured. No significant difference in the total number of ovules per ear was noted between different treatments (Figure 4A). However, the pollination rate was significantly enhanced by 7.2% with all slurry or mineral Zn applications (Figure 4B). The thousand kernels dry weight significantly improved; Zn or slurry rate of 68 t ha^{-1} induced an increase of 15.1% compared to no Zn or slurry applications while high slurry rate of 255 t ha^{-1} resulted in the highest rise of 25.7% (Figure 4C).

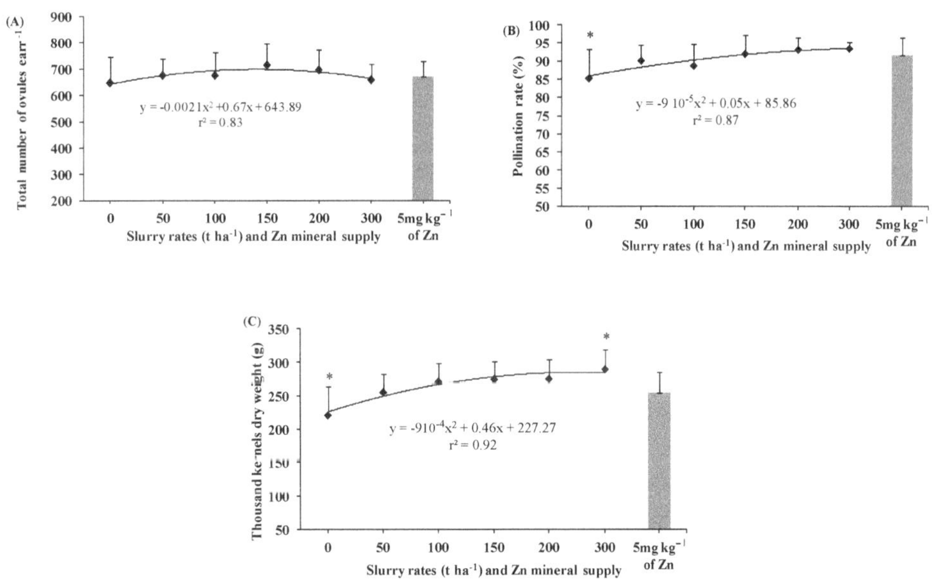

Figure 4. Number of ovules per ear (A), pollination rate (B) and thousand kernels dry weight (C) of corn silage following application of dairy cattle slurry rates or mineral Zn. Vertical bars denote standard deviation. * The value differed from Zn treatment using Dunett test ($p \leq 0.05$)

3.4 Mineral Shoot Content at Harvest

3.4.1 Shoot Zn Content

Zn deficient plants (no Zn or slurry applications) had low shoot Zn content at harvest (16 mg kg^{-1}). This content increased significantly either with slurry or Zn applications (Figure 5). We recorded the highest concentration of 31.67 mg kg^{-1} with Zn supply. Slurry rate of 63 t ha^{-1}, which induced similar yield to Zn fertilization, had a content of 18.60 mg kg^{-1} while a high amount of slurry 200-300 t ha^{-1}, which maximized yield, had a Zn content of around 22 mg kg^{-1}.

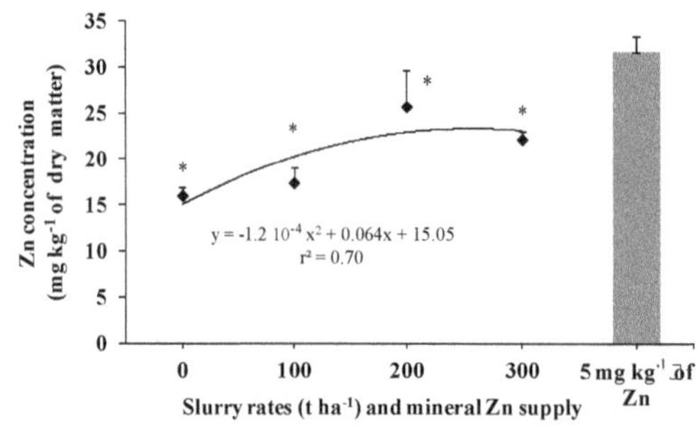

Figure 5. Shoot Zn concentration in corn silage at harvest following application of dairy cattle slurry or mineral Zn supply. Vertical bars denote standard deviation. * The value differed from Zn treatment using Dunett test ($p \leq 0.05$)

3.4.2 Shoot Content on N, P, K, Ca and Mg

High slurry rates (200-300 t ha^{-1}) resulted in a slightly significant increase in N and k shoot content compared to other treatments. They had N and K shoot contents around 1.1% and 0.92%, respectively. Under low slurry rates or Zn supply, these contents were around 0.96% and 0.70%, respectively. The P, Ca, and Mg shoot concentration, even if they were significantly enhanced with slurry rates, stay around 0.24%, 0.13% and 0.09%, respectively, for all treatments (Table 4).

Table 4. Macronutrients concentration (% of shoot dry matter) on corn silage at harvest following applications of dairy cattle slurry spread (t ha^{-1}) or mineral Zn supply

	N	P	K	Ca	Mg
0	0.96±0.05	0.25±0.01*	0.71±0.03	0.07±0.02*	0.14±0.01
100	0.99±0.01	0.24±0.00*	0.78±0.01*	0.07±0.02*	0.13±0.01*
200	1.14±0.01*	0.23±0.00*	0.91±0.01*	0.09±0.02	0.11±0.00*
300	1.13±0.11*	0.24±0.01*	0.93±0.01*	0.10±0.02	0.12±0.01*
5 mg kg^{-1} of Zn	0.95±0.00	0.27±0.00	0.66±0.03	0.13±0.01	0.16±0.01
F test	+	+	+	+	+

Note. Data are the means± standard deviation. * The value differed from Zn treatment using Dunett test ($p \leq 0.05$).

3.5 Soil Extractable Residual Zn

Soil analysis at the next cropping season showed a significant increase in residual DTPA extractable Zn either with addition of slurry or Zn (Figure 6). Slurry application of 300 t ha^{-1} induced the highest DTPA extractable Zn of 0.96 mg kg^{-1} while mineral Zn supply provided an amount of 0.26 mg kg^{-1}.

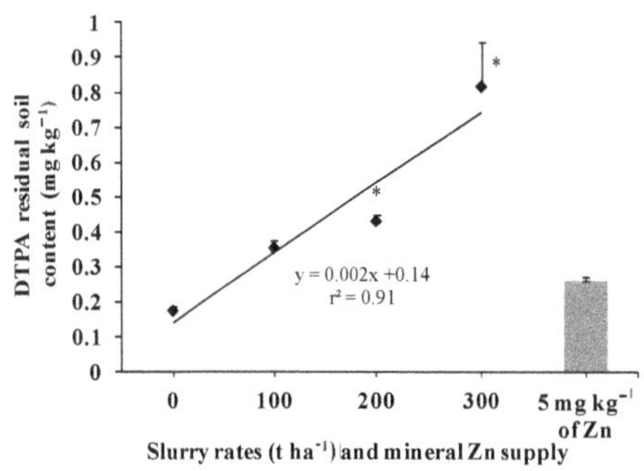

Figure 6. Residual soil DTPA extractable Zn, following application of dairy slurry or mineral Zn, in the next cropping season. Vertical bars denote standard deviation. * The value differed from Zn treatment using Dunett test ($p \leq 0.05$)

4. Discussion

Based on this investigation, slurry spread can substitute Zn mineral supply to prevent Zn deficiency on corn silage grown on sandy soil. Slurry rates between 50 and 60 t ha^{-1} showed similar growth and yield to the adequate Zn mineral supply.

4.1 Zn Deficiency Symptoms

The appearance of Zn deficiency symptoms is especially due to the low Zn content of soil 0.13 mg kg^{-1}, which was below the critical level required for corn production 0.8 mg kg^{-1} (Lindsay & Norvel, 1977). Their appearance at an earlier stage was already reported by other authors (Wang & Jin, 2007; Kuldeep & Banerjee, 1986). The obvious decline of Zn deficiency symptoms with slurry application, as well as mineral Zn supply, proved the role of dairy slurry in preventing Zn deficiency.

4.2 Plant Mineral Content and Growth

Shoot Zn content was enhanced with the addition of slurry as well as Zn supply. Sufficient Zn content, 22 mg kg^{-1} (Singh et al., 2005), was recorded either with slurry rates up or equal to 150 t ha^{-1} or with mineral Zn supply. Zn concentration didn't reach Zn phytotoxic concentration, above 95 mg kg^{-1} (Gupta, Kening, & Siyuan, 2008), despite high slurry rates. On the other hand, even if slurry rates around 50 t ha^{-1} didn't increase shoot Zn content to reach sufficient level, we observed an obvious correction of Zn deficiency symptoms. The beneficial effects of other nutrients supplied by slurry cannot be excluded even if all treatments had received adequate mineral fertilization of macro and micro nutrients. But, such effects remain weak compared to our limited and controlled nutrient factor of Zn. In fact, plants with Zn mineral supply had approximately similar shoot content of N, P, Mg, and Ca as slurry treatments. For all these treatments, we recorded that shoot N content ($\approx$ 0.95%) was approximately equal to the required level of corn at maturity (1%) (Fageria, Baligar, & Jones, 2011). Also, shoot P content ($\approx$ 0.24%) was around the required level ranging between 0.1% and 0.5% (Arnon, 1975), and the shoot Mg concentration ($\approx$ 0.13%) was around the adequate concentration of 0.2% (Fox & Piekielek, 1984). Although the increase of shoot K content was recorded with high slurry rates (200-300 t ha^{-1}), all treatments had contents around the required level (between 0.7% and 0.9%) (Arnon, 1975), and no K deficiency symptoms were noticed. Plant growth was enhanced either with slurry or Zn application. The effect of Zn on stem stretching can be explained by the role of Zn in the metabolism of indole-acetic acid (IAA) as a growth regulator enzyme (Cakmak, Marshner, & Bangerth, 1988). Also, the leaf area and stem diameter were also enhanced. A significant increase of leaf area due to Zn fertilization was noted on corn by Chaab, Savaghebi, Gh, and Motesharezadeh (2011) and on wheat by Khan, Fuller, and Baloch (2008).

4.3 Silage Yield and Dry Matter Partitioning

Silage yield improved either with slurry or Zn applications. 63 t ha^{-1} of slurry resulted in similar yield to the adequate Zn mineral supply. Such slurry rate brought a Zn amount of 1.40 kg ha^{-1} which was lower than the adequate Zn mineral supply (20.25 kg ha^{-1}). This result proved that the effectiveness of Zn contained in slurry is higher than the mineral one. We may explain that by the fact that Zn organic forms, contained in slurry, are more efficient than Zn mineral compounds (Ortega-Blu & Molina-Roco, 2007; Chatterjee & Mandal, 1985; Chand et al., 1981). This also can be explained by the enhancement of soil born Zn availability through the action of organic matter (Nikoli & Matsi, 2011). On the other hand, maximum silage yield was recorded with high slurry rate of 292 t ha^{-1}. This result is consistent with earlier studies of Sutton, Nelson, Kelly and Hill (1986), who reported that liquid dairy manure rate of 224 t ha^{-1} maximized corn yields. Also, Schröder and Dilz (1987) showed that maize forage yield was promoted with cattle slurry application up to 200-300 t ha^{-1}. However, such rate brought an over amount of N (1171 kg ha^{-1}) which may threaten the environment through nitrate leaching. In this context, the nitrate directive of the European Union stated that animal manure application, for a given year, must be around an equivalent amount of 170 -210 kg on N ha^{-1} (European Union, 1991). The biomass weight decline under Zn deficiency (no slurry or Zn applications) can partially be attributed to limited photosynthesis rate (Hansch & Mendel, 2009) due to both disruption in carbonic anhydrase enzyme activity (Sasaki, Hirose, Watanabe, & Ohsugi, 1998) and chlorophyll synthesis (Cakmak & Marshner, 1993).

The dry matter partitioning analysis showed that enhancement of silage yield due to addition of slurry or Zn was mainly attributed to the increase of stem and kernels dry weights. Lithourgidis, Matsi, Barbayiannis and Doras (2007) reported that liquid cattle manure enhanced silage and kernels maize yields. The kernels dry matter, which represented about 50% of shoot dry weight (Figure 2), was mainly enhanced by two components: pollination rate and thousand kernels dry weight. A similar result was reported under Zn foliar spray on corn silage grown on Zn deficient sandy soil (Drissi, Aït Houssa, Bamouh, & Benbella, 2015). Under Zn deficiency, the pollination rate decline can be explained by mal sterility (Sharma, Chatterjee, & Saharma, 1990) while the decrease of thousand kernels weight could be attributed to the photosynthesis decline mentioned above.

4.4 Residual Soil DTPA Extractable Zn

Soil analysis at the next cropping season showed that slurry improved residual DTPA extractable Zn, but only high rate (300 t ha^{-1}) resulted in sufficient residual amount of 0.92 mg kg^{-1}, which was above the critical level of 0.83 mg kg^{-1} required for corn (Lindsay & Norvell, 1977). The Zn soil enrichment due to slurry spread can be explained by its role as a source of Zn, and by its organic matter's role in releasing native soil Zn (Nikoli & Matsi, 2011).

5. Conclusion

A dairy cattle slurry rate ranging between 50 and 60 t ha^{-1} prevented Zn deficiency and showed similar plant growth and silage yield as the adequate mineral Zn supply (5 mg kg^{-1} of Zn as $ZnSO_4{\cdot}7H_2O$). High dairy slurry rates (200 and 300 t ha^{-1}) maximized yield but brought an over amount of N which may pose a threat to the environment.

Acknowledgements

The authors would like to think Mr. Loultiti M. M. and Mr. Coquant J. M. for their financial support. We are grateful to Pr. Maataoui A. and Pr. El Midaoui M. for their valuable help during manuscript preparation. We acknowledge the contribution of Mrs. Mouhtadi R., Mrs. Zehri B., Mr. Ziti S., Mr. Soulaimani A., and Mr. Touhami D. during laboratory analysis. Special appreciation is expressed to Mr. Belbassri M. for his cooperation in conducting the experiment. Also, we address our acknowledgements to Dr. Yachoulti M. and Mr. El Gharradi Y. for language correction.

References

Alloway, J. B. (2008). *Zinc in soils and crop nutrition* (2nd ed.). International Zinc Association Communications. IZA Publications, Brussel. Retrieved from http://www.zinc.org/general/Zinc_in_Soils_and_Crop_Nutrition_ALLOWAY.pdf

Arnon, I. (1975). *Mineral nutrition of maize*. International Potash Institute, Bern. Retrieved from http://www.ipipotash.org/udocs/mineral_ nutrition_ of_ maize.pdf

Beckwith, C. P., Cooper, J., Smith, K. A., Shepherd, M. A. (1998). Nitrate leaching loss following application of organic manures to sandy soils in arable cropping. I. Effects of application time, manure type, overwinter crop cover and nitrification inhibition. *Soil Use and Management, 14*, 123-130. http://dx.doi.org10.1111/j.1475-2743.1998.tb00135.x

Berenguer, P., Cela, S., Santiveri, F., Boixadera, J., & Lloveras, J. (2008). Copper and Zinc Soil Accumulation and Plant Concentration in Irrigated Maize Fertilized with Liquid Swine Manure. *Agronomy Journal, 100*, 1056-1061. http://dx.doi.org/10.2134/agronj2007.0321

Bhupinder, S., Senthil, K. A. N., Singh, B. K., & Usha, K. (2005). Improving zinc efficiency of cereals under zinc deficiency. *Current Science, 88*, 36-44. Retrieved from http://eprints.icrisat.ac.in/46/

Cakmak, I. (2008). Enrichment of cereal grains with zinc: Agronomic or genetic biofortification? *Plant and Soil, 302*, 1-17. http://dx.doi.org/10.1007/s11104-007-9466-3

Cakmak, I., & Marshner, H. (1993). Effect of zinc nutritional status on activities of superoxide radical and hydrogen peroxide scavenging enzymes in bean leaves. *Plant and Soil, 155-156*, 1127-1130. http://dx.doi.org/10.1007/BF00025000

Cakmak, I., Marshner, H., & Bangerth, F. (1988). Effect of zinc nutritional status on growth, protein metabolism and levels of indole 3 acetique acid and other phytohormones in bean (*Phaseolus vulgaris*). *Journal of Experimental Botany, 40*, 405-412. http://dx.doi.org/10.1093/jxb/40.3.405

Chaab, A., Savaghebi, Gh. R., & Motesharezadeh, B. (2011). Differences in the zinc efficiency among and within maize cultivars in a calcareous soil. *Asian Journal of Agricultural Sciences, 3*, 26-31. Retrieved from http://www.maxwellsci.com/print/ajas/v3-26-31.pdf

Chand, M., Randhawa, N. S., & Bhumbla, D. R. (1981). Effectiveness of zinc chelates in zinc nutrition of greenhouse rice crop in a saline–sodic soil. *Plant and Soil, 59*, 217-225. http://dx.doi.org/10.1007/BF02184195

Chatterjee, A. K., & Mandal, L. N. (1985). Zinc sources for rice in soil at different moisture regimes and organic matter levels. *Plant and Soil, 87*, 393-404. http://dx.doi.org/10.1007/BF02184195

Chintala, R., McDonald, L. M., & Bryan, W. B. (2012a). Effect of soil water and nutrients on productivity of Kentucky blue grass systems in acidic soils. *Journal of Plant Nutrition, 35*, 288-303. http://dx.doi.org/10.1080/01904167.2012.636131

Chintala, R., McDonald, L. M., & Bryan, W. B. (2012b). Optimization of water potential and nutrient levels for Kentucky bluegrass-white clover mixture on acidic soils. *Biotechnology, Agronomy, Society and Environment, 16*, 167-177. Retrieved from http://www.popups.ulg.ac.be/1780-4507/index.php?id=8619

Chintala, R., Mollinedo, J., Schumacher, T. E., Malo, D. D., Papiernik, S., Clay, D. E., ... Gulbrandson, D. W. (2013). Nitrate sorption and desorption in biochars from fast pyrolysis. *Microporous and Mesoporous Materials, 179*, 250-257. http://dx.doi.org/10.1016/j.micromeso.2013.05.023

Chintala, R., Schumacher, T. E., McDonald, L. M., Clay, D. E., Malo, D. D., Clay, S. A., ... Julson, J. L. (2014). Phosphorus sorption and availability in biochars and soil biochar mixtures. *CLEAN-Soil Air Water, 42*(5), 626-634. http://dx.doi.org/10.1002/clen.201300089

De Vasconcelos, A. C. F., Clístenes, W. A. N., & Fernando, F. C. F. (2011). Distribution of zinc in maize plants as a function of soil and foliar Zn supply. *International Research Journal of Agricultural Science, 1*, 1-5. Retrieved from http://interesjournal.org/IRJAS/Pdf/2011/March/de%20Vasconcelos%20et%20al.pdf

Drissi, S., Aït Houssa, A., Bamouh, A., & Benbella, M. (2015). Corn silage (*Zea mays* L.) response to zinc foliar spray concentration when grown on sandy soil. *Journal of Agricultural Science, 7*, 68-79 http://dx.doi.org/10.5539/jas.v7n2p68

European Union. (1991). Council directive 91/676/EEC of 12 December concerning the protection of waters against pollution caused by nitrates from agricultural sources. *Official Journal of the European Communities*, L375. Brussels, Belgium. Retrieved from http://eur-lex.europa.eu/legal-content/EN/TXT/PDF/?uri=OJ:L:1991:375:FULL&from=EN

Fageria, N. K., Baligar, V. C., & Jones C. A. (Eds.). (2011). *Growth and mineral nutrition of field crops* (3rd ed.). Taylor and Francis Group.

Fox, R. H., & Piekielek, W. P. (1984). Soil magnesium level, corn (*Zea mays* L.) yield, and magnesium uptake. *Communications in Soil Science and Plant Analysis, 15*, 109-123. http://dx.doi.org/10.1080/00103628409367459

Gupta, U. C., Kening, W. U., & Siyuan, L. (2008). Micronutrients in soils, crops, and livestock. *Earth Science Frontiers, 15*, 110-125. http://dx.doi.org/10.1016/S1872-5791(09)60003-8

Hansch, R., & Mendel, R. R. (2009). Physiological functions of mineral micronutrients (Cu, Zn, Mn, Fe, Ni, Mo, B, Cl). *Current Opinion in Plant Biology, 12*, 259-266. http://dx.doi.org/10.1016/j.pbi.2009.05.006

Khan, M. A., Fuller, M. P., & Baloch, F. S. (2008). Effect of soil applied zinc sulphate on wheat (*Triticum aestivum* L.) grown on a calcareous soil in Pakistan. *Cereal Research Communications, 36*, 571-582. http://dx.doi.org/10.1556/CRC.36.2008.4.6

Kuldeep, S., & Banerjee, N. K. (1986). Growth and zinc content of maize (*Zea mays* L.) as related to soil-applied zinc. *Field Crops Research, 13*, 55-61. http://dx.doi.org/10.1016/0378-4290(86)90007-9

Lindsay, W. L., & Norvell, W. A. (1977). Development of a DTPA soil test for zinc, iron, manganese, and copper. *Soil Science Society of America Journal, 42*, 421-428. http://dx.doi.org/%2010.2136/sssaj1978.03615995004200030009x

Lithourgidis, A. S., Matsi, T., Barbayiannis, N., & Doras, C. A. (2007). Effect of liquid cattle manure on corn yield, composition, and soil properties. *Agronomy Journal, 99*, 1041-1047. http://dx.doi.org/10.2134/agronj2006.0332

Matsi, T., Lithourgidis, A. S., & Gagianas, A. A. (2003). Effects of injected liquid cattle manure on growth and yield of winter wheat and soil characteristics. *Agronomy Journal, 95*, 592-596. http://dx.doi.org/10.2134/agronj2003.0592

Mokhtarpour, H., Teh, C. B. S., Saleh, G., Selamat, A. B., Asadi, M. E., & Kamkar, B. (2010). Non destructive estimation of maize leaf area, fresh weight, and dry weight using leaf length and leaf width. *Communications in Biometry and Crop Science, 5*, 19-26. Retrieved from http://agrobiol.sggw.waw.pl/~cbcs/articles/CBCS_5_1_4.pdf

Murphy, L. S., & Walsh, L. M. (1972). Correction of micronutrients deficiencies with fertilizers. In J. J. Mortvedt, P. M. Giordano & W. L. Lindsay (Eds.), *Micronutrients in Agriculture*, Soil Science Society of

America (pp. 371-381). Retrieved from https://dl.sciencesocieties.org/publications/books/pdfs/sssabookseries/micronutrientsi2/frontmatter

Nikoli, T., & Matsi, T. (2011). Influence of liquid cattle manure on micronutrients content and uptake by corn and their availability in a calcareous soil. *Agronomy Journal, 103*, 113-118. http://dx.doi.org/10.2134/agronj2010.0273

Ortega-Blu, R., & Molina-Roco, M. (2007). Comparison between sulfates and chelated compounds as sources of zinc and iron in calcareous soils. *Agrociencia, 41*, 491-502. Retrieved from http://www.colpos.mx/agrocien/Bimestral/2007/jul-ago/art-1.pdf

Sasaki, H., Hirose, T., Watanabe, Y., & Ohsugi, R. (1998). Carbonic anhydrase activity and CO_2 transfer resistance in Zn deficient rice leaves. *Plant Physiology, 118*, 929-934. http://dx.doi.org/10.1104/pp.118.3.929

Schröder, J. J., Ten Holte, L., & Brourwer, G. (1997). Response of silage maize to placement of cattle slurry. *Netherlands Journal of Agricultural Science, 45*, 249-261. Retrieved from https://library.wur.nl/ojs/index.php/njas/article/view/516/230

Schröder, J., & Dilz, K. (1987). Cattle slurry and farmyard manure as fertilizers for forage maize. In H. G.Van Der Meer, R. J. Unwin, T. A. Van Dijk & G. C. Ennik (Eds.), Animal manure on grassland and fodder crops: Fertilizer or waste? *Developments in Plant and Soil Sciences* (Vol. 30, pp. 137-156). http://dx.doi.org/10.1007/978-94-009-3659-1_9

Sharma, P. N., Chatterjee, A. S. C., & Saharma, C. P. (1990). Zinc deficiency and pollen fertility in maize (*Zea mays*). *Plant and Soil, 124*, 221-225. http://dx.doi.org/10.1007/BF00009263

Sinclair, A. H., & Edwards, A. C. (2008). Micronutrient deficiency problems in agricultural crops in Europe. In B. J. Alloway (Ed.), *Micronutrients deficiencies in global crop production* (pp. 225-244). http://dx.doi.org/10.1007/978-1-4020-6860-7_9

Singh, B., Natesan, S. K. A., Singh, B. K., & Usha, K. (2005). Improving zinc efficiency of cereals under zinc deficiency. *Current Science, 88*, 36-44. Retrieved from http://www.iisc.ernet.in/currsci/jan102005/36.pdf

Sutton, A. L., Nelson, D. W., Kelly, D. T., & Hill, D. L. (1986). Comparison of solid vs. liquid dairy manure applications on corn yield and soil composition. *Journal of Environmental Quality, 15*, 370-375. http://dx.doi.org/10.2134/jeq1986.00472425001500040010x

Van Biljon, J. J., Wright, C. A., Fouche, D. S., & Botha, A. D. P. (2010). A new optimum value for zinc in the main maize producing sandy soils of South Africa. *South African Journal of Plant and Soil, 27*, 252-255. http://dx.doi.org/10.1080/02571862.2010.10639994

Walkley, A., & Black, I. A. (1934). An examination of the Degtjareff method for determining soil organic matter, and a proposed modification of chromic acid titration method. *Soil Science, 37*, 29-38. http://dx.doi.org/10.1097/00010694-193401000-00003

Wallingford, G. W., Murphy, L. S., Powers, W. L., & Manges, H. L. (1975). Effects of beef-feedlot manure and lagoon water on iron, zinc, manganese and copper content in corn and in DTPA soil extracts. *Soil Science Society of America Journal, 39*, 482-487. http://dx.doi.org/10.2136/sssaj1975.03615995003900030031x

Wang, H., & Jin, Ji-Y. (2007). Effects of zinc deficiency and drought on plant growth and metabolism of reactive oxygen species in maize (*Zea mays* L.). *Agricultural Sciences in China, 6*, 988-995. http://dx.doi.org/10.1016/S1671-2927(07)60138-2

Analysis of Agricultural Policies Affecting Medjool Date Palm Cultivation in Jordan

Ebraheem Suliman Altahat[1]

[1] Department of Agricultural Economics and Extension, Faculty of Agriculture, Jerash University, Jerash, Jordan

Correspondence: Ebraheem Suliman Altahat, Department of Agricultural Economics and Extension, Faculty of Agriculture, Jerash University, Jerash 26150, Jordan. E-mail: i.altahat@jpu.edu.jo

Abstract

The objective of this study was analyzing the agricultural policies which affecting Medjool date palm cultivation in Middle Ghors region, Jordan. In this study, Primary data were collected by personal interviews to farmers in the target area while secondary data were collected from different governmental offices. According to the Policy Analysis Matrix (PAM) approach, results showed that the Domestic Resource Cost Ratio (DRC) was less than one 0.27, which means that there was a comparative advantage in producing Medjool date palms. The measures of economic protection showed that there was a subsidy for agricultural producer in the Middle Ghors region because of the domestic resource price of this crop was less than the world price. Therefore, encouraging planting Medjool date palm will enhance the economic efficiency of agricultural resources in the Middle Ghors region in Jordan.

Keywords: medjool, Policy Analysis Matrix (PAM), measures of protection

1. Introduction

Jordan is located in the eastern Mediterranean region, between latitudes 29°30′N and 32°31′N, and covers about 89,287 square Kilometers (Km^2). The climate of Jordan is predominantly of the Mediterranean type, characterized by mild wet winter and extreme variability in rainfall within and among years, and a dry hot summer with short transitional periods of spring and autumn (RJGC, 2000). The agricultural sector is considered to be one of the most important economic sectors in Jordan; however, it has a fair contribution to the total domestic product where the contribution was 3.2% of the value of total domestic product and with a fixed basic prices (Ministry of Agriculture, 2012). Moreover, the sector is closely related to local economic activities which contributed with about 27% of GDP (Ministry of Agriculture, 2009).

The palm cultivation in Jordan is relatively new. The last fifteen years have seen a significant expansion in the palm cultivation, this is mainly because palm trees are more adaptable to dry environment due to unsustainable high degrees of heat, drought and salinity (Alawaideh et al., 2008). That is besides the availability of appropriate environmental and climatic and investment conditions in the field of date palm cultivation in Jordan (Ministry of Agriculture, 2012). Despite the scarcity of private water resources, the area planted with palm trees has increased from 1082.9 to 21519.8 dunum. This is accompanied by the evolution in the production of 892.5 tons to 11980.7 tons during the period from 1994 to 2013. In the context of consumption patterns prevailing in Jordan, and high quality types of dates such as Medjool, Sukari and Albrahi, which forced farmers to expand the cultivation of those varieties; the number of palm trees has increased from 14857 trees to 304735 during the period between 1994 to 2013 (Department of Statistics, 2014) .The importance of this study came from the importance of the role played by agriculture sector in the study area in terms of the limited basic agricultural resources such as land, water, and capital and the high contribution of the region in the value of agricultural output (GDP). Particularly, the economic importance to the area in the production of Medjool date. This study will also provide the necessary economic information to make decisions that would raise the economic efficiency of agricultural resources available, and provide information for decision-makers to take appropriate decisions related to this important product.

1.1 Objectives of the Study

The central goal of this study is to analyze the agricultural policies affecting Medjool date palm cultivation in

Jordan. To that end, this study also has other objectives:

1) Estimation of returns and costs using market prices and world prices for Medjool date crop.

2) Computing various measures of economic protection.

3) Agricultural policy analysis and policy formulation for the Medjool date cultivation in Jordan.

2. Review of Literature

Nothing was found in literature review regarding the policies affecting on date palm cultivation in Jordan using policy analysis matrix (PAM) Approach.

El-Habbab and Jabarin (1991) studied the impact of government price policy on the production of wheat in the rainfed areas in Jordan. The results of the study indicated that the Nominal Protection Coefficient (NPCo) of wheat using traditional and modern methods was (1.01), this means that with policy, the price of wheat received by farmers increased by (1%). Using the Nominal Protection Coefficient (NPCi) for production inputs for both production techniques was (1.06), indicating an increase in the costs paid by farmers as fees (6%). The Effective Protection. Coefficient (EPC) was one. This means that the final result of the price policy was almost zero. Also, the results of the mentioned study indicated that there was a comparative advantage in producing wheat.

El-Habbab and Jabarin (1991) studied the impact of the government price policy related to irrigated wheat on water in the Jordan Valley area, on the efficiency input use, self-sufficiency ratio and the possibility of producing fodder crops in the southern areas of Jordan. The results of the study indicated that wheat production in the irrigated areas uses resources efficiently if it is compared to tomato production in the Jordan Valley or alfalfa production in the southern desert area. Also, the results indicated that the Domestic Resource Cost (DRC) was less than one in the case that the price of water is 6 fils/cm in the Jordan Valley and 27 fils/cm in the southern desert.

Kubursi (2002) studied the Lebanon's Agricultural Potential: A Policy Analysis Matrix Approach. The results of the study suggested that there were a number of agricultural products where Lebanon has a major comparative advantage. These include apples, tomatoes, pears, oranges, cucumber, watermelons, cantaloupes, etc. There were a few products where the subsidies sustained economically (profit-wise) inefficient outputs. This was particularly true for tobacco, wheat and sugar beets. Olives appeared to be marginal and in some cases inefficiently produced.

Salem (1996) studied the impact of wheat policy on dry land wheat production in the Republic of Yemen. The results of the study indicated that dry land wheat production is economically feasible, and Yemen has a comparative advantage of producing it. The Nominal Protection Coefficient (NPC) of tradable inputs (NPCI) for dry land wheat was 59% and the value of the subsidy for the tradable inputs and domestic factors for dry land wheat was 3098 Yrs/ton.

Wiendiyati and Umbu Reku Raya (2003) studied the Policy Analysis Matrix (PAM) on the impact of government policy toward the comparative and competitive advantage of soybean farming in Ngada Regency, Nusa Tenggara. The results of the study indicated some important findings, those were the followings: Imposing of import tariff on soybean increased the price of private output, and increased the competitive advantages of soybean farming system in Ngada.

Altahat (2010) studied the impact of governmental subsidy policy and risk analysis in vegetables production In the Yarmouk Basin (Syria) and in the Jordan Valley (Jordan). The results of the study showed that the Domestic Resource Ratio (DRC) for summer tomato, spring potato in Yarmouk basin in Dra'a and for spring potato in Deir Alla district in the Jordan Valley was less than one, which means that there was a comparative advantage in producing those crops in the irrigated areas. There were incentives to encourage farmers to produce these irrigated crops. While production of summer tomatoes in Deir Alla district in the Jordan Valley was not feasible. The measures of production indicated that the presence of taxing for output and subsidization for one or more input for all kinds of irrigated vegetables.

3. Methodology and Procedure

3.1 Primary and Secondary Data

This study based upon primary data from "Econometric Analysis of Medjool Date Production Costs in Jordan" (Altahat et al., 2013). In addition to primary data, Secondary data have been collected from different sources, governmental officials, research, literatures, and other sources that have been collected and used in the analysis. Two kinds of budgets were estimated: private and social budget. Private budget was that with prevailing market prices. Social budget was that estimated with adjusted prices (social prices or efficiency prices). The enterprise budget was then used to build accounting matrix. The accounting matrix is called a Policy Analysis Matrix (PAM). Measures of protection and comparative advantage measures abstracted from PAM to determine if the production makes efficient use of domestic factors.

3.2 Study Area

The study was carried out in the Middle Ghors (Deir Alla); one of the four regions in the Jordan valley, Jordan. It covers about 83 thousand dunums, 2548 agricultural units (each unit about 30 dunums), 10 basins. The area is several degrees warmer in winter than the rest of the country. Agriculture is the major economic activity in the area. Palm cultivated area covered in the study is 6468.0 dunums, which is 29.4% of the total cultivated palm area in Jordan (21519.8 dunums). The number of palm trees in the study area is 84,877 trees; 27.9% of the total number of palm trees planted in Jordan (304,735 trees). The number of palm fruitful trees in Jordan 222340 trees, of which 28.4% (63178 trees) in the study area. The production of palm trees in the study area reaches up to 3664.3 tons; 30.6% of the total production on the level of Jordan (11980.7 tons) (Department of Statistics, 2014).

3.3 Policy Analysis Matrix (PAM) Method

Governments often intervene in the national economy to achieve certain objectives. These objectives includes; improving the distribution and stability levels of the real income, increasing production of food commodities, and improving the efficiency of resource allocation. To reach these objectives in the agricultural sector, a government uses different policy tools, such as (a) intervention in project plans; (b) fixing input prices; (c) subsidization of output prices; (d) quotas on imports competing with domestic products; and (e) multiple exchange rates for import and export. Markets play a crucial role in allocating resources and direct agricultural production. However, the public sector plays an important role in supporting markets. Whereas, the public sector providing the necessary infrastructure (such as roads), providing sound macroeconomic policies that avoid high rates of inflation and appreciated exchange rates, providing agricultural market information, research and extension, and facilitating the creation of an environment conducive to the development of competitive markets (ESCWA, 1995).

This study will be presented as an evaluation of the current production system of the Medjool date palm produced in the Middle Ghors region by means of the Policy Analysis Matrix (PAM) methodology. PAM method will be used to determine the impact of Jordanian Governmental policies, including output and input prices, credit subsides and exchange rates, on the efficiency of production. The principles of Policy Analysis Matrix (PAM) developed by Pearson and Monke in 1989 were used. The PAM method serves both as a logical framework for thinking about the effects of agricultural policy and as empirical analytical tool for measuring the policy impacts. The information used to feed PAM has been obtained by primary and secondary data. The PAM model is portrayed in Table 1. Private profits were defined in the first row as D = A - B - C, social profits were defined in the second row as H = E - F - G, and the policy effect were defined in the third row as L = I - J - K. The third row of the matrix shows the divergences or differences between the first row (private valuation) and the second row (social valuation). If market failure does not exist in the product market, distorting policies causes all divergences between private and social prices of tradable output and input. Policies, which may cause divergence, includes subsidies, taxes and quantitative controls applied to domestic production or trade of the commodity, price policies may also cause distortions. The first column in the matrix is labeled revenues. Costs are divided into two components, costs of tradable inputs (inputs which are traded in the world markets) such as fertilizer and pesticides, area included in the second column. The third column of the matrix includes domestic factors (non-tradable inputs because there is no international markets for these inputs) such as land water, labor, and capital. Column four in the matrix is labeled profits. Economic profit is the fundamental part of the PAM analysis. Profit is defined as the difference between the value of outputs (revenues) and the costs of all inputs (costs).

Table 1. The Policy Analysis Matrix (PAM)

Revenues	Costs		Profits	
	Tradable inputs		Domestic factors	
Private prices	A	B	C	D
Social prices	E	F	G	H
Policy effect	I	J	K	L

Source: Monke and Pearson (1989).

Note. The symbols (capital letters) stand for:

A: Total revenues in private price (market prevailing prices some times called financial prices).

B: Cost of tradable inputs (such as fertilizer and Chemicals) in private prices.

C: Cost of domestic factors (such as labor and capital) in private prices.

D: Private profit.

E: Total revenues in social price (prices that are adjusted to reflect government intervention).

F: Cost of tradable inputs (such as fertilizer and Chemicals) in social prices.

G: Cost of domestic factors (such as labor and capital) in social prices.

H: Social profit;

Private profit: (D) = A - B - C;

Social profit: (H) = E - F - G;

Output transfer: (I) = A - E;

Input transfer: (J) = B - F;

Factor transfer: (K) = C - G;

Net transfer: (L) = D - H.

4. Result and Discussion

Table 2 contains a summary for the policy analysis results of Medjool date palm cultivation in Middle Ghors region. The results were reported on a per dunum basis. Which have been dependable to get to the impact of agricultural policy. It was found that revenue for Market prices (local) was higher than the revenue for social prices (global) and this means that there was a subsidy that equivalent to 38.2% of the social price to harvested dates. As it turns out that the tradable inputs received a financial support equal to 9.9% of the cost of the tradable economic production price, while there was a financial support for local resources equal to a rate of 38.4% of the cost for local resources and economic price. Therefore, the overall impact of the policy was in the favor of the producers.

Table 3 shows the measures of economic protection for Medjool date palm cultivation The DRC or comparative advantage indicator is less than one, which means that the production of Medjool date palm makes an efficient use of domestic resources and Jordan has a comparative advantage in this production. Also, it was found that the Nominal Protection Coefficient for tradable output (NPCo) was 1.38 which means that there is a subsidy for agricultural producers and that the adopted agricultural

Table 2. Policy Analysis Matrix for Medjool date palm in the Middle Ghors

	Revenues	Cost of tradable inputs	Cost of domestic factors	Profits
Private prices	6289	411	675	5203
Social prices	4550	456	1096	2999
Effect of divergence & policy	1738	(45)	(421)	2204

Note. Figures in brackets ( ) are negative numbers.

Source: Calculated and tabulated from primary and secondary data.

Policy allows local market price to be larger than the international price. As for the Nominal Protection Coefficient for tradable inputs (NPCi), it was found that there is a decline in the costs paid by the producer (farmer) of 0.90 which indicates that the government has a high financial support to those farmers. In other words, domestic prices of tradable inputs were less than social prices. As far as that the Effective Protection Coefficient (EPC) was greater than one, which means there are incentives for farmers as a result of the positive impact for the policy of subsidy input and output. The Private Cost of Resources (PCR) was less than one, which indicates that local costs or money invested were less than the added value achieved, and therefore farmers reached high profitability. Using the Profitability Coefficient (PC) that is greater than zero means that there is profitability in the production of this crop. Subsidy Ratio to Producers (SRP) was greater than zero which indicates that there was a financial support for agricultural producer.

Table 3. Indicators of competitiveness, efficiency and policy impacts calculated for Medjool date palm in the Middle Ghors

Items	Value
PCR(C/A-B) =Private Cost of Resources	0.12
DRC(G/E-F) = Domestic Resource Ratio	0.27
NPCo(A/E) = Nominal Protection Coefficient for Output	1.38
NPCi (B/F) = Nominal Protection Coefficient for Input	0.90
PC (D/H) = Profitability Coefficient	1.74
EPC{(A-B)/(E-F)} = Effective Protection Coefficient	1.44
SRP(D-H/E) = Subsidy Ratio to Producers	48%

Source: Calculated from Table 2.

5. Recommendations

The study recommended the importance of encouraging planting Medjool date palm since there was a comparative advantage and led to enhance the economic efficiency of agricultural resources in the Middle Ghors region in Jordan.

References

Alawaideh, M. A., Fida, A., Najah, M., Samer, A., Moustafa, A., & Aminh, A. (2008). *The reality of palm cultivation in Jordan*. Agricultural credit corporation, Amman, Jordan.

Altahat. (2010). *The impact of governmental subsidy policy and risk analysis in vegetables production in the Yarmouk Basin (Syria) and in the Jordan Valley (Jordan)* (Ph.D. Thesis, unpublished). Aleppo University, Agricultural Economic Department.

Altarawneh, M., & Ebraheem, A. (2013). Econometric analysis of Medjool date production costs in Jordan. *Scientific Journal of King Faisal University, Basic and Applied Sciences, 14*(1).

Department of Statistics. (2011-2013). *External Trade Report*. Jordan.

Department of Statistics. (2014). *Crops statistics*. Jordan.

Economic and Social Commission for Western Asia (ESCWA). (1995). *Evaluation of Agricultural Policies in the Syrian Arab Republic (Policy Analysis Matrix)*. New York.

El-Habbab, M. S., & Jabarin, A. S. (1991). The impact of wheat policy on traditional and modern rainfed wheat production in Jordan. *Dirasat, 18B*(3).

El-Habbab, M. S., & Jabarin, A. S. (1991). The impact of wheat policy on irrigated wheat production in Jordan. *Dirasat, 18B*(4).

Kubursi, A. A. (2002). *Lebanon's Agricultural Potential: A Policy Analysis Matrix Approach*. McMaster University and Econometric Research Limited.

Ministry of Agricultural. (2009). *Annual Report*.

Ministry of Agriculture. (2012). *Annual Report*.

Monke, E. A., & Scott, R. P. (1989). *The Policy Analysis Matrix for agricultural development*. Cornell, USA: Cornell University Press.

Royal Jordanian Geographic Center (RJGC). (2000). *Monthly Reports*. Amman, Jordan.

Salem, M. A. (1996). The impact of wheat policy on dry land wheat production in the republic of Yemen. *Dirasat, Agricultural Sciences, 23*(1).

Wiendiyati and Umbu Reku Raya. (2003). *Policy Analysis Matrix (PAM) on the impact of government policy toward the comparative and competitive advantage of soybean fanning in Ngada regency*. Nusa Tenggara, Nusa Cendana University in West Timor, Indonesia.

Appendix

Appendix 1. Enterprise budget of Medjool date palm in the Middle Ghors (in JD's per Dunum)

Item	Market prices	Social prices
Tradable Outputs	6288.9	4550.4
- Crop produce (output) (JD/du.)	6099	4360.5
- By-product (JD/du.)	189.9	189.9
Tradable Inputs	410.8	455.8
- Cuttings Palm (JD/du.)	312.5	329.8
- Chemical Fertilizer (JD/du.)	42.4	69.5
- Plant Protection Chemicals(JD/du.)	11.8	12.5
- Machinery labor (JD/du.)	36.6	36.6
-Threads (JD/du.)	7.4	7.4
Non-Tradable Inputs	675.2	1095.7
- Water Requirement (JD/du.)	185.3	520.0
- Manure (JD/du.)	9.0	9.0
- Manual Labor (JD/du.)	244.5	244.5
- Interest (JD/du.)	80.4	166.2
- Land Rent (JD/du.)	150	150
-Maintenance (JD/du.)	6.0	6.0

Source: Calculated and tabulated from primary and secondary data.

15

Economic Effects of Climate Change on Maize Production and Farmers' Adaptation Strategies in Nigeria: A Ricardian Approach

A. S. Coster[1] & A. I. Adeoti[1]

[1] Department of Agricultural Economics, University of Ibadan, Nigeria

Correspondence: A. S. Coster, Department of Agricultural Economics, University of Ibadan, Nigeria. E-mail: lekecoster@yahoo.co.uk

Abstract

Increasing threat of climate change is aggravating the problem of declining agricultural productivity in the face of rapid population growth. This implies that rural sustainance and food security is under threat and stress. Given that crops differ in climate requirements and economic importance, analysis of the attendant effect of climate change on specific crop remains to be adequately explored. This study uses Ricardian approach to examine the effects of climate change on maize production in Nigeria.

Multistage sampling technique was employed for the study. Data were collected on 346 maize-based farming households in three different agro-ecological zones of Nigeria.

Average age of the farmers and household size were 45years and 8 persons respectively. Average years of farming experience and years of schooling were 25.6 and 6.5 years respectively. The empirical results showed that maize net revenue is sensitive to climate change. Seasonal marginal impact analysis showed that increase in rainfall during rainy season increased Maize Net Revenue (MNR) in rainforest, guinea and montane savanna respectively. Marginal increase in rainfall during the dry season increased MNR in rainforest while it decreased MNR in guinea and montane savanna respectively. However, marginal increase in temperature during dry season has positive impacts on MNR in all AEZs. The predicted results using a range of climate scenarios confirm that climate change will have negative impact on maize net revenue in the future. Maize Farmers have taken adaptive measures against climate change which are changing the planting dates, changed land-use practices, mixed cropping and mixed farming. The major barriers to adaptation are inadequate credit or saving and inadequate knowledge of appropriate adaptation strategies suited for the local climate conditions.

Keywords: climate change, maize production, adaptation strategies, Ricardian analysis

1. Introduction

Climate change as defined by United Nation Framework Convention on Climate Change (UNFCC) refers to a change of climate that is attributed directly or indirectly to human activity that alters the composition of the global atmosphere and that is in addition to natural climate variability observed over comparable time periods. Succinctly put, this effect of the pursuit of livelihood and comfort leads to emission of Greenhouse gases (GHG). These gases are mainly carbon dioxide (CO_2), methane (CH_4) and nitrous oxides (NO_2) (Molua, 2002). It is now well known that climate is changing worldwide. The past two decades have witnessed globally a rapid increase in the awareness about climatic changes. This triggered widespread apprehension among scientists and governments about their global implications (Gadgil, 1996). The major environmental concern regarding increased concentration of CO_2 and other trace gases is their greenhouse potential that is ability to trap solar energy in the atmosphere. This trapped solar energy causes global warming and other changes in the world's climate (Houghton et al., 1996).

Climate change is primarily caused by the developed countries. However, it is the bitter irony of destiny that Africa contributes least (920,000 t of CO_2 each year, less than 4% of the global production) of all the continents to the climate change, but will probably suffer most from its consequences. Economists refer to this as a typical case of negative external effects, an externalisation of costs: A non-involved party bears the costs of a third party's actions (Medugu, 2008). FAO (2007) reported that up to11% of arable land could be highly affected by climatic change in the developing world. There will be a reduction of cereal production in 65 countries and retardation of about 16% of agricultural GDP. A decrease of up to 30% in world food production due to effects of climate change on agriculture is generally predicted (IPCC 2007).

In Africa estimates indicate that about 60-70 percent of the population is dependent on the agricultural sector for the employment, and the sector contributes on average nearly 34 percent to gross domestic product (GDP) per country (Adam et al., 1993). In Nigeria, Agriculture contributed 41.25 of GDP in 2005 almost the same as in 2004 (CBN 2005). Over 2006, 2007, 2008, 2009, 2010, 2011,2012 and 2013 the share of agriculture in overall real GDP amounted to 41.72%, 42.01%, 42.13%, 41.70%, 40.76 %, 40.11%, 39.26% and 38.56% respectively (NBS, 2010, 2014).

More than 60% of the working adult populations of Nigeria are employed in the agricultural sector directly and indirectly. Over 90% of Nigeria's agricultural output comes from peasant farmers who dwell in the rural area where 60% of the population live (Agricultural Report, 2007). Despite these contributions, Crosson (1997) pointed out that considering the lower technological and capital stocks, the agricultural sector in developing countries is unlikely to withstand the additional pressures imposed by climate change without a concerted response strategy.

Nigeria is experiencing adverse climatic conditions with negative impacts on the welfare of millions of people. Persistent droughts and flooding, off season rains and dry spells have sent growing seasons out of orbit, on a country dependent on a rain fed agriculture (Medugu, 2008). Nigerian agriculture is almost entirely rain-fed hence inherently susceptible to the vagaries of weather. Only about a million hectare is currently irrigated in Nigeria out of the total 30.5 millions arable hectares of land (Madu et al., 2010). As global warming accelerates, it is expected that agricultural adaptation to climate change can only be meaningful, if irrigated agriculture gains prominence. Agriculture in Nigeria is therefore particularly vulnerable to the impacts of climate change (FAO, 2008; Medugu, 2008; IFAD, 2007). The consequences are that the increasing frequency and severity of droughts are likely to cause: crop failure; high and rising food prices; distress sale of animals; de-capitalization, impoverishment, hunger, and eventually famine.

Many people and most households in Nigeria depend on cereals (most especially, maize) as a contributing, if not principal, source of food and nutrition (CBN, 2005). Maize is one of the important grains in Nigeria, not only on the basis of the number of farmers that engaged in its cultivation, but also in its economic value. Maize is a major important cereal crop being cultivated in the rainforest and the derived savanna zones of Nigeria (Iken & Amusa, 2004). Despite its high yield potential, maize production is however faced with numerous constraints. One of the major constraints is intermittent drought during the growing season, which, significantly reduce maize yield (Ayanlade & Odekunle, 2006).

Declining agricultural productivity in the face of rapid population growth as a result of climate change is worrisome and cause for a great concerned. Militating against the climate change requires understanding the impact of the change and effectiveness of the coping strategies. Given that different crops have different climate requirements necessitate the need for specific crop analysis which has not been adequately explored in Nigeria. Hence the effect of climate change on notable food crop like maize which has long term implication for food security was therefore investigated. Specifically, the study sought to analyze the effects of climate change on maize farmers' income; assess farmers' perception and adaptations to climate change and identify the barriers to adaptation to climate change.

2. Literature Review/Theoretical Framework

Previous studies that examined the impact of climate change on agriculture made use of agronomic (otherwise known as production function) studies (Mearns et al., 1997; Downing, 1992; Schulze et al., 1993; Adams et al., 1990, 1993; Parry et al., 1999) and Ricardian approach (Mendelsohn et al., 1994; Deressa & Hassan, 2009; Gbetibouo & Hassan, 2005; Sanghi, 1998; Sanghi et al., 1998; Kumar & Parikh, 1998). The agronomic models simulate a laboratory type set up and provide data on climatic factors and crop growth. Although the agronomic models provide a controlled and randomized application of environmental conditions, it does not take adaptive behavior of an optimizing farmer into account. Ricardian model on the other hands measure the impact of climatic factors through their contribution to farmland-prices and have been extensively used for incorporating farm level adaptation (Mendelsohn et al., 1994). However, availability of land prices as well as non-existence of efficient land markets are two major obstacles in applying the Ricardian method to most of the developing countries, hence Semi-Ricardian model using data on annual net revenue per hectare instead of land prices, since land value is the present value of a future stream of net revenue (Seo & Mendelsohn, 2007).

The Ricardian approach estimates the importance of climate and other variables on the capitalized value of farmland. The method utilized the typical economic measure of farm performance: net revenue or net farm income. The approach has been used to evaluate the contribution of environmental measures to farm income. By regressing land value on a set of environmental inputs, one can measure the marginal contribution of each input to farm income. Ricardian model has been extensively criticized on the ground that; it fails to fully control for the impact of variables that could also explain the variation in farm incomes, it assumes prices will remain constant, failure to take account of water supply and failure to account for the effect of factors that do not vary across space such as carbon dioxide concentrations that can be beneficial to crops.

Despite its shortcomings, Ricardian model has been extensively applied to a number of countries such as United States (Mendelsohn et al., 1994, 1999), Brazil (Sanghi, 1998) China (Liu et al., 2004) and India (Sanghi et al., 1998; Kumar & Parikh, 1998, 2001). Similar approach has been applied in Africa: Cameroon (Molua, 2002, Molua & Lambi, 2006), Senegal (Sene et al., 2006), Kenya (Kabubo & Karanja, 2007), Srilanka (Seo et al., 2005; Kurukulasuriya & Ajwad, 2004), Ethiopia (Deressal et al., 2005) South Africa (Gbetibou & Hassan, 2005), Egypt (Eid et al., 2005) and Nigeria (Ajetomobi et al., 2011; Fonta et al., 2011).

2.1 Theoretical Framework

The theoretical basis of the Ricardian method is deeply rooted in the famous theory of economic rents by David Ricardo (1817). However, much of its application to climate-land value analysis, draws extensively from the work of Mendelsohn et al. (1994).

Mendelsohn et al. (1994) propose an alternative economic approach, which makes use of cross-sectional data to capture the influence of climatic as well aseconomic and other factors on land values (or farm income). The technique has been named the Ricardian method because it is based on the observation made by Ricardo (1817), that land values would reflect land productivity at a site under perfect competition. It is possible to account for the direct impact of climate on yields of different crops as well as the indirect substitution among different inputs including the introduction of different activities, and other potential adaptations to different climates by directly measuring farm prices or revenues by using the Ricardian model. The Ricardian technique captures the flexibility of farmers better than the agronomic method. The method examines how land values (or rents) shift with climate and other control variables. Because farmer adaptations are reflected in land values, the approach accounts for the costs and benefits of adaptation.

To measure the economic effect of Climate change on maize crop in Nigeria, the standard Ricardian method (Mendelsohn et al., 1994) was adopted. However, the specification follows Seo et al. (2005) due to its simplicity.

Assuming that the revenue maximizing function of maize farmers in Nigeria is derived from the cost function, production (output function) and the cost of land as follow.

$$C_i = C_i(Q_i, P_x, E) \tag{1}$$

Where Q_i represents the quantity of maize, $C_i(.)$ is the relevant cost function associated with maize production, P_x represents the vector of prices of inputs associated with maize production except land, and E reflects a vector of environmental characteristics of the farmer's land including climate (i.e., temperature and precipitations). Given the cost function in Equation (1), under the assumption of perfect competition in the market for maize crop production, the farmer will maximize net revenue as:

$$Max\ NR = P_i * Q_i(R, E) = C_i(Q_i P_x, E) - P_L L_i = 0 \tag{2}$$

Where NR represents the net revenue per hectare proxies for farm land value, R is a vector of inputs, P_L is the rent and L_i the land. If we assume that a maize farmer chooses inputs, R, to maximize NR, then we can express the resulting outcome of NR in terms of E alone as:

$$NR = f(E) \tag{3}$$

And, the resulting welfare value of a change in the environment from state A to B is:

$$W = \sum f(E_{iB}) * L_i - \sum f(E_{iA}) \times L_i \tag{4}$$

Where, L_i is the amounts of land of type i (Seo et al., 2005). Equation (4) indicates that the welfare value of change in environment is equal to the difference in the net revenue given the two states of nature. However, since maize crop grow and develop very well under preferred temperatures and rainfall, thus, levels far above or below the optimal ranges would obviously reduce productivity. This suggests that the relationship between NR and these climate variables should be hill-shaped as has been extensively discussed in the literature (Mendelsohn et al., 1994; Kurukulasuriya & Mendelsohn, 2006; Seo et al., 2005). To capture this hill-shaped relationship, NR for maize crop production in Nigeria is specifying using the model of Equation (3) as:

$$NR = \sum(\beta_0 + \beta_1 T_1 + \beta_2 T_2^2 + \beta_3 R_3 + \beta_4 R_4^2) + \sum \gamma_i S_i + \sum \alpha_i X_i + \varepsilon \tag{5}$$

Where T_i and R_i represent normal temperature and rainfall in each season, S is the set of soil variables, X is the set of economic and social variables like access to market and capital and other relevant social characteristics and erepresents the error term. Equation (5) represents the empirical model to be estimated for Nigerian maize farmlands

2.2 Marginal Impact Analysis

The marginal impact analysis was undertaken to observe the effect of an infinitesimal change in temperature and rainfall on net maize revenue.

From Equation (5), the mean marginal impact of a climate variable on maize farm revenue will be derived. The expected marginal impact of a climate variable on net revenue, the proxy of farm value, evaluated at the mean is:

$$E\left[\frac{\partial NR}{\partial T_i}\right] = \beta_1 + 2 * \beta_2 T_2 * E[T_i] \quad (6)$$

$$E\left[\frac{\partial NR}{\partial R_i}\right] = \beta_3 + 2 * \beta_4 R_4 * E[R_i] \quad (7)$$

3. Methodology

3.1 Study Area

The study area is Nigeria. Nigeria is one of the sub-Saharan African nations in the western part of the Africa and shares land border with the republic of Benin to the west, Chad and Cameroon to the east, Niger republic to the north, and its coast lies on the gulf of Guinea (Wikipedia, 2009).

The total land area of Nigeria (923,766 km^2) is divided into seven broad ecological or land resource zones namely mangrove swampy forest, Rainforest, Montane forest/grassland, Derived savanna, Guinea savanna, Sudan savanna, and Sahel savanna (Olufemi & Ameh, 1999). The categorization is based on the similarity of climate and vegetation cover as well as the type of crops that are adapted to each land area. With the exception of the montane region, the length of wet season (days) and temperature increase from the coast to the hitherland. In this categorization no state of the federation can boast of one ecological zone. A state may have up to three ecological zones. All these zones support maize production.

3.2 Sampling Procedure

Multistage sampling technique was employed for the study. Three major maize producing states were purposively selected to represent agro-ecological zones (Niger: guinea savannah, Taraba: montane savanna and Oyo: rainforest). The second stage was the purposive selection of two Local Government Areas with records of highest maize production in each of the states. Thirdly, 5 villages were randomly selected from each LGA and lastly 400 maize-based farming households were randomly selected from the list of maize producing farmers obtained from the ADP of each zone in a proportionate sampling method.

Although a total of 400 questionnaires were administered on the respondents, 54 of these were found unsuitable for analysis and consequently, data from 346 questionnaires were analyzed for the study.

3.3 Type and Sources of Data

The study employed both primary and secondary data. Data were collected on households demographic and farm characteristics, input and output in maize production, institutional factors, perception of climate change and adaptation strategies using structured questionnaire.The questionnaire instrument was adapted and modified from the Global Environmental Fund (GEF) Regional Climate, Water and Agriculture Project of the Centre of Environmental Economics and Policy in Africa (CEEPA), University of Pretoria, South Africa (See, http://www.ceepa.co.za/climate_change/index.html).

3.4 Secondary Data

Data on climate variables that is temperature and rainfall for 41 years (1970-2011) was employed in the study. The data was obtained from the Nigerian Metrological Agency in Oshodi, Lagos State. Averages of temperature and rainfall data for the 41 years from six weather stations (Yola, Ibi, Minna, Bida, Ibadan and Saki) in the three agro-ecological zones were pooled to allow enough variation in the data set. Their averages for the two predominant seasons (Dry and Rainy season) in the country were estimated for the 41 years period. The dry season ranges from November – March while the rainy season range from April – October. It is assumed that farms within the same location experiences the same weather conditions hence the same climate data are matched to farms that are within the neighbourhood of each station.

3.5 Soil Data

The soil data for the three selected states representing each maize producing agro-ecological zone were obtained from the Food and Agricultural Organization (FAO, 2003). The FAO provides information about the major and minor soils in each location, including the slope and texture. In all, there exists 8 types of soil (Arenisols, Vertisols, Luvisols, Nitrosols, Lexisols, Fluvisols, Leptosols and Plinthosols) in the three states and the percentage share of each particular dominant soil was obtained.

3.6 Net Revenue

Maize production performance was measured using average yield as well as using gross revenue calculations based on the market valuation of output and all variable input costs including cost of labour. The real net revenue that was used for this study defined as gross revenue minus the cost of fertilizer and pesticides, hired labour (valued of median market wage rate), light farm tools (such as files, axes, cutlass, machets etc), rental on heavy machinery (tractors, ploughs, threshers etc) and household labour (valued at own wage rate).

4. Results and Discussion

4.1 Socio-Economic Characteristics of Maize Farmers

The distribution of the farmers by agro-ecological zone as presented in Table 1 shows that between 77% and 97% of maize farmers were male across the three agro-ecologies with an overall male dominance of 90.8%. Male dominance has severally been attributed to the laborious nature of peasant farming due to high dependence on manual labour. Also limited access of women to productive inputs has also made men the major actors. The comparison of socio-demographic characteristics of maize farmers across the agro-ecological zones revealed that there were significant differences in all the estimated variables, at the one percent level of significance. Farmers in the rainforest had the highest mean age of 51.54 years, followed by guinea savanna (43.48 years) and montane savanna (41.61 years). Farmers' experience in farming showed that maize farmers in guinea savanna had higher average experience of 30.64 years, which is significantly higher than rainforest (24.93 years) which in turn is higher than montane savanna (15.81 years). This implies that with the level of experience of the sampled farmers they are expected to have more knowledge and information about climate change and agronomic practices that they can use in response to climate change. According to Maddison (2006) and Nhemachena and Hassan (2007) experience in farming increases the probability of uptake of adaptation measures to climate change. The year of education attainment was found to be significantly low across the three agro-ecological zones with mean of 9.16 years, 4.58 years and 7.64 years for rainforest, guinea savanna and montane savanna respectively. The implication of this is that there would be low level of technological adoption to improve maize farming and limited exposure of farmers to warning system of climate information that can facilitate effective adaptation strategies. The household size also showed a significant difference across the three agro-ecological zones. Maize farmers in the guinea savanna had an average household size of 10 persons, montane savanna (8 persons) and rainforest (5 persons). The average distance of product and input market place to the farms varied significantly across the agro-ecological zones. In all, the average distance of product and input market to farms was about 7.90 km and 7.45 km. The number of extension visits varied among the farmers in each agro-ecological zone. The average number of visit was 2.26, 2.3 and 1.47 times in rainforest, guinea and montane savanna respectively. On the average, about 61.7%, 54.9% and 34.2% of maize farmers had access to credit in rainforest, guinea and montane savanna respectively. In terms of agricultural subsidy, average subsidy received by farmers in all the zones was N5, 837 in a year. About 67%, 89.2% and 68.4% of the maize farmers in rainforest, guinea and montane savanna respectively reported to have property right on their farmland which was acquired through inheritance and outright purchased.

4.2 Farm Characteristics

The land holding and proportion of land area allocated to maize production varied significantly across the agro-ecological zones (Table 1). The total farm size holding in guinea savanna is significant higher (6.57 ha) than total farm size holding in montane savanna (4.66 ha) which in turn is higher than total farm size holding in rainforest (3.49 ha). The proportion of land area allocated to maize production also varied significantly across the three agro-ecological zones. The average land area allocated to maize production was significantly higher in guinea savanna (1.89 ha) than montane savanna (1.56 ha) and rainforest agro-ecological zones (0.84 ha) respectively. This suggests that maize farming is still predominantly on small scale level. Maize yield in the montane savanna (4.01 ton/ha) was significantly higher than its yield in guinea savanna (3.40 ton/ha) and rainforest (2.54 ton/ha) respectively. The result is consistent with previous studies (Abiodun et al., 2011; Madiyazhagan et al., 2004) that highest maize production occurs in the savanna because these zones have the most favourable climate and soil for maize production. The distribution of the farmers by their use of inputs across agro-ecological zones shows maize farmers in guinea savanna used an average of 300 kg/ha of fertilizer in maize production. This is significantly higher than average fertilizer used by rainforest (153.56 kg/ha) and montane savanna (125.75 kg/ha) respectively. The yearly average of the sample was about 272.6 kg/ha which is lower than the recommended usage of 400 kg/ha application level for effective optimum growth and yield of maize (Law-Ogbomo, 2009).The result also shows that maize farmers in the guinea savanna used an average 12.6litres/ha of pesticides which is significantly higher than average pesticides used by montane savanna (2.98 lt/ha) and rainforest (2.87 lt/ha) respectively. The use of irrigation for maize crop production shows that majority (77.7%) of the farmers did not use irrigation for maize production while about 22.3% made use of irrigation to grow maize. Montane savanna had the highest percentage (39.2%) of farmers using irrigation than guinea (20.2%) and rainforest (11.7%) respectively.

4.3 Ricardian Analysis

4.3.1 Descriptive Statistics of Relevant Variables Used in Ricardian Analysis

1) Net Revenue

Net maize revenue varied across the three agro-ecological zones. It was found to be significantly higher for maize farmers in montane savanna (₦158,554.60) than maize farmers in rainforest (₦140,081.90) which in turn is higher than guinea savanna (₦109,237.15). The variability in maize net revenue across the agro-ecological zones can be explained by the difference in climatic condition and the level of technology.The empirical analysis therefore tried to find the climatic, soil, and socioeconomic variables that would help explain this variability.

2) Climate

The study considered two climate data namely temperature and rainfall. This is because their data were readily available in the climate model. The averages for the two predominant seasons (Dry and Rainy season) in the country were estimated for the 41 years period (1970-2011). The dry season ranges from November – March while the rainy season range from April – October. Result in Table 1 showed that rainforest agro-ecological zone recorded the highest monthly average rainfall (176.3 mm) during the rainy season followed by guinea savanna (163.4 mm) and montane savanna (147.1 mm). Rainfall pattern during the dry season was found to be very low in guinea (3.8 mm) and montane (6.8 mm) savanna than rainforest zone (24.1 mm). Results showed that Montane savanna had higher monthly average temperature during the rainy and dry season (31 °C and 33.4 °C) than guinea savanna (30.3 °C and 32.9 °C) and rainforest (29.5 °C and 32 °C).The average annual rainfall for the 41 years period of study was estimated at about 1262 mm, 1166 mm and 1062 mm for rainforest, guinea and montane savanna respectively while average annual temperature for the same period was estimated at about 31.7 °C, 32.7 °C and 32.9 °C for rainforest, guinea and montane savanna respectively.

3) Soil

In Table 1, it showed that there are eight soil samples that were dominant in the study areas based on FAO (2003) soil data classification. The percentage shares of each particular dominant soil in the soil mapping were used in the regression. Across the agro-ecological zones, rainforest farmlands were covered by three dominant soil types in respect of lixisol (50%), luvisol (50%) and leptosol (34%). Guinea savanna farmlands were located on six dominant soil types which includes; lixisol (90%), luvisol (50%), leptosol (34%), fluvisol (35%), pinthosol (70%) and nitrosol (70%). Montane savanna agro-ecological zone were located on seven dominant soil types. They are; arenisol (40%), lixisol (50%), leptosol (50%), fluvisol (35%), pinthosol (70%), veritosol (60%), nitrosol (70%). This implies that endowment of soil resources differed from one agro-ecological zone to another.

Table 1. Descriptive statistics of variables used in Ricardian Regression

Variables	Rainforest	Guinea	Montane	All zones	F-stat.
Net revenue (₦)	140,081.90	109,237.15	158,554.60	128.878.87	3.95***
Dry season rainfall (mm)	24.1(11.4)	3.83(4.05)	6.8(8.91)	11.43(4.94)	
Rainy season rainfall (mm)	176.3(10.30)	163.4(19.85)	147.1(37.30)	156.7(20.75)	
Annual rainfall l(mm)	1262(573.45)	1166.7(19.85)	1062.9(254.27)	1035.21(214.25)	
Dry season temperature (°C)	32.02(2.47)	32.9(7.11)	33.4(6.54)	32.8(5.87)	
Rainy season temperature (°C)	29.5(2.24)	31.3(5.17)	31.0(5.09)	30.6(1.53)	
Annual temperature (°C)	31.7(4.04)	32.7(5.92)	32.9(5.45)	32.2(2.05)	
Arenisol (%)			40	0.09(0.17)	
Fluvisol (%)		35	35	0.25(0.16)	
Lixisol (%)	50	90	50	0.70(0.20)	
Luvisol (%)	50	50		0.39(0.21)	
Leptosol (%)	34	34	50	0.38(0.07)	
Nitrosol (%)		70	70	0.49(0.30)	
Pinthosol (%)			70	0.51(0.31)	
Veritosol (%)			60	0.14(0.10)	
Farm size (ha)	3.49(3.16)	6.57(5.38)	4.66(3.86)	5.31(5.52)	28.02***
Land area formaize (ha)	0.84(0.51)	1.89(1.12)	1.56(0.97)	1.54(0.81)	6.83***
Maize yield (t/ha)	2.54(4.64)	3.40(3.29)	4.01(2.82)	3.31(3.64)	3.69**
Fertilizer (kg/ha)	153.56(152.7)	404.6(373.7)	125.37(75.36)	272.7(284.04)	47.72***
Pesticides (lt/ha)	2.87(4.16)	12.61(20.07)	2.98(2.15)	7.76(15.12)	14.61***
Irrigation (%)	11.7	20.2	39.2	22.3	
Gender (% male)	90.4	97.1	77.2	90.8	
Age of household head (years)	51.54(11.20)	43.48(7.38)	46.61(11.91)	45.44(10.45)	29.81***
Education (years)	9.16(6.17)	4.58(6.15)	7.64(6.24)	6.46(6.47)	17.29***
Household size (number)	5.08(3.38)	10.26(3.57)	7.52(4.24)	8.33(4.27)	55.85***
Farming experience (years)	24.39(12.85)	30.64(10.79)	15.89(10.16)	25.56(12.69)	47.49***
Distance to output market (km)	7.16(8.23)	10.26(5.42)	3.61(1.89)	7.90(6.39)	36.42***
Distance to input market (km)	8.01(9.26)	8.58(5.52)	4.28(3.02)	7.45(6.59)	12.84***
Extension contact (number)	2.26(0.75)	2.5(2.13)	1.47(1.02)	2.19(1.68)	10.76***
Access to credit (%)	61.7	54.9	34.2	52	
Land ownership (%)	67	89.1	68.4	78.3	
Subsidy (₦)	4581(1760.4)	6423(8567.7)	11354(15684)	5837.28(9575)	

Note. * **Significant at 1%; **Significant at 5%; Values in parenthesis are standard deviation.

Source: Computed from field data, 2012.

4.3.2 Model Estimation

Three models were estimated for the regression analysis. The first model estimates the response of maize net revenue to climate variables only. The second model integrates the soil variables into the first model and the third model added the socio-economic characteristics of the household to the second model.

1) Relationship between Maize Net Revenue and Climate Variables (Mode 1)

Model 1 shows the response of maize net revenue with climate variables only. Results (Table 2) showed that rainfall during the dry season, temperature during the rainy season and square of temperature during the dry season exhibits strong collinearity and hence were dropped from the model. The co-efficient of determination (R^2) of 0.254 indicate that 25.4% of the variations in net revenue were explained by the climate variables. The adjusted R^2 (0.243) is a little lower than R-squared but not too much suggesting that the model do not have a serious over fitting problem. The F-statistic (F = 23.137***) indicates that the overall model is significant at I percent,

Result shows that dry season rainfall has a negative relationship with net revenue and is statistically significant ($p < 0.1$). The reason for this is that low rainfall associated with dry season period reduced maize yield which decreased maize net revenue, thus weakening the purchasing power of the farmers.

The quadratic term of dry season rainfall exhibits a U-shaped relationship with maize net revenue implying that increased rainfall during the dry season may be associated with higher productivity. Adequate rain during dry season is capable of lengthening growing season for maize with greater output possible per hectare.

Rainy season rainfall, however, has a positive relationship with maize net revenue and exhibits a U-shaped relationship with it. This implies that increased rainfall during the rainy season is beneficial for maize production.

The quadratic term for rainy season temperature have a U-shaped relationship with net revenue, implying that temperature during the rainy season could be beneficial for maize production. This results show that climate exhibits a non-linear relationship with net revenue which is consistent with findings in literature (Mendelsohn et al., 1994; Kurukulasuriya & Mendelsohn, 2006; Kabubo & Karanja, 2007; Ajetomobi et al., 2011).

2) Relationship between Maize Net Revenue, Climate and Soil Variables (Model 2)

Model 2 (Table 2) shows the response of maize net revenue with climate and soil variables. The inclusion of soil variables was to capture the spatial heterogeneity across the agro-ecological zones sampled. The soil variables introduced did not in any way improved the model as the R^2 (0.254) and F- statistics (F = 23.137***) remain the same as in the first model although all the climate variables and soil variables were negatively signed and significant at $p < 0.01$, $p < 0.05$ and $p < 0.1$ probabilities level.

Results show that only two of the soil variables (luvisol and pinthosol) among the eight soil types introduced into the model are significant while the other soil types dropped because of strong collinearity. When the soil variable is negatively significant, it means the soil is poor and its spatial extension will cause a decrease in farm net revenues (Sene et al., 2006). This indicates that the dominant soil types in the three agro-ecological zones are not suitable for maize production in their present state. The reason attributed for this may be due to continuous cropping of the same land over the years for maize production with its attendant fertility depletion especially in a cropping system dominated by maize which is a high nutrient feeder.

3) Relationship between Maize Net Revenue, Climate Variables, Soil Factors and Household's Socio-Economic Variables (Model 3)

Model 3 (Table 2) show the results of maize net revenue regression with socio-economic variables combined with the first and second model. Introduction of these variables raises the co-efficient of determination (R^2) from 25.4% to 36.3%. The adjusted R^2 (0.323) suggested that the model was not over-fitted. The F- statistics (9.158; $p < 0.01$) and Durbin Watson (1.928) indicates that the overall model is significant and well behaved. The result also confirms the quadratic relationship between net farm revenue and climate variables.

Results showed an indirect relationship between dry season rainfall and net revenue and this is statistically significant ($p < 0.05$). This implies that low rainfall during the dry season tends to reduce net revenue in maize production. Also, dry season rainfall intensity is capable of increasing humidity which hinders proper drying of maize cobs on the field thereby making storage difficult. This in turn affects grain quality and consequently market price for the product. However, the second order effect of dry season rainfall exhibits a U-shaped relationship with net revenue. This implies increased rainfall during the dry season positively relates to increased maize net revenue as farmers will take advantage of rain during this period for dry season maize cropping. By implication, increased dry season rainfall is beneficial to maize if it falls between the critical growth period but could be harmful if erratic or inadequate during this period and intense at later period when higher humidity could affect the drying process.

The quadratic term of rainy season rainfall exhibits a U-shaped relationship with net revenue and is statistically significant ($p < 0.01$). This suggests that increased rainfall during the rainy season will increase maize productivity.

Also, dry season temperature has a positive and significant relationship with net revenue ($p < 0.05$). Such relationship points to the possibility of farmers capitalizing on the adequate sunlight intensity to dry their grains on the field and thus made them store better for greater market value. This explains the reason why some farmers preferred to leave the maize to dry properly on the field. The maize has to be dried until it attains a moisture level of 12% which is the recommended level for long term storage. Such maize cobs are then shelled and sold at the grain market. Dried maize, command high price than green maize and are bought in large quantity by agro-allied industries for further processing into useful bye products and livestock's feeds.

Rainy season temperature has a U-shaped relationship with net revenue and statistically significant ($p < 0.05$), implying that temperature during the rainy season could be beneficial for maize production. This is because rainy season temperature is required for seed germination, formative crop growth and enhances physiological and grain maturity (Abiodun et al., 2011).

The dominant soil variable (luvisol) has a negative relationship with net revenue and is statistically significant ($p < 0.05$). This implies that the dominant soil type in the three agro-ecological zones on which maize is grown is not suitable for increased maize production at its present state. Its continued cropping will reduce net revenue of maize annually.

Most of the households' variables have a significant impact on maize net revenue per hectare.

Gender of the household head has a positive relationship with net revenue and statistically significant ($p < 0.05$). This indicates that maize net revenue may increase with male headed household than the female headed household. This may be because of the considerable physical and tedious task involved in maize farming or because of possible discrimination faced by women when selling farm produce; which limited their access to productive input resources. This result corroborates the findings by Kurukulasuriya and Ajwad (2004), Kabubo and Karanja (2007), and Ajetomobi et al. (2011).

Household size has a positive and significant impact on net revenue ($p < 0.10$). This indicates that household members in this study were not majorly dependents and were productive people. A large household improves the potential labour position on the farm. An increase in the number of productive dependant in a household will increase net farm revenue.

Education level of the head has a positive relationship with net revenue and statistically significant ($p < 0.05$). This implies that higher levels of education attainment generally leads to higher net revenue per hectare of maize. By implication, an educated household head possessed the ability to adopt new and improved technologies and ability to better optimize on farming practices.

Based on a priori expectation, co-efficient of irrigation is expected to have a positive impact on maize net revenue because irrigation controls for rain fluctuation most especially in the dry season. However, results showed that irrigation has a negative relationship with net revenue and is statistically significant ($p < 0.05$). This result is not surprising given that maize production is rain-fed in the country and that very few farmers in the study areas cultivated maize during the dry season that required irrigation. As reported by Kurukulasuriya and Ajwad (2004) rain-fed agriculture is more profitable than irrigated agriculture because there is a cost associated with the latter than the former.

Distance to input market has a negative relationship with net revenue and statistically significant ($p < 0.05$). This indicates that farmers incurred more cost in terms of money and time if the market place becomes farther away from their farms. This result supports the finding of Fonta et al. (2011) and Deressa (2007) that market distance has a negative impact on net revenue per hectare.

Contrary to a priori expectation, extension agents' contact with farmers is negatively related to net revenue and statistically significant ($p < 0.05$). This implies that number of visit by extension agents to keep the farmers informed about new technologies and modern farming technique was inadequate. This is not surprising giving that in our sample the average number of contact between extension agent and farmer was twice per month. This result is in agreement with findings by Adeogun et al. (2008); Sabo and Zira (2009) and Saka (2011) that the frequency of contact by extension agents was found to be low among Nigerian farmers.

Household access to subsidy (cash or kind) has a positive relationship with net revenue and statistically significant ($p < 0.05$). This implies that access to farm subsidy have a significant impact in increasing farm revenue.

Land area has inverse relationship with net revenue and statistically significant ($p < 0.01$). This indicates that small farms are more productive on a per hectare basis than large farms. The possible reason for this observation is that small farms use fixed resources such as household labour and other inputs over a smaller area than large farms.

Similar results by Kabubo and Karanja (2007) and Ajetomobi et al. (2011) showed that there is inverse relationship between farm size and productivity indicating the reductive importance of small farm holding.

Table 2. Ricardian regression estimates of maize net revenue model

	Model 1		Model 2		Model 3	
	Co-efficient	T-value	Co-efficient	T-value	Co-efficient	T-value
Dry season rainfall	-0.051*	-1.537	-0.043**	-2.121	-0.065**	-2.011
Dry season rainfall square	0.002**	2.213			0.002**	2.232
Rainy season rainfall square	1.164e-5*	1.549	-9.28e-6	-1.249	2.57e-5***	3.806
Dry season temperature	0.049	0.602	-0.120*	-1.903	0.013**	2.125
Rainy season temperature square	0.003**	2.808			0.003**	2.835
Soil: Luvisol			-1.668***	-4.369	-0.588**	-2.674
Pinthosol			-1.893**	-2.841		
Gender					0.160**	2.030
Age						
Household size					0.002	0.529
Years of schooling					0.009*	1.688
Farming experience					0.001	0.545
Irrigation					-0.174**	-2.154
Ownership of land					-0.010	-0.142
Distance to output market					0.004	0.668
Distance to input market					-0.013**	2.078
Livestock ownership					-0.024	-0.425
Extension contact					-0.023*	-1.639
Received subsidy					6.237e-6**	2.618
Access to credit					0.044	1.007
Land area for maize					-0.316***	-3.744
Land area for maize square					0.091***	3.759
Constant	-0.372	-0.197	11.345***	3.734	-0.531	0.270
R – Square	0.254		0.254		0.363	
Adj – R square	0.243		0.243		0.323	
F – Statistics	23.137***		23.137***		9.158***	
Durbin Watson	1.731		1.731		1.928	

Note. Legend: * Significant at 10%; ** Significant at 5%; *** Significant at 1%.

Source: Computed from field data, 2012.

4.3.3 Marginal Impacts of Climate Variables on Maize Net Revenue

In this subsection, marginal impacts of climate on maize agriculture (Table 3) were estimated using the regression results for model 3. The marginal impact analysis was undertaken to observe the effect of small changes in temperature and rainfall on maize net revenue. Results showed that marginal increase in rainfall during the dry season significantly increases net revenue per hectare for maize farms in the rainforest agro-ecological zone by ₦ 4,398.57 while it reduces net revenue per hectare for maize farms in guinea savanna and montane savanna by ₦ 5,429.09 and ₦5,593.36 respectively. The reason may be that majority of the farmers in guinea and montane savanna cultivated maize once per year because of the unimodal pattern of rainfall in these agro-ecological zones thus leaves the maize on the field to dry properly before harvesting and storage. Any slight increase in rainfall during this period may result in diseases and pests build-up which may cause economic damage for the maize farmers. In addition, the drying process could be disturbed thereby promoting grain deterioration and consequent reduction in market value of the product. While in the rainforest agro-ecological zone maize is cultivated twice per year (early and late maize) because of bimodal rainfall distribution hence increase in rainfall during the dry season will be beneficial for maize farmers in this zone as it increased net revenue.

The rainy season rainfall has a positive impact on maize net revenue in all the agro-ecological zones. Increasing rainfall in the rainy season increases net revenue per hectare for maize farms in rainforest, guinea and montane savanna by ₦1,274.75, ₦917.59 and ₦1,205.01 respectively. During the rainy season, increasing rainfall by 1mm increases net revenue in all farms by ₦1,043.92. As noted by Durand (2006), precipitation is the most important driver of maize production. It is therefore important that farmers be encouraged to irrigate their maize farms to mitigate the impact of rainfall deficiency, all other things being equal.

Also, Table 3 shows that temperature during the dry season increase net revenue per hectare for maize farms in rainforest by ₦1,821.06, guinea savanna by ₦1,420.08 and montane savanna by ₦2,061.21. The scenario above shows that, even though maize production is rain-fed, intense sunlight is required to dry the maize cobs after it has completed its growth processes before harvesting and storage for further processing. Sowunmi and Akintola (2009) reported that savanna agro-ecological zones are found to be suitable for maize production in terms of good soil and temperature. Also, small changes in temperature during the rainy season increases maize net revenue in rainforest, guinea and montane savanna by ₦24,794.43, ₦20,514.34, ₦29,491.16.

As indicated from the result of this study, temperature is not harmful for maize production. This result did not agreed with the findings from many studies in literature (Mendelsohn et al., 1994; Kurukulasuriya & Mendelsohn, 2006; Kabubo & Karanja, 2007) who argued that temperature is harmful for crop production. The distinct point of disagreement is that these studies used the combination of different crops and aggregated revenues which are different from this study that used single crop. Moreover, one should be mindful that each crop have known climate on which it can grow best which emanated from the result of this study.

Table 3. Marginal impact of climate change on maize net revenue across the agro-ecological zones

	Rainforest (₦)	Guinea (₦)	Montane (₦)	All zones (₦)
Dry season rainfall	4.398.57	-5,429.09	-5,933.36	-3,221.97
Rainy season rainfall	1,274.75	917.59	1,205.01	1,043.92
Dry season temperature	1,821.06	1,420.08	2,061.21	1,678.65
Rainy season temperature	24,794.43	20,514.74	29,491.16	23,682.18

Source: Computed from field data, 2012.

4.3.4 Predicting the Impact of Climate Change on Maize Agriculture in Nigeria

The study simulated the impact of future climate change scenarios on maize crop across the three agro-ecological zones namely rainforest, guinea savanna and montane savanna. In this simulation, the only variable subject to change were the climate variables while other factors were constant. However, in the long run, technology, capital, consumption etc. are bound to change and these will have tremendous impacts on future maize net revenue. Therefore, we are not making a forecast of how net revenue of maize will actually change but simply isolating the effect of climate change on net revenue of maize. In this section, the regression result for model 3 was used to project the impact of global warming on maize agriculture in Nigeria. To assess the impact of different climate scenarios three General Circulation Models (GCMs) scenarios were used. They are Canadian General Circulation Model (CGCM2), Hadley Centre for Climate Prediction and Research (HADCM3), and the Parallel Climate Model (PCM). The three models predict a wide range of climate outcomes. The emphasis of the present study is to develop scenarios of near-term climate change based on the existing trend and knowledge of how climate may change in the near future. To make prediction for the impact of climate on maize agriculture in the year 2030, three scenarios were highlighted and used. These are (a) increase in temperature only by 1.5 °C (b) decrease in rainfall by 7% (c) increase in temperature by 1.5 °C and decrease in rainfall by 7%. The scenarios were applied separately for rainforest, guinea savanna and montane savanna respectively.

Results in Table 4 show that if rainfall decline by 7% it would result in 35.7% loss in maize net revenue in all the zones. However, there is a marked difference in the predicted loss across the agro-ecological zones. In rainforest, 18.3% (-₦1,056.61/ha) loss in maize net revenue, 20.2% (-₦10,923.72/ha) loss in guinea savanna and 49.7% (-₦ 12,414.83/ha) loss in montane savanna agro-ecological zones respectively were predicted. This gives an indication that maize production is rain dependent and that further decline in rainfall in the near future would result in substantial loss in maize revenue. Studies (CIMMYT, 1988; Diallo et al., 1989) reported that availability of adequate rainfall is by far the most limiting factor in maize production in sub-Saharan Africa. Kurukulasuriya and Ajwad (2004) in their study highlight the importance of precipitation relative to temperature for agriculture in tropical countries. The results from the four climate models used in their study reveal that change in precipitation levels, not temperature, drives productivity. Table 4 revealed further that increase in temperature by 1.5 °C would result in 4.6% (₦1,260.74/ha) gain in maize net revenue in rainforest zone, 4.4% (₦983.14/ha) gain in guinea savanna and 4.3% (₦1,426.99/ha) gain in montane savanna zones respectively. In all zones, a gain of 4.4% in maize net revenue was predicted. The result showed that rainforest, guinea and montane savanna agro-ecological zones are more likely to suffer economic loss in maize production from declining rainfall than from rising temperature. This is because maize is a hot season crop, with optimum productivity in the temperature range of 21-30 °C (Abiodun et al., 2011; Sowunmi & Akintola, 2009). This finding is therefore consistent with the conclusion in Mendelsohn et al. (1994) that climate change could be beneficial under some conditions. The cumulative effect of the two factors will be devastating for the maize farmers in the three agro-ecological zones under study as they were going to suffer losses in maize net revenue in year 2030. Estimated losses were 2.5%, 35.5% and 28.1% for rainforest, guinea and montane savanna agro-ecological zones respectively. In all zones, estimated loss in maize net revenue is 15.5%. This implies that climate change (rainfall and temperature) will have harmful effect on maize revenue in the future. Madiyazhagan et al. (2004) observed that high temperature (greater than 30 °C) compounded by water stress occurring at the same time decreases maize kernel planted in dry land environments. Also, Akpalu et al. (2008) reported that the impact of precipitation on maize yield is stronger than that of temperature, meaning that the impact of climate variability on maize yield could be negative if the change increases temperature but reduces precipitation at the same rate and simultaneously. This result indicate that climate change could have negative impacts on maize net revenue in the future which may pose a serious threat to household food security since maize is a staple food and widely consumed in Nigeria. Thus, effective and efficient adaptation measures should be promoted to prepare stakeholders in maize production systems to enhance their resilience and flexibility when facing adverse climatic change.

Table 4. Predicted impacts of different climate scenarios by agro-ecological zones

	Rainforest (₦)		Guinea (₦)		Montane (₦)		All Zones (₦)	
Climate scenarios	ΔNR	%Δ	ΔNR	%Δ	ΔNR	%Δ	ΔNR	%Δ
-7% rainfall	-2,050.6	-18.3%	-10,923	-20.2%	-12,414	-49.7%	-5,661.9	-35.7%
+1.5 °C	1,260.74	4.6%	983.13	4.4%	1,426.9	4.3%	1,147.01	4.4%
(-7% +1.5 °C)	-789.86	-2.5%	-9,940	-35.5%	-10,987	-28.1%	-4,513.7	-15.5%

Note. *** ΔNR = Change in net revenue per ha; %Δ = percentage change in net revenue/ha.

Source: Computed from field data, 2012.

4.4 Perception of Climate Change by Agro-Ecological Zones

Table 5 presents the farmers' perceptions of long term climate change by agro-ecological zones. Results indicate that significant number of farmers in the three agro-ecological zones; rainforest (35.1%), guinea savanna (53.2%) and montane savanna (75.9%) said that temperature has increased over the years. By contrast, 12.8%, 15% and 7.6% of farmers in rainforest, guinea and montane savanna respectively said that temperature has decreased. Farmers in montane savanna recorded the lowest percentage in terms of decreased temperature (7.6%), altered temperature change (3.8%) and no change in temperature (2.5%) than their counterpart in other agro-ecological zones.

The result for rainfall (Table 5) shows a similar pattern across the three agro-ecological zones. Majority of the farmers in the three agro-ecological zones; rainforest (59.4%), guinea savanna (68.6%) and montane savanna (77.1%) said that rainfall had decreased over the years. Also, 5.3% and 1.7% of the farmers perceived increase in rainfall for rainforest and guinea savanna respectively while none of those sampled farmers in montane savanna said that rainfall had increased. Also, 25.6% of the farmers in the rainforest and guinea savanna (22.5%) said that there had been change in the timing of rains while only 10.1% of the respondents in the montane savanna said the same. In addition, 11.7%, 8.2% and 13.9% of the farmers in rainforest, guinea and montane respectively said they had lived through a change in the frequently of droughts while 11.7%, 8.2% and 1.3% of farmers in rainforest, guinea and montane respectively said that there is no change in rainfall pattern. From the above it suggests that the farmers are perceptive to climate change which is a basic precondition for adaptation.

Table 5. Perception of climate change by agro-ecological zones (% of respondents)

Perception	Rainforest	Guinea	Montane
Increased temperature	35.1	53.2	75.9
Decreased temperature	12.8	15.6	7.6
Altered temperature change	12.8	15.0	3.8
No change in temperature	8.5	9.2	2.5
Increased rainfall	5.3	1.7	0.0
Decreased rainfall	59.4	68.6	77.1
Change in timing of rains	25.6	22.5	10.1
Change in frequency of drought	11.7	8.2	13.9
No change in rainfall	11.7	3.5	1.3

Source: Computed from field data, 2012.

4.5 Short-Term Adaptation to Climate Change

Households reported that they have used more than one type of adaptation strategies. This implies that a single strategy is not adequate in adapting to the impact of climate change as combination of several strategies is likely to be more effective than a single strategy. Table 6 reveals the analysis of adaptation made by respondents across the agro-ecological zones. Results showed that the major adaptation method adopted by maize farmers across the agro-ecological zones was changing the planting dates of maize. This was observed to have the highest percentage in all the three agro-ecological zones, that is, Rainforest (77.7%), Guinea savanna (83.8%) and Montane savanna (69.5%). Changed land-use practices such as crop rotation, shifting cultivation etc was shown to be important across the agro- ecological zones; rainforest(75.5%), guinea (65.3%) and montane savanna (48.1%) respectively. Adopting mixed cropping or multi cropping is widely practiced in rainforest zone (78.7%) compared to farmers in guinea savanna (35.8%) while it is less adopted by farmers in montane savanna (19%). Adopting mixed farming i.e planting crops and rearing livestock are more common in montane savanna (69.2%) and guinea savanna (57.6%) than rainforest (27.7%). Uses of improved maize seed are adopted by 52.8%, 48.1% and 42.7% of the farmers in rainforest, guinea and montane savanna respectively. The use of irrigation as an adaptation strategy to cushion the adverse effect of climate change was commonly practiced in montane savanna (35.3%) than guinea (10.6%) and rainforest (2.6%) agro-ecological zones. Adopting terrace or contour across the slope was universally practiced in the rainforest zone (31.7%) to guide against soil erosion. 43.2% and 25.6% of the maize farmers in montane and guinea savanna change from farming to non-farming during the hot climate but this practice is almost irrelevant in rainforest (4.3%). Changing the use of chemicals and fertilizers as an adaptation strategy was more common among maize farmers in montane savanna (43.2%) than other maize farmers from guinea (36.8%) and rainforest zones (26.6%). Prayer or ritual offering are made by farmers across the agro-ecological zones but it is more pronounced in guinea savanna (42.2%) and montane savanna zones (38.3%). Few of the maize farmers report no changes in agricultural practice across the three agro-ecological zones; rainforest (10.6%), montane savanna (7.6%) and guinea savanna (4.6%). This implies that every respondent across the agro-ecological zones have made at least one adaptation

Table 6. Short term Adaptation to climate change by agro-ecological zones (% of respondents)

Adaptation	Rainforest	Guinea savanna	Montane savanna
Mixed/Multi cropping	78.7	35.8	19.0
Changed land-use practices	75.5	65.3	48.1
Increased use of irrigation	2.6	10.6	35.3
Mixed farming	27.7	57.6	69.2
Changed time of planting	77.7	83.8	69.5
Terrace/Contour construction	31.7	1.7	16.5
Use of improved seed varieties	52.8	48.1	42.7
Change from farming to non-farming	4.3	25.6	43.2
Change use of chemicals and fertilizers	26.6	36.8	43.2
Prayer or ritual offering	11.4	42.2	38.3
No adaptation	10.6	4.6	7.6

Source: Computed from field data, 2012.

4.6 Barriers to Adaptation by Agro-Ecological Zones

The analysis of barriers to adaptation in Nigeria maize agriculture by agro-ecological zones revealed some important differences in the extent and prevalence of different adaptation measures. The possibility of these differences may be due to differences in the perception of climate change across the agro-ecological zones or due to institutional differences between the agro-ecological zones. Six major constraints to adaptation were identified by maize farmers across the agro-ecological zones. These are inadequate information on climate change, inadequate knowledge of appropriate adaptation strategies, inadequate credit or saving, no access to water/river/stream, inadequate access to improved seed variety and land tenure problem.

Results (Table 7) show that majority (67.9%) of the sampled farmers lacked appropriate knowledge of adaptation strategies. This is closely followed by inadequate credits or saving to invest in appropriate adaptation strategies (64.2%), inadequate information on climate change (63.9%), inadequate access to improved seed (18.8%) and inaccessibility to water or stream to facilitate irrigation (11.8%). However there is a marked difference in the barriers to adaptation across the agro-ecological zones. Larger percentage of farmers in rainforest (61.7%), guinea savanna (57.8%) and montane savanna (79.7%) were impeded by inadequate information on climate change. This could be attributed to the fact that research on climate change and indigenous adaptation options are still at lower ebb in the country and thus information is lacking in this area. A large number of maize farmers in montane savanna (84.8%), rainforest (62.8%) and guinea savanna (55.5%) opined that inadequate credit or saving hindered them from possessing the necessary resources and technologies that can assist them to adapt to climate change. Also, inadequate knowledge of appropriate adaptation strategies that suited the local climate conditions are considered a major barrier to adaptation in guinea savanna (75.1%), rainforest (61.7%) and montane savanna (59.5%). inaccessibility to water/river/stream were major adaptation constraint in guinea savanna (46.8%) and rainforest (30.9%) while few farmers in montane savanna (15.2%) felt they were impeded. Inadequate access to improved seed variety was anticipated to be a major barrier in adaptation but was perceived not to be a barrier except in montane savanna agro-ecological zone. Land tenure problem is a major barrier to adaptation because of the high population pressure which forces farmers to cultivate a small portion of land over the years and makes them unable to conserve from further damages. Nearly half of the maize farmers in montane savanna claimed to have been impeded by land tenure problem while very few were constrained in rainforest (27.7%) and guinea savanna (4.0%) respectively. This lack of security of property right is a major problem in technology adoption.

Table 7. Barriers to adaptation by agro-ecological zones

	Rainforest	Guinea	Montane	Pooled
Inadequate information on CC	58 (61.7)	100 (57.8)	63 (79.7)	229 (63.9)
Inadequate knowledge of appropriate adaptation strategies	58 (61.7)	130 (75.1)	47 (59.5)	235 (67.9)
Inadequate credit or savings	59 (62.8)	96 (55.5)	67 (84.8)	222 (64.2)
No access to water/river/stream	29 (30.9)	81 (46.8)	12 (15.2)	41 (11.8)
No access to improve seed	9 (9.6)	9 (5.2)	47 (59.5)	65 (18.8)
Land tenure problem	26 (27.7)	7 (4.0)	39 (49.4)	72 (20.8)

Source: Computed from field data, 2012.

5. Conclusion

The paper assesses the economic effects of climate change on maize farmer net revenue per hectare using the Ricardian estimation technique. A number of interesting findings emanating from this study showed that there is a non-linear relationship between net revenue and climate variables suggesting that climate change affect maize productivity. The study revealed further that gender, household size, education; subsidy has a positive relationship with maize net revenue while luvisols, irrigation, distant to inputs market, and land area has inverse relationship with maize net revenue. The marginal impact analysis showed that increasing rainfall marginally during dry season increases maize net revenue in rainforest zone while it decreases maize net revenue in guinea and montane savanna agro-ecological zones. However, increasing temperature marginally during dry and rainy season increases maize net revenue in the three agro-ecological zones under study. The prediction results confirm that climate change will have a substantial impact on maize net revenue, and that the impact will be more pronounced in guinea and montane savanna than rainforest zone. The results suggest that farmers in Nigeria were aware of climate change, that most of them have noticed an increase in temperatures and decrease in rainfall and that some have taken adaptive measures.

6. Policy Implication and Recommendation

Evidence from this study and available literatures has shown that climate had changed and is still changing while we continued to live with it. The climate change phenomenon is an environmentally induced factor which cannot be controlled but rather to adjust and cope with it. To ensure livelihood sustainability and food security which is being threatened on daily basis by climatic variation and changes there is need for policies framework to be developed at local farm level to counter this adverse effect.

- The study revealed that climate change affects maize production and that farmers were going to suffer losses from declining rainfall than rising temperature. It is suggested that government and private organization should assist in the provision of irrigation facilities to local farmers so as to cope with rainfall deficiency.

- Low revenue of the maize farmers significantly weaken their purchasing power hence makes it difficult for investing in irrigation technology. Policy measures directed towards strengthening financial institutions i.e Microfinance banks, Nigerian Agricultural Bank, Bank of Industry etc should be put in place to encourage them to give credit facilities to farmers at low interest rate.

- The dominant soil types in the study area are not suitable for improved maize production in their present state. Policy directed towards intensification of extension agents' technical service on soil management and conservation techniques that will help in improving soil structure and fertility for increased maize productivity should be focused. Also, government should support maize farmers in the provision of fertilizers at subsidized rate.

- Climate change information is a necessary perquisite for adapting to climate change. Nigerian meteorological agency and other meteorological agencies should be encouraged and strengthened to provide farmers with early warning signal through extension agents to enable them make informed decisions and allow them to better prepare for adverse weather conditions.

- There is preference for specific adaptation strategies among farmers based on the prevailing climatic conditions in their respective zone. Adaptation policies by government should target different agro-ecological zones based on the constraints and potentials of each agro-ecological zone instead of recommending uniform interventions.

References

Abdullahi, A. B., Iheanacho, A. C., & Ibrahim, A. (2006). Econometric Analysis of the Relationship between Drought and Millet Production in the Arid Zone of Nigeria: A Case Study of Borno and Yobe States. *Journal of Agriculture & Social Science, 3*, 170-174.

Abiodun, B. J., Salami, A. T., & Tadross, M. (2011). Climate Change Scenarios for Nigeria: Understanding Biophysical Impacts. *Climate Systems Analysis Group, Cape Town, for Building Nigeria's Response to Climate Change Project*. Ibadan, Nigeria: Nigerian Environmental Study/Action Team (NEST).

Adeogun, A. O., Ajana, A. M., Ayinla, O. A., Yarhere, M. T., & Adeogun, M. O. (2008). Application of Logit Model in Adoption Decision.A Study of Hybrid Clarias in Lagos State, Nigeria. *American-Eurasian J. Agric & Environ. Sci., 4*(4) 468-472.

Ajetomobi, J. O. (2010). Supply Response, Risk, and Institutional change in Nigerian Agriculture. *AERC Research Paper 197*.

Ajetomobi, J., Abiodun, A., & Rashid, H. (2011). Impacts of Climate Change on Rice Agriculture in Nigeria. *Tropical and Subtropical Agroecosystems, 14*(2011), 613-622.

Ayanlade, S., & Odekunle, T. O. (2006). *Assessing Rainfall Variability Impacts on Maize Yield in Guinean Savanna Part of Nigeria, Using GIS Technique*.

Central Bank of Nigeria. (2005). *Annual Reports and Statement of Account*.

CIMMYT. (1988). *Maize Production Regions in Developing Countries*. CIMMYT. Mexico: El-Batan.

Cline, W. (1996). The impact of global warming on Agriculture: Comment. *The American Economic Review, 86*(5), 1309-1311.

Coster, A. S. (2015). *Economic Effects of Climate Change on Maize Production and Adaptation Strategies in Nigeria* (Unpublished Ph.D Thesis). Department of Agricultural Economics, University of Ibadan, Nigeria.

Crosson, P. (1997). Impacts of Climate Change on Agriculture. *Climate Issues Brief 4*. Resources for the Future, Washington, D.C.

Deressa T. (2007). Measuring the Economic Impact of Climate Change on Ethiopian Agriculture: Ricardian Approach. *World Bank Policy Research Paper No. 4342*. World Bank, Washington DC.

Deressa, T., Hassan, R., & Poonyth, D. (2005). Measuring the economic impact of climate change on South Africa's sugarcane growing regions. *Agrekon, 44*(4), 524-542. http://dx.doi.org/10.1080/03031853.2005.9523726

Diallo, A. O., Edmeades, G. O., & Johnson, E. C. (1989). Breeding strategies to overcome constraints and increase maize producitivity in Sub-Saharan Africa. In B. Gebrekidan (Ed.), *Maize Improvement, Production and Protection in Eastern and Southern Africa*. Proceedings of the Third Eastern and Southern Africa Regional Maize Workshop held 18-22 September at Nairo-Ketale, Kenya.

Durand, W. (2006). Assessing the impact of climate change on crop water use in South Africa. *CEEPA Discussion Paper #28*. University of Pretoria, South Africa.

Eid, H. M., El-Marsafawy, S. M., & Ouda, S. A. (2006). Assessing the Economic Impacts of Climate Change on Agriculture in Egypt: A Ricardian Approach. *CEEPA Discussion Paper No. 16*.

FAO. (2003). *The Digital Soil Map of the World (DSMW) CD-ROM*. Food and Agriculture Organization, Italy, Rome. Retrieved from http://www.fao.org/AG/agl/agll/dsmw.stm

FAO. (2007). *Climate change and food security*. FAO, Rome, Italy.

FAO. (2008). *Hunger on the Rise*. Food and Agriculture Organisation. Retrieved August 7, 2010, from http://www.fao.org/newsroom/EN/news/

Fonta, W. M., Ichoku, H. E., & Urama, N. E. (2011). Climate Change and Plantation Agriculture: A Ricardian Analysis of Farmlands in Nigeria. *Journal of Economics and Sustainable Development, 2*(4).

Gadgil, S. (1996). Climate change and agriculture – An Indian perspective. In Y. P. Abrol, et al. (Eds.), *Climate Variability and Agriculture* (pp. 1-18). Narosa Publishing House, New Delhi.

Gbetibouo, G., & Hassan, R. (2005). Economic impact of climate change on major South African field crops: A Ricardian approach. *Global and Planetary Change, 47*, 143-152. http://dx.doi.org/10.1016/j.gloplacha.2004.10.009

Houghton, J. T., MeiraFilbo, L. G., Callander, B. A., Harris, N., Kattenberg, A., & Maskell, K. (1996). Climate Change 1995. *The Science of Climate Change*. Cambridge University, UK.

IFAD. (2007). *Rural Poverty in Nigeria: Agriculture in the Federal Republic of Nigeria*. International Fund for Agricultural Development. Retrieved August 7, 2010, from http://www.ruralpovertyportal.org

Iken, J. E., & Amusa, A. (2004). Maize Research and Production in Nigeria. *African Journal of Biotechnology, 3*, 302-307. http://dx.doi.org/10.5897/AJB2004.000-2056

IPCC. (2007). Summary for Policymakers. In M. L. Parry, O. F. Canziani, J. P. Palutikof, P. J. van der Linden & C. E. Hanson (Eds.), *Climate Change 2007: Impacts, Adaptation and Vulnerability* (pp. 7-22). Contribution of Working Group II to the Fourth Assessment Report of the Intergovernmental Panel on Climate Change, Cambridge University Press, Cambridge, UK.

Kabubo-Mariara, J., & Karanja, F. K. (2007). The economic impact of climate change on Kenyan crop Agriculture: A Ricardian Approach. *Global Planet Change, 57*, 319-330. http://dx.doi.org/10.1016/j.gloplacha.2007.01.002

Kumar, K., & Parikh, J. (1998). Climate change impacts on Indian agriculture: The Ricardian approach. In A. Dinar et al. (Eds.), *Measuring the Impact of Climate Change on Indian Agriculture* (World Bank Technical paper 402). Washington, DC.

Kumar, K., & Parikh, J. (2001). Indian agriculture and climate sensitivity. *Global Environmental Change, 11*, 147-154. http://dx.doi.org/10.1016/S0959-3780(01)00004-8

Kurukulasuriya, P., & Ajwad, M. I. (2004). Ricardian Study of Sri Lankan Farmers. In R. Mendelsohn, A. Dinar, A. Baist, P. Kurukulasuriya, M. I. Ajwad, F. Kogan & C. Williams (Eds.), *Cross sectional Analyses of Climate Change Impacts* (World Bank Policy Research Working Paper 3350).

Kurukulasuriya, P., & Mendelsohn, R. (2006). A Ricardian analysis of the impact of climate change on African crop land. *CEEPA Discussion Paper No. 8*. Centre for Environmental Economics and Policy in Africa. Pretoria, South Africa: University of Pretoria.

Kurukulasuriya, P., & Rosenthal, S. (2003). Climate change and agriculture: A review of impacts and adaptations. *Paper No. 91 in Climate Change Series*. Agriculture and Rural Development Department and Environment Department, World Bank, Washington, DC.

Law-Ogbomo, K. E., & Law-Ogbomo, J. E. (2009). The Performance of *Zea mays* as Influenced by NPK Fertilizer Application. *Not. Sci. Biol., 1*(1), 59-62.

Liu, H., Li, X., Fischer, G., & Sun, L. (2004). Study on the impacts of climate change on China's agriculture. *Climate Change, 65*, 125-148. http://dx.doi.org/10.1023/B:CLIM.0000037490.17099.97

Maddison, D. (2006). The perception of and adaptation to climate change in Africa. *CEEPA Discussion Paper No. 10*. Centre for Environmental Economics and Policy in Africa. Pretoria, South Africa: University of Pretoria.

Madiyazhagan, R., Birch, C. J., Carberry, P. S., & Micheal, R. (2004). *Water and high temperature stress effects on maize production*. Paper Presented at the 4th International Crop Science Congress.

Madu, D. B., Gundiri, M. N., Musa, I. K., Nwa, E. U., Adamu, M. S., Bala, M. S., ... Ogunniyi, E. O. (2010). *Country Profile – Nigeria*. Abuja: Ministry of Water Resources.

Medugu, N. I. (2008). *Nigeria: Climate Change – A Threat to the Country's Development*. Retrieved from http://www.allafrica.com/nigeria/

Mendelsohn, R., & Neumann, J. E. (Eds.). (1999). *The Economic Impact of Climate Change on the Economy of the United States*. Cambridge, UK: Cambridge University Press. http://dx.doi.org/10.1017/CBO9780511573149

Mendelsohn, R., Nordhaus, W., & Shaw, D. (1994). The Impact of Global Warming on Agriculture: A Ricardian Analysis. *American Economic Review, 84*, 753-771.

Molua, E. L. (2002). Climate variability, vulnerability and effectiveness of farm-level adaptation options: The challenges and implications for food security in southwestern Cameroon. *Environment and Development Economics, 7*, 529-545. http://dx.doi.org/10.1017/S1355770X02000311

Molua, E. L., & Lambi, C. M. (2007). The economic impact of climate change on agriculture in Cameroon. *The World Bank Policy Research Working Paper 4364*.

National Bureau of Statistics. (2007). *Agricultural survey Report.*

National Bureau of Statistics. (2010). *Annual Abstract of Statistics.*

National Bureau of Statistics. (2014). *DLM Research Estimates.*

Nhemachena, C., & Hassan, R. (2007). Micro-Level Analysis of Farmers' Adaptation to Climate Change in Southern Africa. *IFPRI Discussion Paper No. 00714.* International Food Policy Research Institute. Washington DC.

Nigeria Institute of Social and Economic Research. (2007). *Agricultural Report.*

Odjugo, P. A. O. (2010). General Overview of Climate Change Impacts in Nigeria. *Journal of Hum.Ecol., 29*(1), 47-55.

Ricardo, D. (1817). *On the principles of political economy and taxation.* John Murray, London, England.

Sabo, E., & Zira, D. Y. (2009). Awareness and Effectiveness of vegetable technology information packages by vegetables farmers in Adamawa State, Nigeria. *African Journal of Agricultural Research, 4*(2), 65-70.

Saka, J. O. (2011). *Land Use Intensification and Efficiency of Food Crop Production in South-western Nigeria* (Unpublished PhD Thesis). Department of Agricultural Economics, University of Ibadan, Nigeria.

Sanghi, A. (1998). *Global warming and climate sensitivity: Brazilian and Indian agriculture* (PhD Dissertation). Department of Economics, University of Chicago.

Sanghi, A., Mendelsohn, R., & Dinar, A. (1998). The climate sensitivity of Indian agriculture. In A. Dinar et al. (Eds.), *Measuring the impact of climate change on Indian Agriculture* (World Bank Technical Paper 402). World Bank, Washington, DC.

Sene, I. M., Diop, M., & Dieng, A. (2006). Impacts of Climate Change on the Revenues and Adaptation of Farmers in Senegal. *CEEPA Discussion Paper No. 20.*

Seo, S. N., & Mendelsohn, R. (2007). An integrated farm model of crops and livestock: Modeling Latin. American agricultural impacts and adaptations to climate change. *World Bank Policy Research Series Working Paper 4161* (p. 41) World Bank, Washington DC., USA.

Seo, S. N., Mendelsohn, R., & Munasinghe, M. (2005). Climate change and agriculture in Sri Lanka: A Ricardian valuation. *Environment and Development Economics, 10,* 581-596. http://dx.doi.org/10.1017/S1355770X05002044

Sowunmi, F. A., & Akintola, J. O. (2010). Effect of Climatic Variability on Maize Production in Nigeria. *Research Journal of Environmental and Earth Sciences, 2*(1), 19-30.

16

Biology of the Root-Attacking Moth, *Isoceras sibirica* Alpheraky

Zhixiong Liu[1], Hongxia Liu[1], Jintong Zhang[1] & Xupeng Chen[1]

[1] Institute of Chemical Ecology, Shanxi Agricultural University, Taigu, Shanxi, China

Correspondence: Jintong Zhang, Institute of Chemical Ecology, Shanxi Agricultural University, Taigu, Shanxi, China. E-mail: zhangjintong@126.com

Abstract

The root-attacking moth, *Isoceras sibirica* Alpheraky (Lepidoptera: Cossidae), is a pest of the food crop *Asparagus officinalis* Linn, and can seriously affect yield. Data for this study were collected from both the laboratory and in the field over a 3-year period (2009 - 2011). Periods for egg incubation and pupal stage ranged from 24.49 ± 4.92 to 28.82 ± 4.35, 19.79 ± 4.11 to 21.32 ± 3.46 days, respectively. Larval weight during the developmental period was described by the equation: $y = 0.0109e^{0.1888x}$. An average longevity of 6.3 days was observed for male and 5.1 days for female adults. A forewing wingspan of 52.4 ± 0.83 mm and 45.2 ± 0.89 mm was observed for females and males, respectively. This study provides a basis for biological pest control of *Isoceras sibirica*.

Keywords: biology, *Isoceras sibirica*, life cycle, asparagus

1. Introduction

Due to the excellent conditions for the cultivation of asparagus, China was the largest producer of asparagus in the world in 2010, followed by Peru and Germany (National Agricultural Statistics Service, 2006). Asparagus farming therefore contributes significantly to the local economy. *Isoceras sibirica* Alpheraky (Lepidoptera: Cossidae), a destructive pest, is widely distributed in asparagus farms in China. *I. sibirica* larvae burrow into the root crowns, causing destruction of the freshly planted asparagus. This damage seriously affects growth and total asparagus yield. The larvae of *I. sibirica* grow within the roots of asparagus, making control difficult. At present, the main control method is the broad use of chemical pesticides. Year by year, long-term and heavy application of pesticides has resulted in serious environmental pollution, which can potentially compromise human health. Although strenuous efforts have been made to control the moths (Guoqi et al., 2010; Hongxia, Yujie, & Jintong, 2010; Gaochuan et al., 2013); however, their biology has been studied little. Such biological insight would facilitate the formulation of effective management strategies (Kioko, Raina, & Mueke, 2000), and is therefore the basis of this study.

2. Methods

This study was conducted both in the laboratory and in the field of Shanxi province, North China, which lies between latitudes 34°42'N and 40°42'N, longitudes 110°24'E and 114°42'E altitude, and is 1000 m above sea level. Experiments conducted in the laboratory focused on the morphological structure of eggs, larvae, pupae and adults of *I. sibirica*. The field experiments were carried out during 2009, 2010 and 2011 in Shanxi.

2.1 Egg Clusters and Incubation Period

To determine egg size, 1000 eggs from six *I. sibirica* female moths were measured under a stereomicroscope using an ocular micrometer (Mbahin, Raina, Kioko, & Mueke, 2012). The average length of 100 freshly laid egg clusters was determined using a Vernier calliper. An electronic balance was used to determine egg weight. Incubation periods were determined for 100 egg clusters, which were laid daily and marked. Eggs were monitored daily over 40 days and hatching times were recorded for each egg.

2.2 Larval Period

After hatching, larvae were transferred to separate asparagus plants and reared. The larval period was determined by observing the first through seventh instars of 100 cohorts of *I. sibirica* larvae until they emerged. The number of instars was calculated by visual observation of discarded exuviae (Mbahin et al., 2007; Schmidt, Campbell, & Trotter, 1977). Larval weight during the developmental period is presented in Figure 1.

2.3 Pupal Period

To measure the pupal period, spinning dates of 125 cocoon nests were recorded prior to transfer to an open sleeve (50×45×45 cm) until the adults emerged.

2.4 Adults

The life span of male and female moths was measured from the point at which larvae became viable. The moths were deposited into a wooden cage (100×100×100 cm) with nylon screens mounted on all four sides and a door on one side (Gothilf, 1968). Wingspan, as determined by the maximum distance between the tips of each forewing, was measured using a Vernier caliper.

2.5 Data analysis

Mean values were compared using a t-test (Peacock, & Worner, 2008) or chi-square (Moses, Emerson, & Hosseini, 1984). The SPSS 17.0 software package was used for data analysis. The degree of significance is shown as follows:

*: significant ($P < 0.05$);

**: highly significant ($P < 0.01$);

ns: not significant ($P > 0.05$).

3. Results

3.1 Egg Clusters

The egg of *I. sibirica* is oval in shape, measures 1.96 ± 0.02 mm at the major axis and 0.73 ± 0.02 mm at the minor axis (n = 79). Eggs weigh an average of 0.84 ± 0.003 mg (n = 68), with regular hexagon ridges projecting upward. They are white in color, but become brown prior to hatching. The eggs are usually laid in clusters on the epidermis of asparagus stems. The mean number of freshly laid eggs in a cluster was 204 ± 6 (n = 25). The incubation period ranged from 24 to 28 days. Mean incubation periods are shown in Table 1.

Table 1. Mean (±SD) incubation period (days) of *I. sibirica* in the asparagus field (2009-2011)

Year	*Isoceras sibirica* Alpheraky	
	n	Incubation period (days)
2009	17	28.82 ± 4.35
2010	23	26.30 ± 5.18
2011	29	24.49 ± 4.92

n: number of egg clusters.

3.2 Larvae

Newly emerged larva chew a small cavity, approximately 0.5 mm in diameter, into the stem and crawl down the stem to the root, which is typically 2-3 cm below the soil surface. Based on our observations, the larvae are gregarious from the first to the seventh instars. Direct observation of expelled exuviae by larvae reared through the larval stage indicated that there are six molts in the period from hatch to pupation. From the first larval instar until pre-pupa, young larvae grew from 1.8 mg to more than 4,600 mg. Developmental data are presented in Figure 1. The equation used to determine the curve is showed as $y = 0.0109e^{0.1888x}$.

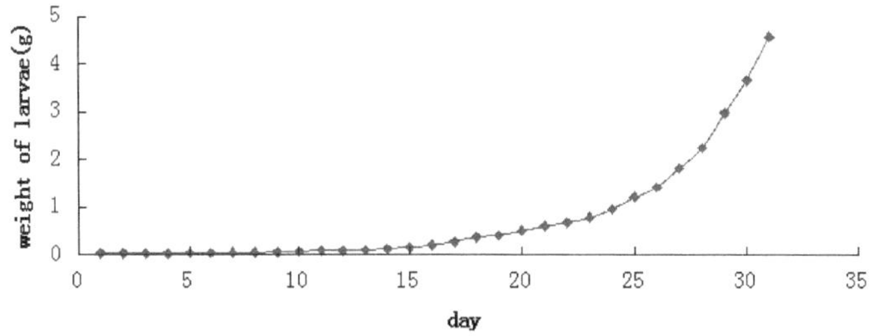

Figure 1.The developmental tendency of weight of larvae

3.3 Pupa

After the larvae period is complete, mature larvae leave the asparagus in search of a convenient place to pupate. The cocoons averaged 8 cm in length and 1.5 cm in diameter and were gray in color. Within the cocoon, the larvae remained quiescent in a pre-pupa stage for 3 to 5 days before pupating. Cocoon construction and the pre-pupa stage required between 5 to 10 days, depending upon the ambient temperature. The length and width of male pupae ranged from 1.7 to 2.8 cm (mean 2.2 ± 0.27) and 0.6 to 1 cm (mean 0.77 ± 0.10) (n = 43), while the length and width of female pupae ranged from 2.2 to 3.2 (mean 2.6 ± 0.28) and 0.6 to 1.2 cm (mean 0.84 ± 0.12) (n = 38), respectively. There was a highly significant difference between the lengths and widths of the two sexes. The weight of the male pupae ranged from 0.23 to 0.47 g (mean 0.35± 0.06), whereas female pupae weight ranged from 0.27 to 0.56 g (mean 0.43 ± 0.07). There were also highly significant differences between the weights of the pupae of the two sexes. The mean (±SD) pupal period (days) of *I. sibirica* in the asparagus field is presented in Table 2.

Table 2. Mean (±SD) pupal period (days) of *I. sibirica* in the asparagus field (2009-2011)

Year	*Isoceras sibirica* Alpheraky	
	n	Pupal period (days)
2009	10	20.60 ± 3.02
2010	14	19.79 ± 4.11
2011	22	21.32 ± 3.46

n: number of cocoon nests.

3.4 Adults

Moths emerged from April through June. On average, mean forewing length was 28.4 ± 0.80 mm for females (Figure 2.4) (n = 187), compared to 24.3 ± 0.67 mm for males (n = 198). Forewing wingspan was 52.4 ± 0.83 mm and 45.2 ± 0.89 mm for females and males, respectively. Highly significant differences were found between the mean length of forewing and the wingspan of the male and the female moths. Mean life span of *I. sibirica* in the asparagus field is presented in Table 3.

Table 3. Adult lifespan (days) of *I. sibirica* in the asparagus field (2009-2011)

Year	*Isoceras sibirica* Alpheraky				t-test
	n	Male	n	Famale	
2009	69	5.72 ± 2.02	57	4.86 ± 1.56	ns
2010	57	6.22 ± 1.74	67	5.07 ± 1.34	**
2011	78	6.95 ± 2.18	88	5.23 ± 1.43	**

n: number of males or females moths; ns: not significant difference between male and female; **: highly significant difference between male and female.

3.5 Occurrence of I. sibirica

Time of occurrence of *I. sibirica* is listed in Table 4.

Table 4. Time of occurrence of the various stages of *I. sibirica* in the asparagus field

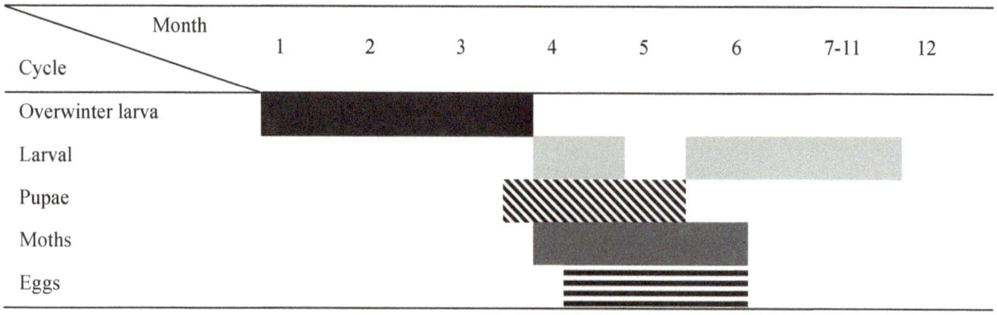

4. Discussion

Experimental results presented herein have resolved much of the confusion over the timing of *I. sibirica* development (Al-Mehmmady, 2000). Total number of eggs laid varied by a few hundred or more, however, no significant differences ($P < 0.01$) were found between numbers of eggs from different egg clusters after mating (Figure 2.3). Our results correlate with the work of Guoqi (2008), who reported an average of 226 and noted that incubation periods are dependent on climate. Oviposition can be influenced by many factors, including plant volatiles, secondary chemicals, temperature and waxes on leaf surfaces (Asian Vegetable Research and Development Center, 1986; Lu, Chen, & Zhu, 1988; Talekar & Shelton, 1993), however, eggs continue to be laid so long as the temperature is favorable (Gothilf, 1970).

It appears that variations in larvae periods correlate with the availability of food and favorable climatic conditions. The newly emerged larvae are active and roll in the soil looking for food. When they encounter asparagus, the larvae first consume the stem, then the roots, which renders the crop unfit for human consumption. As a result, major reductions in asparagus yield can occur (Guoqi et al., 2008).

Other factors affecting pupation include the thickness of the soil. Previous studies of the time of pupation reported that the average date for eclosion was May 21, but this varied depending upon the temperature (Markin, Nagata, & Taniguchi, 1989).

Moths remain quiescent due to an aversion to light. The life span of males is longer than that of females. It is possible that females consume too much energy before dying. In addition, some morphological differences exist between the forewing and anterior wing (Figures 2.1 and 2.2).

Additional studies should fully elucidate the biology of *I. sibirica* and provide information to facilitate the development of more effective control strategies for this pest. Accurate and detailed information regarding the biology of *I. sibirica* is necessary for development of an effective program for management of this important and widespread pest.

Initial data suggest that the application of sex pheromones may be an effective strategy to reduce attacks by *I. sibirica* (Alpheraky), and temper crop damage (C. J. DeLoach, A. D. DeLoach, & Cordo, 1979). Future work focused on ultrastructural findings of *I. sibirica,* in relation to the underlying biochemical processes of sex pheromone biosynthesis or plant volatiles, may also prove informative (Yokoyama et al., 2003).

2.1 2.2 2.3 2.4

Figure 2. 2.1 forewing; 2.2 interior wing; 2.3 mating; 2.4 female

Acknowledgements

This research was supported by the National 12th Five-year Science and Technology Support Plan of China (Grant no.2012BAD19B07) and the Research Foundation of the Introduction of Talents of Shanxi Agricultural University (Grant no.2014ZZ09).

References

Al-Mehmmady, R. M. (2000). Biological studies on the okra moth, *Earias vittella* (F.)(Lepidoptera: Noctuidae) in Jeddah, Saudi Arabia. *Research Bulletin, 96*, 5-18. Retrieved from http://hdl.handle.net/123456789/7916

Asian Vegetable Research and Development Center. (1986). *AVRDC Progress Report Summaries 1985* (p. 96). Shanhua, Taiwan: Asian.

DeLoach, C. J., DeLoach, A. D., & Cordo, H. A. (1979). Observations on the biology of the moth, *Samea multiplicalis*, on waterlettuce in Argentina. *J. Aquat. Plant Manage, 17*, 42-44.

Gaochuan, S., Hongli, Q., Zhanbei, Z., et al. (2013). Main asparagus diseases and pests and key control methods of Pollution-free production in Shanxi Province. *Journal of Shanxi Agricultural Sciences, 4*(4), 377-380. http://dx.doi.org/10.3969/j.issn.1002-2481.2013.04.22

Gothilf, S. (1968). The biology of the carob moth *Ectoniyelois ceraioniae* (Zell.) in Israel. III. Mass culture on artificial diet. *Israel J. Ent., 3*(2), 109-118.

Gothilf, S. (1970). The biology of the carob moth *Ectomyelois ceratoniae* (Zell.) in Israel. III. Phenology on various hosts. *Israel J. Ent., 5*, 161-175.

Guoqi, D., Zhanbei, Z., et al. (2008). The bionomics of *Isoceras sibirica. Chinese Bulletin of Entomology, 45*(3), 397-400. http://dx.doi.org/10.3969/j.issn.0452-8255.2008.03.012

Guoqi, D., Zhanbei, Z., et al. (2010). Preference of *Isoceras sibirica* to storage root and stem of asparagus and effects of humidity on development of its pupae. *Journal of Environmental Entomology, 32*(2), 152-157. http://dx.doi.org/10.3969/j.issn.1674-0858.2010.02.002

Hongxia, L., Yujie, H., & Jintong, Z. (2010). Study on the Occurrence Regularity and Control Techniques of *Isoceras sibirica* Alpheraky and *Crioceris quatuordecimpunctata* Scopolic, *Xiandai Nongye Keji, 14*, 145-146. http://dx.doi.org/10.3969/j.issn.1007-5739.2010.14.095

Jump-up USDA. (2006). *Vegetables 2005 Summary*. National Agricultural Statistics Service.

Kioko, E. N., Raina, S. K., & Mueke, J. M. (2000). Survey on diversity of wild silk moths species in East Africa. *East African Journal of Science, 2*(1), 1-6.

Lu, Z. Q., Chen, L. F., & Zhu, S. D. (1988). Studies on the effect of temperature on the development, fecundity and multiplication of *Plutella xylostella* L. *Insect Knowl., 25*, 147-149.

Markin, G. P., Nagata, R. F., & Taniguchi, G. (1989). Biology and behavior of the South American moth, *Cyanotricha necyria* (Felder and Rogenhofer) (Lepidoptera: Notodontidae), a potential biocontrol agent in Hawaii of the forest weed, *Passiflora mollissima* (HBK) Bailey. *Proc Hawaiian Entomol Soc., 29*, 115-123. http://dx.doi.org/ 10.1303/jjacz.27.17

Mbahin, N, Raina, S. K., Kioko, E. N., et al. (2007). Spatial distribution of cocoon nests and egg clusters of the silkmoth *Anaphe panda* (Lepidoptera: Thaumetopoeidae) and its host plant *Bridelia micrantha* (Euphorbiaceae) in the Kakamega Forest of western Kenya. *International Journal of Tropical Insect Science, 27*(3-4), 138-144. http://dx.doi.org/10.1017/S1742758407859662

Mbahin, N., Raina, S. K., Kioko, E. N., & Mueke, J. M. (2012). Biology of the wild silkmoth *Anaphe panda* (Boisduval) in the Kakamega forest of Western Kenya. *International Journal of Forestry Research, 2012, 186549*. http://dx.doi.org/10.1155/2012/186549

Moses, L. E., Emerson, J. D., & Hosseini, H. (1984). Analyzing data from ordered categories. *The New England Journal of Medicine, 311*(7), 442-448. http://dx.doi.org/10.1056/NEJM198408163110705

Peacock, L., & Worner, S. P. (2008). Biological and ecological traits that assist establishment of alien invasive insects. *New Zealand Plant Protection, 61*, 1-7. Retrieved from http://www.nzpps.org/terms_of_use.html

Schmidt, F. H., Campbell, R. K., & Trotter, S. J. (1977). Errors in determining instar numbers through head capsule measurements of a lepidoptera – A laboratory study and critique. *Annals of the Entomological Society of America, 70*(5), 750-756.

Talekar, N. S., & Shelton, A. M. (1993). Biology, ecology, and management of the diamondback moth. *Annual Review of Entomology, 38*(1), 275-301. http://dx.doi.org/10.1146/annurev.en.38.010193.001423

Yokoyama, N., Fónagy, A., Tatsuki, S., et al. (2003). Ultrastructural studies on the pheromone-producing cells in the silkmoth, *Bombyx mori*: formation of cytoplasmic lipid droplets before adult eclosion. *Acta Biologica Hungarica, 54*(3), 299-312. http://dx.doi.org/10.1556/ABiol.54.2003.3- 4.8

17

Rearing Larvae of Different Strains of *Coregonus lavaretus* under Cold Water Conditions: Comparison of a Special Cold-Water Line with a Standard Agglomerated Microdiet

Stephanie Leithner[1] & Josef Wanzenböck[2]

[1] Organismic Biology, University of Salzburg, Salzburg, Austria

[2] Research Institute for Limnology, University of Innsbruck, Mondsee, Austria

Correspondence: Stephanie Leithner, Organismic Biology, University of Salzburg, Hellbrunnerstraße 34, 5010 Salzburg, Austria. E-mail: stephanie.leithner@stud.sbg.ac.at

Abstract

We studied the performance of a commercial, agglomerated larval microdiet in terms of survival, growth and feed acceptance and compared a dedicated "cold water line" of the product to the standard product involving a Baltic and an Alpine strain of whitefish. Experiments were conducted under cold water conditions < 8 °C (average 5.1 °C) for 60 days after hatching. Replicate tanks of 400 and 200 liter volume received filtered lake water disinfected with ozone and UV light and were stocked with 15 larvae per liter. Daily rations of 4% body weight were manually delivered in six portions every two hours and group samples were taken weekly to determine length, weight and feed acceptance. Dead larvae were removed and counted daily. Before our experiment was influenced by an outbreak of an unknown disease during the last two weeks, we observed similar growth, survival and feed acceptance for each of the feed types with low variability within each of the two strains. Strains showed characteristic differences in growth and survival related to lower feed acceptance during the first four weeks of the Alpine strain. We concluded that the specific cold water line of the commercial microdiet did not provide any advantage compared to the standard product under low temperature conditions. However, differential performance might exist under different temperature regimes.

Keywords: European whitefish, growth, survival, feed acceptance, AgloNorse®

1. Introduction

The industrialization and intensification of fish production in Europe (mainly trout and salmon) started in the 1960s when the industry demanded better and more balanced food products (Goddard, 1996). However, for sensitive fish species, such as European whitefish, the developed feeds seemed to be of limited suitability as shown in many studies during the 1970s, 1980s and 1990s (Eckmann & Kausch, 1976; Rösch & Dabrowski, 1986; Rösch, 1988; Koskela & Eskelinen, 1992). These studies were rarely satisfactory because at this time available dry feeds (extruded feeds) often included insufficient qualitative and quantitative nutrient composition for whitefish larvae and they starved to death (Kummer & De Verga, 1992). More recently the industry increasingly uses the process of agglomeration to improve performance of starter feeds (microdiets) for small larvae of mainly marine species. Through this process, individual particles are enlarged in order to facilitate handling and to avoid dust formation (Langdon & Barrows, 2011; Holt et al., 2011). However, the suitability of these agglomerated dry feeds has not been extensively tested for whitefish respectively published.

Previous studies showed that rearing of European whitefish under cold-water conditions (< 8 °C) was difficult and survival rate increased with increasing water temperature (Rösch, 1988; Koskela & Eskelinen, 1992; Rojas Beltran et al., 1992). However, in most hatcheries the ambient water temperatures during early spring, when whitefish hatch, are low and water warming takes considerable energy and thus may not be economically viable. At the same time, producers of microdiets for marine fish larvae propagated special "cold-water" lines of their microdiets (e.g. AgloNorse® distributed by Ewos in 2006/2007) for which improved success was reported with cod compared to the standard product (pers. communication). This cold-water line differed from the standard product in proportion of highly unsaturated fatty acids (for details see Material & Methods). Therefore, the question arose to which extent special "cold-water" microdiets show improved performance also for whitefish larvae under ambient low water temperatures (< 8 °C).

Rearing of whitefish larvae is of variable success and a large part of this variability depends on the strain used for experiments (Rösch & Dabrowski, 1986; Mahmoudzadeh et al., 2009). Consequently results cannot reliably be transferred to other strains (Rösch, 1992) or species (the taxonomy of European whitefish is a matter of intensive debate) e.g. from Baltic/Scandinavian strains to those of the central European region of the Alps. Potentially the latter strains might have advantages in marketing as regional special products. However, aquaculture of Alpine whitefish is considerably less developed compared to Baltic and Scandinavian strains (Siikavuopio et al., 2012).

The aim of our study was to compare rearing success in terms of growth, survival and feed acceptance of two whitefish strains (Baltic vs. Alpine) using a special "cold-water" and a standard agglomerated commercial microdiet under ambient cold-water conditions.

2. Materials and Methods

2.1 Experimental Design and Conditions

In winter 2011/2012 eggs of European whitefish (*Coregonus lavaretus*), from an existing breeding line originating from whitefish of the southern Baltic region (sometimes designated *Coregonus maraena*) and reared since the 19th century in carp-pond aquaculture in Bohemia and northern Austria (Susta, 1888; Winkler et al., 2011) and bred in our own hatchery in the fourth generation were stripped and artificially fertilized. The offspring of these fish is thereafter called strain 1. Native whitefish (*Coregonus lavaretus*) from the Austrian lake "Hallstättersee" (designated *Coregonus renke* by Kottelat & Freyhof, 2007) were caught in early December 2011 and wild fish were stripped to form the first hatchery generation of what is referred to in the following as strain 2. The fertilized eggs were incubated in incubation vessels (Zuger jars) at a temperature of 4 – 5 °C in our hatchery in Mondsee, Austria. The experiment started a few days after hatching on 08.02.2012 and encompassed the first 60 days of development which covered the most sensitive phase of whitefish larvae when changes from endogenous to exogenous feeding occurs. Because of technical restrictions we had to use tanks of different sizes: Replicate groups of 6000 larvae were reared in round PVC tanks of 400 liters volume or 3000 larvae in tanks of 200 liters volume. All test tanks received water pumped from the lake at 10 m depth, filtered by a sand filter and treated with Ozone and UV light. For strain 1 four tanks were used (two big and two small tanks) and for strain 2 six tanks were used (three big and three small tanks). The exchange rate of the water per tank was set to 1.0 times per hour. Within the first 3 - 4 weeks feeding was started with 100 to 200 µm particles, and during the next 3 - 4 weeks feeding continued with 200 to 300 µm particles maintaining a daily ration of 4% body weight per day. Feed was delivered manually in six portions every two hours. Tanks were siphoned daily to remove faeces, food residues and eventually dead larvae. Motionless larvae lying on the bottom were considered dead, siphoned off and counted. Water quality parameters (temperature, pH, redox potential) were recorded by means of probes in intervals of 30 minutes and logged. A sample of 30 individuals per tank was collected weekly to determine fresh weights to the nearest 0.1 mg. For this purpose, larvae were sacrificed with an overdose of the anesthetic phenoxyethanol, blotted of paper towels and then weighed using an analytical balance. Subsequently they were placed under a dissecting microscope to determine if food particles were visible in the gut of the larvae.

2.2 Feed Types

A special agglomerated larval starter feed, originally developed for marine fish species called "AgloNorse®", i.e. the standard product and "AgloNorse® Extra", replacing the "cold-water" version of 2006/2007 (Tromsø Fiskeindustri AS, Norway) was used in our experiment. The composition of the dry feed was: 58-60% protein, 17–20% fat, 10% ash, 1% fibre and 7–9% moisture. The only difference between the two types of feed was in the percentage of the n-3 PUFA and n-6 PUFA ("AgloNorse®": 2.1–2.6% PUFA n-3 and 2.5–2.6% PUFA n-6; "AgloNorse® Extra": 3–3.5% PUFA n-3 and 1.3% PUFA n-6).

2.3 Statistical Analysis

Numbers of survivors per tank were calculated daily by subtracting dead individuals from the initial number minus those collected for measurements or after extensive cleaning. Before any statistical analysis, the metric data were checked for normal distribution and for equal variance, and optionally log transformation (log (data)) was applied. Analysis of Covariance (ANCOVA) was used as a test procedure for the significance of the growth data including time as a co-variable. As a second test for significance, student's t-test was chosen. If data were not normally distributed or equal variance was not given, despite transformation, a nonparametric test, i.e. the Mann-Whitney Rank Sum test, was used.

3. Results

The initial lengths of the larvae were significantly different (12.37 ± 0.34 mm for strain 1 and 10.89 ± 0.76 mm for strain 2, Mann-Whitney test: P = <0.001, Strain 1 n = 90 and strain 2 n = 87). Also the initial weights of the two whitefish strains at the start of the experiment were significantly different (7.9 ± 1.1 mg for strain 1 and 6.5 ± 1.0 mg for strain 2, Mann-Whitney test: P = <0.001, Strain 1 n = 120 and strain 2 n = 180). Water temperature was very low at the beginning of the experiment, at about 3 °C (Table 1), due to an exceptionally cold weather period. Subsequently, a slowly rising trend in temperatures was observed but never reached 7°C. The average temperature during the experiment was 5.1 ± 1.2 °C.

Table 1. Weekly average temperature and relative growth rate (RGR)

			Strain 1		Strain 2	
			AgloNorse®	AgloNorse® Extra	AgloNorse®	AgloNorse® Extra
Date	**Week**	**Weekly average temperature**	**RGR**	**RGR**	**RGR**	**RGR**
8 – 15 February	1	3.24 ± 0.37	0.006	0.006	0.019	0.032
16 – 23 February	2	3.93 ± 0.22	0.011	0.011	0.000	-0.015
24 – 29 February	3	4.36 ± 0.14	0.019	0.018	-0.001	-0.004
01 – 07 March	4	5.00 ± 0.14	0.014	0.015	0.000	0.006
08 – 14 March	5	5.11 ± 0.19	0.022	0.024	0.028	0.025
15 – 21 March	6	6.16 ± 0.28	0.026	0.018	0.025	0.020
22 – 28 March	7	6.45 ± 0.20	-0.007	0.016	0.028	0.029
29 March – 04 April	8	6.37 ± 0.46	0.002	0.027	0.030	0.004

Note. Weekly average temperature and relative growth rate (RGR) of the different strains and feed during the experiment. Data are presented as weekly means ± standard deviations.

The major differences between strains and feed types are summarized in Figures 1a and 1b representing measured mean weights of all replicate tanks: For strain 1 we observed mean constant growth for the entire experimental period for "AgloNorse® Extra" (high standard deviation occurred because larvae of one of the replicate tanks grew considerably less from day 40 onwards) and declining weights for "AgloNorse®" during the last two weeks. For strain 2, stagnant average weights were evident for both feed types until day 30, followed by continuous mean growth in the "AgloNorse®" groups. A halt in growth was observed in strain 2 for "AgloNorse® Extra" during the last week. For the entire experimental period, the measured mean fresh weight did not differ significantly between groups receiving different feed types in strain 2, however, it differed in strain 1 (data were transformed and tested using ANCOVA. Strain 1: n = 960, P = 0.004 and strain 2: n = 1440, P = 0.541). However, also in strain 1, fresh weights were similar in each of the first six weeks (Student's t-test, p-values between 0.672 and 0.985, n = 30 for "AgloNorse®" and n = 30 for "AgloNorse® Extra"). In strain 2 fresh weights were similar in each of the first seven weeks (Student's t-test and Mann-Whitney test, p-values between 0.031 and 0.887, n = 30 for "AgloNorse®" and n = 30 for "AgloNorse® Extra").

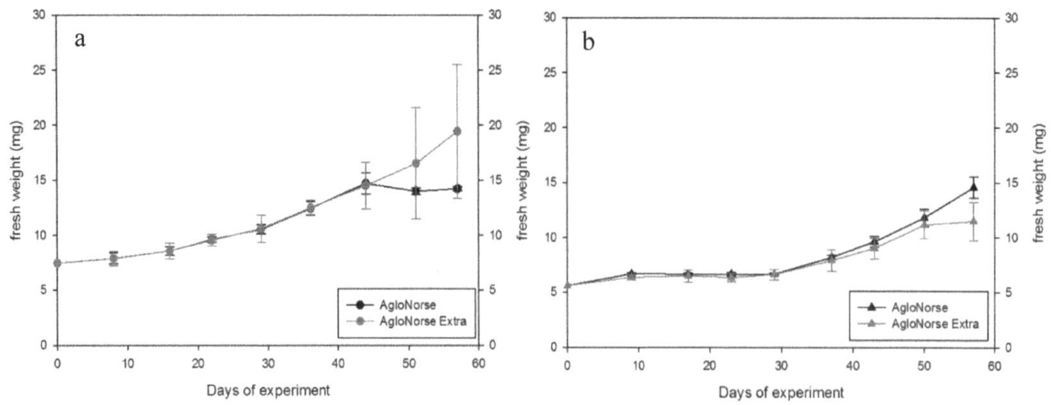

Figure 1. Growth curves

Note. Growth curves for different feed types and different fish strains including mean weights ± standard deviation of all replicate tanks: a) Strain 1: "AgloNorse®" and "Aglo Norse® Extra". b) Strain 2: "AgloNorse®" and "AgloNorse® Extra".

The final weight of strain 1 differed significantly between the two feed types (P =< 0.001, Mann-Whitney test on transformed data) and was 14.2 ± 0.2 mg (n = 60) for "AgloNorse®" and 19.4 ± 6.6 mg (n = 60) for "AgloNorse® Extra". Also the final weights of strain 2 differed significantly between the two feed types (P =< 0.001, Mann-Whitney test on transformed data). The final weight for "AgloNorse®" was 14.6 ± 4.2 mg (n = 90) and for "AgloNorse ®Extra" was 11.5 ± 3.8 mg (n = 90).

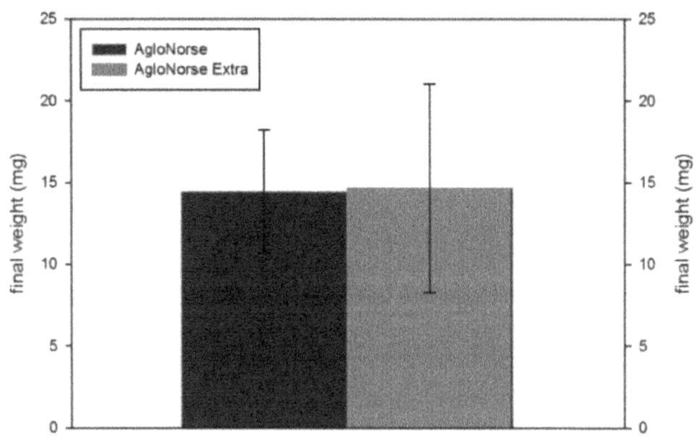

Figure 2. Comparison of feed types

Note. Comparison of the two types of feed. Bars illustrate mean ± standard deviation of final fresh weight (mg) of all larvae regardless to the strain.

For "AgloNorse®", there was no statistically significant difference (P = 0.772, Mann-Whitney test on transformed data, strain 1: n = 60 and strain 2: n = 90) between final weights of strain 1 and strain 2. However, for "AgloNorse® Extra", there was a statistically significant difference between final weights of strain 1 and strain 2 (P =< 0.001, student's t-test on transformed data, strain 1: n = 60 and strain 2: n = 90). Comparing the two types of feed, combining all fish without regard to the strain, no statistically significant difference was found in final weights (Figure 2, P = 0.283, Mann-Whitney test on transformed data, "AgloNorse®": n = 150 and "AgloNorse® Extra": n = 150).

In terms of relative growth rate, there was no statistically significant difference within strain 1 and strain 2 (Table 1), based on the different types of feed (ANCOVA: strain 1: P = 0.174, "AgloNorse®" n = 14 and for "AgloNorse® Extra" n = 14. Strain 2: P = 0.126, "AgloNorse®" n = 21 and for "AgloNorse® Extra" n = 21). Within the two feed types, there were also no statistically significant differences in relative growth rates between strain 1 and strain 2 (ANCOVA: "AgloNorse®": P = 0.480, strain 1 n = 14 and strain 2 n = 21. "AgloNorse® Extra": P = 0.076, strain 1 n = 14 and strain 2 n = 21).

Comparing the two types of feed in the relative growth rates, combining all fish without regard to strain, no statistically significant difference was found (ANCOVA: P = 0.645, "AgloNorse®" n = 35 and "AgloNorse® Extra" n = 35). Comparing the two strains in the relative growth rates, combining all fish without regard to feed type, no statistically significant difference was found (ANCOVA: P = 0.423, strain 1 n = 28 and strain 2 n = 42).

The survival rates were different for the two strains and partly also for groups receiving different feed types (Figure 3). In strain 1 the survival rates remained very high for six weeks for both groups and started to decrease around day 40. The group receiving "AgloNorse®" showed strongly declining survival with lower variability as both replicate tanks were affected with high mortality. The group receiving "AgloNorse® Extra" showed high variability as only in one replicate tank elevated mortality was observed. This resulted in a final average survival rate of 47.5%. There was a statistically significant difference in average survival between groups (ANCOVA: P = 0.003, n = 120 for both feed types). For strain 2 the survival rate started to decrease much earlier, at day 30 of the experiment. There was noticeable increase of mortality at this time in all replicate tanks. However, unlike strain 1, the decline in survival stopped around day 40 and remained constant for about ten days before the two groups diverged. We observed a second decline in survival in one group ("AgloNorse® Extra") but not in the other group ("AgloNorse®"). The average final survival rate of strain 2 was 45.58 % for "AgloNorse®" and 37.32 % for "AgloNorse® Extra", with relatively low variability. There was no statistically significant difference in survival of strain 2 (ANCOVA: P = 0.277, n = 190 for both feed types) between the two types of feed.

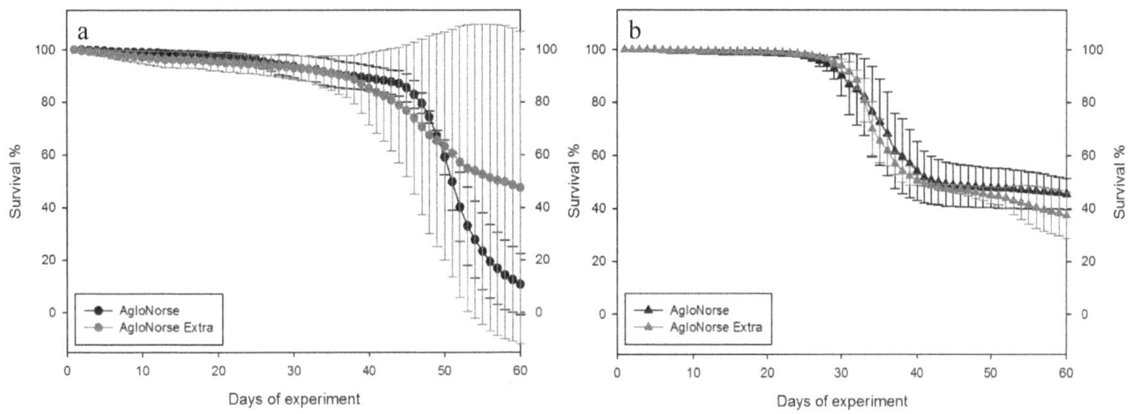

Figure 3. Survival rate

Note. a) Survival rate (%) of strain 1. b) Survival rate (%) of strain 2. Data are presented as daily average of survival rate + standard deviation.

Comparing the two different feed types without regard to the strain, there was no statistically significant difference (ANCOVA: P = 0.123, n = 300 for both feed types). However comparing the strains without regard to the different types of feed, there was a statistically significant difference (ANCOVA: P = 0.001, n = 240 for strain 1 and n = 360 for strain 2), i.e. strain 1 showing lower survival.

At the beginning, strain 1 accepted the offered feed faster than strain 2. Towards the end of the experiment differences disappeared and reached almost 100% (Figure 4). There was a statistically significant difference between the strains (ANCOVA: P = 0.000; strain 1 n = 29 and strain 2 n = 48). But between the types of feed there was no statistically significant difference in strain 1 nor in strain 2 (ANCOVA: Strain 1: P = 0.699, "AgloNorse®" n = 16 and "AgloNorse® Extra" n = 16, Strain 2: P = 0.971, "AgloNorse®" n = 24 and "AgloNorse® Extra" n = 24).

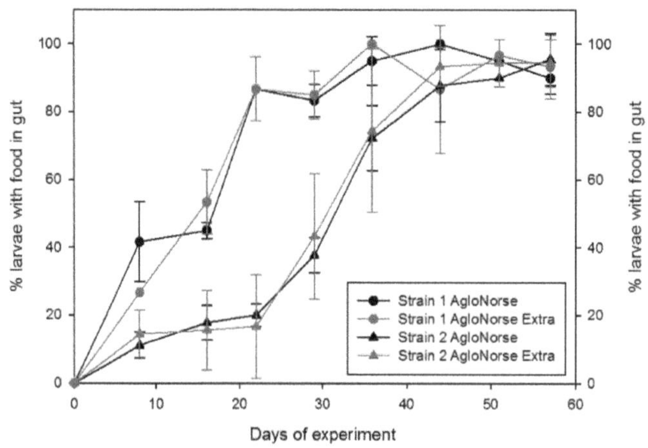

Figure 4. Feed acceptance

Note. Feed acceptance of strains. Data are represented as mean ± standard deviation.

4. Discussion

4.1 Identifying the Time Period for Valid Comparisons

Our results were influenced by an unidentified disease towards the end of the experiment. Therefore, the potential influence has to be assessed and separated from other factors to be investigated, i.e. strain and feed type.

For strain 1, the experiments were clearly influenced by disease during the last two weeks as seen in a complete halt of growth in the test group receiving feed "AgloNorse®" (in which all replicate tanks were affected, Figure 1a) and in the distinct increase of variance in the group receiving feed "AgloNorse® Extra" (in which one replicate tank was affected, Figure 1a). Up until day 45 the growth curves do not show any sign of disease influence, i.e. low variance, and the survival curves (Figure 3a) also show high survival of 80% until day 40, before plunging down. Therefore, we are confident to make valid comparisons in this strain during the first six weeks of the experiment.

For strain 2, we observed almost identical and low-variance growth curves until day 50 and diverging average weights only during the last week (Figure 1b). This corresponds to almost identical survival curves with noticeable differences appearing during the last ten days (Figure 3b). The group receiving "AgloNorse®" was not affected by the disease in none of the three replicate tanks and variances in weights and survival remained low (Figures 1b and 3b). For the group receiving "AgloNorse® Extra" one of three replicate tanks was affected which resulted in lower average weight and elevated variability compared to the first group of strain 2 (Figure 1b). For this strain we are convinced to make valid comparisons for the first 7 weeks of the experiment. Although the number of replicates of the performed experiment was relatively small, the variation among replicates was extremely low before disease outbreak. Therefore, the number of replicates used was considered sufficient.

4.2 Comparisons of Feed Types

During six weeks before disease outbreak, a steadily increasing growth curve with very low variance was registered in strain 1 and with both feed types (Figure 1a). The average weights in each of the six weeks were almost identical among different feed types and the relative growth rates were largely similar as well (Table 1). Therefore, it seems evident that feed type did not lead to any observable differences between groups of strain 1, neither in growth (Figure 1a) nor in survival (Figure 3a) or feed acceptance (Figure 4). Divergence of the two groups receiving different feed types was observed during the last two weeks of our experiment and was clearly associated with disease outbreak. This affected the two groups differently (all replicate tanks in one group vs. one replicate tank in the other group), however, we suspect this was unrelated to feed type.

For strain 2 we observed almost constant average weights over time for the first four weeks followed by steadily increasing weights for the following three weeks before disease outbreak. This pattern was found to be identical for both groups receiving different feed types and average weights were similar in the two groups in each of the seven weeks. Within that time period there was no indication of feed type leading to differences in growth, survival or feed acceptance in strain 2. Divergence of the two groups with regard to growth (Figure 1b, Table 1) and survival (Figure 3b) during the last week were again associated with differential affection of the two groups by disease.

Even if the influence of disease is ignored, no significant difference in final weights of fish receiving the different feed types was found when fish of both strains were pooled (Figure 2). To our knowledge, no previous study has investigated dedicated cold-water feed types and compared their performance to regular feed types. Most previous studies on whitefish compared feed types of vastly varying ingredients in experimental feeds (Luczynski et al., 1986; Dabrowski & Poczyczynski, 1988; Harris & Hulsman, 1991) or commercial microdiets of different manufacturers (Koskela & Eskelinen, 1992, Rojas-Beltran et al., 1992, Rissanen, 1995). In those cases, the observed differences in performance of whitefish larvae might be of little surprise, however, our results provide evidence that one dedicated cold-water microdiet ("AgloNorse® Extra") does not provide advantages compared to the regular product ("AgloNorse®") of the same manufacturer. In that respect, whitefish larvae seem to react differently to the two feed variants compared to cod (pers. € communication).

4.3 Comparison of Strains

For the time period before disease outbreak we observed considerable differences between strains with regard to growth and survival. These differences most probably have their common source in the differences regarding feed acceptance which was clearly delayed in strain 2 compared to strain 1 (similarly in both feed types, Figure 4). The feed acceptance in strain 1 increased quickly for both feed types which lead to increasing weights right from the start of the experiment and to low mortalities. For strain 2 the feed acceptance rates remained low for the first three weeks. Individuals which did not feed and lost weight outbalanced those feeding and gaining weight and therefore, average weights remained almost constant. Individuals refusing to accept the dry feed could probably live on their yolk reserves until day 25 after which survival started to drop in strain 2. Concurrently the rate of feed acceptance apparently rose as "refusers" were dying off. Similarly the average weight started to increase and average growth rates attained the levels of strain 1. This process was nearly finished around day 40 when survival remained constant and feed acceptance rates as well as growth rates had reached the levels of strain 1. The following week strain 1 was partly influenced by the disease and strain 2 was partly influenced one week later making further comparisons difficult. However, when only unaffected tanks are taken into consideration, we did not find noticeable differences among strains. Strain 1 was hit harder and earlier by the disease compared to strain 2 resulting in lower survival and partly losing its initial growth advantage to end up in similar final weights, at least in the "AgloNorse®" group.

Differences in strains also found in previous studies (Rösch & Dabrowski, 1986; Rösch, 1988, 1992). Similarly much of the differences were found to originate from differential feed acceptance. It remains unclear why such differences exist and if domestication might increase feed acceptance in a certain strain. Our results indicate that this might be the case because strain 1 is regarded more domesticated, i.e. adapted to hatchery conditions, compared to strain 2.

4.4 Agglomerated Microdiets Compared to Previous Ones

It seems that agglomerated microdiets can reach at least similar performance as previously used ones (Lucynski et al., 1986; Rösch & Dabrowski, 1986) even under cold water conditions. Survival of around 50% has been reported (Lucynski et al., 1986; Rösch & Dabrowski, 1986) in studies using "old generation" microdiets, however, they were performed under more favorable, higher water temperatures (> 8 °C). As we know from previous studies, food particle must be in an adequate physical structure so that they are readily digestible for whitefish larvae and nutritious enough to provide sufficient energy for growth (Holt et al., 2011; Langdon & Barrow, 2011). Food acceptance is strongly influenced by the aroma of the food (Appelbaum, 1980; Dabrowski et al., 1984) and therefore "AgloNorse®" as well as "AgloNorse® Extra" are suspected to allow the larvae to smell the food particles and to identify them as a tempting meal (Kolkovski et al., 1997; Yúfera et al., 2002). Because the exact composition of the tested feed is a company secret, no detailed information can be given due to the missing data.

4.5 Water Conditions

Cold-water conditions (< 8 °C) seem to be sufficient for the very early period of larval development (2-3 weeks) of whitefish but appear to exert negative influences during later development. If disease outbreak after six, respectively seven weeks was facilitated by cold-water conditions remains unclear, as disease outbreaks have been reported under higher temperatures as well (Dabrowski et al., 1984; Lucynski et al., 1986; Dabrowski & Poczyczynski, 1988). But it is noticeable that disease outbreak was recognized in those studies exclusively at higher water temperatures (> 8 °C). A shift of optimal temperatures during larval/juvenile development is known for many temperate species, e.g. nose carp (Keckeis et al., 1999) and is probably highly pronounced in whitefish.

Anyhow, a test of agglomerated microdiets developed for marine fish larvae under higher water temperature regimes would be the next logical step to be performed using whitefish.

Acknowledgements

This study was funded by the Austrian Research Promotion Agency (FFG), under the "Bridge" program project number 827556. Sincere thanks are given to the Federal Austrian Forest Agency (Österreichische Bundesforste AG) for their cooperation, financial support and the supply with fish from lake Hallstatt. Franz Lahnsteiner and Sabine Wanzenböck provided helpful comments to improve an earlier version of this manuscript, which is appreciated very much.

References

Appelbaum, S. (1980). Versuche zur Geschmacksperzeption einiger Süsswasserfische im larvalen und adulten Stadium. *Arch. Fisch. Wiss., 31*, 105-114.

Dabrowski, K., & Poczyczynski, P. (1988). Laboratory experiment and mass rearing of coregonid fish fed exclusively dry diets. *Aquaculture, 69*, 307-316. http://dx.doi.org/10.1016/0044-8486(88)90338-9

Dabrowski, K., Charlon, N., Bergot, P., & Kaushik, S. (1984). Rearing of coregonid (*Coregonus schinzi palea* Cuv. et Val.) larvae using dry and live food, I. Preliminary data. *Aquaculture, 41*, 11-20. http://dx.doi.org/10.1016/0044-8486(84)90385-5

Eckmann, R., & Kausch, H. (1976). Fütterung von Felchenbrut mit künstlichen Futtermitteln. *In der Fischwirt, 26*, 9-11.

Goddard, S. (1996). *Feed management in intensive aquaculture*. Capman & Hall (Hrsg.). http://dx.doi.org/10.1007/978-1-4613-1173-7

Harris, K. C., & Hulsman, P. F. (1991). Intensive culture of lake whitefish (*Coregonus clupeaformis*) from larvae to yearling size using dry feeds. *Aquaculture, 96*, 255-268. http://dx.doi.org/10.1016/0044-8486(91)90156-2

Holt, G. J., Webb, K. A., & Rust, M. B. (2011). Microparticulate diets: testing and evaluating success. In G. Joan (Ed.), *Holt Larval Fish Nutrition* (pp. 353-372). John Wiley & Sons. http://dx.doi.org/10.1002/9780470959862.ch13

Keckeis, H., Bauer-Nemeschkal, E., Menshutkin, V., Nemeschkal, H. L., & Kamler, E. (1999). Effects of female attributes and egg properties on offspring viability in a rheophilic cyprinid, *Chondrostoma nasus*. *Canadian Journal of Fisheries and Aquatic Sciences, 57*, 789-796. http://dx.doi.org/10.1139/cjfas-57-4-789

Kolkovski, S., Koven, B., & Tandler, A. (1997). The mode of action of *Artemia* in enhancing utilization of microdiet by gilthead seabream *Sparus aurata* larvae. *Aquaculture, 155*, 193-205. http://dx.doi.org/10.1016/s0044-8486(97)00117-8

Koskela, J., & Eskelinen, U. (1992). Growth of larval European whitefish (*Coregonus lavaretus*) at different temperatures. *Polskie Archiwum Hydrobiologi, 39*, 677-682. http://dx.doi.org/10.1016/s0044-8486(03)00275-8

Kottelat, M., & Freyhof, J. (2007). *Handbook of European freshwater fishes* (pp. 1-646). Kottelat, Cornol, Switzerland and Freyhof, Berlin, Germany, Publications Kottelat. http://dx.doi.org/10.1643/ot-08-098a.1

Kummer, H., & De Verga, V. (1992). Vergleich unterschiedlicher Futtermittel bei der Aufzucht von Coregonen- und Seesaiblingsbrut in künstlichen Systemen. *Österreichs Fischerei, 45*, 267-276.

Langdon, C., & Barrow, R. (2011). Microparticulate diets: Technology. In G. Joan (Ed.), *Holt Larval Fish Nutrition* (pp. 305-351). John Wiley & Sons. http://dx.doi.org/10.1002/9780470959862.ch12

Luczynski, M., Majkowski, P., Bardega, R., & Dabrowski, K. (1986). Rearing of larvae of four coregonid species using dry and live food. *Aquaculture, 56*, 179-185. http://dx.doi.org/10.1016/0044-8486(86)90333-9

Mahmoudzadeh, H., Ahmadi, M. R., & Shamsaei, M. (2009). Comparison of rotifer *Brachionus plicatilis* as a choice of live feed with dry feed in rearing *Coregonus lavaretus* fry. *Aquaculture Nutrition, 15*, 129-134. http://dx.doi.org/10.1111/j.1365-2095.2008.00575.x

Rissanen, I. (1995). The favourable feeding rate of whitefish (*Coregonus lavaretus* L. S. L.) in first feeding at 11 °C temperature. *Arch. Hydrobiol. Spec. Issues Advanc. Limnol., 46*, 325-330. http://dx.doi.org/10.1016/0044-8486(82)90075-8

Rojas Beltran, R., Champigneulle, A., Gillet, C., & Le Rouilly, N. (1992). Influence of egg source, initial characteristics, and rearing conditions on the growth, survival and development of *Coregonus lavaretus L.* larvae initially fed on a dry diet. *Polskie Archiwum Hydrobiologi, 39*, 683-691.

Rösch, R. (1988). Mass rearing of *Coregonus lavaretus* on a dry diet. *Finnish Fisheries Research, 9*, 345-351.

Rösch, R. (1992). Food intake and growth of larvae of *Coregonus lavaretus*: Effect of diet and light. *Polskie Archiwum Hydrobiologi, 39*, 671-676.

Rösch, R., & Dabrowski, K. (1986). Tests of artificial food for larvae of *Coregonus lavaretus* from Lake Constance. *Arch. Hydrobiol. Beih. Ergebn. Limnol., 22*, 273-282.

Siikavuopio, S., Knudsen, R., Amundsen, P., & Stainar Saether, B. (2012). Growth performance of European whitefish (*Coregonus lavaretus* (*L.*)) under a constant light and temperature regime. *Aquaculture Research, 43*, 1592-1598. http://dx.doi.org/10.1111/j.1365-2109.2011.02963.x

Susta, J. (1888). Die Ernährung des Karpfens und seiner Teichgenossen. *In neue Grundlagen der Teichwirtschaft* (p. 252). Herrcke & Lebeling.

Winkler, K., Pamminger-Lahnsteiner, B., Wanzenböck, J., & Weiss, S. (2011). Hybridization and restricted gene flow between native and introduced stocks of Alpine whitefish (*Coregonus sp.*) across multiple environments. *Molecular Ecology, 20*, 456-472. http://dx.doi.org/10.1111/j.1365-294x.2010.04961.x

Yúfera, M., Kolkovski, S., Fernández-Diaz, D., et al. (2002). Free amino acid leaching from a protein-walled microencapsulated diet for fish larvae. *Aquaculture, 214*, 273-287. http://dx.doi.org/10.1016/s0044-8486(01)00902-4

18

Crop Evapotranspiration (ET) Estimation Models: A Review and Discussion of the Applicability and Limitations of ET Methods

Abhinaya Subedi[1] & José L. Chávez[1]

[1] Civil and Environmental Engineering Department, Colorado State University, Fort Collins, Colorado, USA

Correspondence: José L. Chávez, Civil and Environmental Engineering Department, Colorado State University, 1372 Campus Delivery, Fort Collins, Colorado 80523-1372, USA. E-mail: jose.chavez@colostate.edu

Abstract

This is a review paper on existing methodologies to calculate crop evapotranspiration (ET_c). We have attempted to present all the important ET estimation procedures to date starting from the simple empirical Blaney Criddle method to the complex Shuttleworth model. The common approach to calculate ET_c is to estimate a reference crop ET rate (ET_{ref}) using weather variables from nearby weather station, and multiplying it by an appropriate crop coefficient (K_c). Recently, there have been attempts to calculate actual crop ET (ET_a) directly without using K_c. The latter method is still in the developmental phase. This study reviews the existing literature on ET estimation and identifies research needs in the current methods and technology. The extension of the Shuttleworth model for hourly time step and the validity of the Irmak and Mutibwaa model at field level for various crops would be a good milestone for the one step ET estimation. Furthermore, there are indications that the development of a new variable canopy surface resistance (r_c) model, which can be applicable for different crops at different climatic conditions, would be a good contribution in this field.

Keywords: ET_c models applicability, crop evapotranspiration (ET_c), ET_c models review, ET_c models limitations

1. Introduction

Water is the basis of life. In the modern world, the demand of water is increasing because of the growing population as well as the increased urbanization and industrialization. As a result, water for agriculture is becoming limited. For this reason, accurate estimation of crop water requirement is very important. The problem of over irrigation or under irrigation will be minimized if we are able to accurately estimate crop water requirement or crop evapotranspiration (ET_c, mm/d). The schematic diagram of the ET_c process is shown in Figure 1. Various methods have been developed so far to estimate the ET_c. John Dalton (1766-1844) was the pioneer in developing an equation for evaporation from large water bodies, such as lakes and reservoirs. In his equation, the evaporation rate was calculated as the product of the vapor pressure deficit and a factor "K" which is dependent on the wind speed. Since then, various ET methods have been developed, which are described in this article.

This paper reviews various ET_c estimation models that have been developed to date. Among these models, the Penman Monteith (PM) equation is found to be more consistent over a wide range of climatic conditions (Allen et al., 2005). The most challenging part in the PM equation is to calculate the canopy surface resistance. The FAO 56 PM equation (Allen et al., 1998) and ASCE-EWRI (Environmental and Water Resources Institute of American Society of Civil Engineers) 2005 Standardized PM equation (ASCE-EWRI, 2005) are based on the fixed canopy surface resistance approach. These two methods calculate the reference crop ET, which along with the crop coefficient (K_c) is used to calculate ET_c or the actual crop ET (or ET_a). Recently, some researchers have pointed out flaws in this technique of estimating ET_a, so there have been attempts to calculate ET_a directly using variable surface resistance values, without requiring crop coefficients. Katerji and Perrier (1983), Todorovic (1999), and Shuttleworth (2006) are notable researchers in variable surface resistance approach, which are described later in the manuscript. The subsequent sections describe different ET estimation techniques in a chronological order.

2. Evolution of Different Methods

2.1 Blaney-Criddle Method

The Blaney-Criddle method was first developed in 1942. It is an empirical equation and very simple to use. They developed a simple mathematical model as given by Equation (1), (Blaney & Criddle, 1962).

$$u = kf \tag{1}$$

$$U = \sum kf = KF \tag{2}$$

where,

u = monthly consumptive use, in inches;

$f = T_F \times p / 100$ is the monthly consumptive use factor;

T_F = mean monthly temperature, in degrees Fahrenheit (°F);

p = monthly percentage of daytime hours of the year;

k = empirical consumptive use crop coefficient for monthly period;

U = seasonal consumptive use (or evapotranspiration), in inches;

F = sum of the monthly consumptive use factors for the period (sum of the products of mean monthly temperature and monthly percentage of daytime hours of the year);

K = empirical consumptive use crop coefficient for irrigation season or growing period.

In metric units,

$$u = kp\left(\frac{45.7T_a + 813}{100}\right) \tag{3}$$

where,

u = monthly consumptive use, in millimeters;

T_a = mean monthly temperature, in degrees Centigrade/Celsius (°C).

Although the method was originally developed to compute ET on a monthly basis, it can be modified to estimate daily values of ET with mean daily temperature (ASCE, 1990). As temperature methods tend to underestimate ET in arid regions while overestimating ET in humid regions, local calibration of the empirical coefficients is required to produce reliable estimates of ET (ASCE, 1990). The advantage of this method is the simplicity and disadvantage is that it underestimates ET grossly compared to the measured ET values (Sammis et al., 2011).

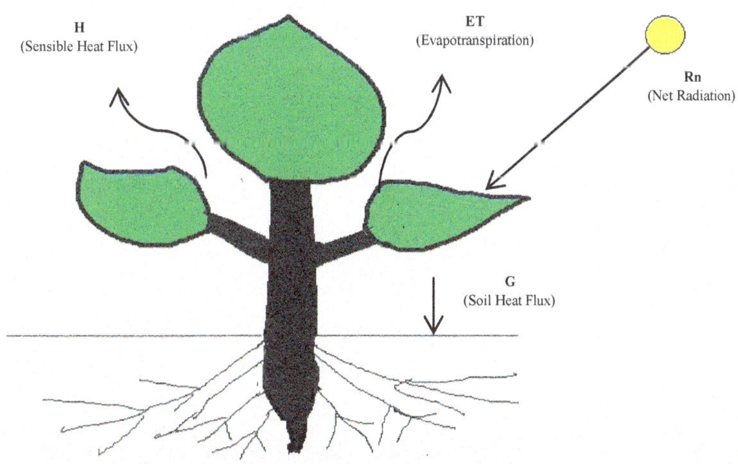

Figure 1. Schematic diagram of the ET process from a crop canopy

2.2 Thornthwaite Method

In 1948, Thornthwaite and Penman both developed potential evapotranspiration equation independently. Potential ET here refers to the maximum ET that can occur from a given crop surface. Penman's equation was more mechanistic while Thornthwaite's equation was more empirical. The Thornthwaite (1948) equation is simpler than Penman's equation because the method requires less climatic data. Thornthwaite's equation is as follows:

$$PET = 16(10T_a / I)^a \quad (4)$$

where,

PET = potential evapotranspiration rate, in mm per month;

T_a = mean monthly air temperature, in degrees Celsius (°C);

I = summation of the 12 monthly heat index i, where i = $(T_a / 5)^{1.514}$;

a = an empirical coefficient, which is calculated using the following equation:

$$a = 0.675*10^{-6} I^3 - 77.1*10^{-6} I^2 + 0.01792I + 0.49239 \quad (5)$$

This method is not based on strong mathematical and physical principles as it is purely empirical. However, as it is simple to use and gives acceptable result, in many parts of the world the method is still used to estimate irrigation water requirement.

Kumar et al. (1987) compared the Thornthwaite and Penman methods, in India, to calculate potential ET. They found that Penman's method seemed more realistic in estimating the mean annual potential evapotranspiration distribution over India. They also reported that Penman's potential evapotranspiration estimates were higher than Thornthwaite's estimates during the winter and pre-monsoon months and lower during the monsoon months, at most of the Indian stations. Pereira and De Camargo (1989) concluded that Thornthwaite's method was not appropriate for estimating ET in advective condition; however, they indicated that the method could be used for irrigation scheduling purposes when the fetch requirement is met. Bautista et al. (2009) concluded that Thornthwaite's method worked very well during the rainy months in both of their research sites; however, for drier months the use of Thornthwaite's method was not recommended without the adjustment of its coefficient "16".

2.3 Hargreaves Equation

Hargreaves (1975) developed an equation for estimating ET which doesn't require wind speed data. His equation is as follows:

$$ET_0 = 0.0075 R_s T_F \quad (6)$$

where,

ET_0 = potential ET for a grass reference surface in the same units as R_s;

R_s = global solar radiation at the surface in equivalent water evaporation, usually mm of evaporation;

T_F = mean air temperature in degrees Fahrenheit (°F).

For degrees Celsius, the equation is modified as:

$$ET_0 = 0.0135 R_s (T_a + 17.8) \quad (7)$$

Hargreaves (1977) developed the equation for R_s as shown below. R_s units as R_a units:

$$R_s = 0.075 R_a S^{0.50} \quad (8)$$

Hargreaves (1977) developed the equation for S to be applicable for Central America as:

$$S = 12.5(100 - RH)^{0.50} \quad (9)$$

where, RH = mean monthly relative humidity, %.

Hargreaves and Samani (1982) developed an equation to determine R_s from extraterrestrial radiation (R_a) and the air temperature range (TD):

$$R_s = K_{RS} R_a TD^{0.50} \quad (10)$$

where, R_s and R_a are in the same units as indicated abvoe, K_{RS} is a calibration coefficient and TD is mean maximum minus mean minimum daily temperatures, in degree Celsius, for a given period (generally one week).

Later on, Hargreaves and Samani (1985) obtained the following equation for ET_0.

$$ET_0 = 0.0022 R_a (T_a + 17.8) TD^{0.50} \quad (11)$$

For months of peak demand, Hargreaves and Samani (1985) recommended that the coefficient (0.0022) be increased to 0.0023. ASCE-EWRI (2005) and Allen et al. (1998) recommended using Equation (11) when solar radiation, relative humidity and/or wind speed data are missing.

Hargreaves equation is also empirical, simple, and easy to use. Bautista et al. (2009), in a study, compared the results of the Hargreaves equation (Equation 11) with those of the FAO 56 PM equation (Allen et al., 1998), considering the latter equation as the standard method (reference). They found that the Hargreaves method compared well with the FAO 56 PM method with a resulting coefficient index (of determination) of 0.82. However, the coefficient index improved to 0.91 after adjusting Hargreaves equation coefficient "0.0022" in Equation (11) from 0.0021 to 0.0024 (based on seasons) for tropical subhumid climate sites and from 0.0022 to 0.0026 for semiarid climate sites. Ravazzani et al. (2012) also compared the Hargreaves and Samani (1985) equation (HS) to FAO 56 PM equation for daily time steps in alpine river basins and found that the HS equation didn't perform well. The HS method showed overestimation of ET_0 at lower elevation sites and underestimation at higher elevation sites. However, after using a correction factor, they found that the HS equation was in very good agreement with the FAO 56 PM equation.

2.4 Christiansen Method

Christiansen (1968) developed a simple method to estimate pan evaporation and crop evapotranspiration. According to Christiansen, the reasons for using pan evaporation data were: they were more consistent, already considerable work had been done to relate pan evaporation data with crop consumptive use and the pan evaporation data were readily available. The mathematical model that he developed was as follows:

$$E = K\,R_a\,C \tag{12}$$

where E is used in a general sense to apply to evaporation or evapotranspiration, K is a dimensionless constant developed empirically from data analysis, and C is a dimensionless coefficient related to climatic parameters, and R_a is the extraterrestrial radiation, expressed as equivalent depth of evaporation in the same units as E. The coefficient C is expressed as the product of any number of subcoefficients that are functions of specific climatic parameters that are found to have a significant effect on the evaporation or evapotranspiration (Christiansen, 1968). Mathematically,

$$C = C_T C_W C_H C_S C_E \tag{13}$$

where, C_T, C_W, C_H, C_S and C_E represent the coefficients for air temperature, wind speed, relative humidity, sunshine percentage and elevation, respectively. The value of K was adjusted so that all coefficients were equal to unity for standard and approximate mean values of the parameter they represent (Christiansen, 1968). Christiansen (1968) described in detail how to calculate the different parameters in Equation (13) using Tables in his article.

This method is purely empirical as it is not based on any physical equation. This method can somehow accurately estimate ET on a monthly basis. However, this method cannot be used to calculate ET_c on a daily basis or for shorter time steps. Wai et al. (2004) evaluated the performance of the Christiansen method and Penman methods with respect to the measured pan evaporation in Malaysia. They found that the Penman method ET results compared better than the Christiansen's method ET results to measured values (Pan) of potential evapotranspiration.

2.5 Penman Related Equations

2.5.1 Original Penman Equation

Penman (1948) developed a mechanistic approach to calculate ET. He used a combination approach by combining the surface energy balance equation with an aerodynamic equation. Several ET estimation models, for example, FAO 56 PM equation, ASCE-EWRI Standardized PM equation, CIMIS Penman method (as described in Temesgen et al., 2005) have been based on Penman equation. The original Penman equation is as follows:

$$ET = \frac{\left(\Delta(R_n - G) + k_w(e_s - e_a) f(u)\gamma\right)}{\lambda(\Delta + \gamma)} \tag{14}$$

$$f(u) = a_w + b_w\,u_2; \tag{15}$$

where,

f(u) = wind speed function

k_w = unit coefficient (6.43 for ET in mm/d or 0.268 for ET in mm/h).

Penman (1948) recommended the value of a_w and b_w as 1.0 and 0.537, respectively, for clipped grass. Doorenbos and Pruitt (1977) in the FAO 24 paper recommended a constant of 6.61 in place of 6.43. They also recommended the values of a_w and b_w as 0.75 and 0.993 for full cover alfalfa. Wright (personal communication, 1987) as cited by Allen et al. (1989) derived an improved form of the Wright (1982) variable wind function by using the normal probability density function equation to approximate the change in a_w and b_w coefficients as a function of the Julian day (J) or day of the year for an alfalfa reference surface at Kimberly, Idaho. The equations for a_w and b_w are:

$$a_w = 0.4 + 1.4 \exp\left[-\left(\frac{[J-173]}{58}\right)^2\right] \tag{16}$$

$$b_w = \left\{0.007 + 0.004 \exp\left[-\left(\frac{[J-243]}{80}\right)^2\right]\right\}(86.4) \tag{17}$$

Equations (14) and (15), with a_w and b_w calculated with Equations (16) and (17) were termed 1982 Kimberly Penman equation by Allen et al. (1989).

Sun and Song (2008) evaluated the performance of the original Penman equation with measured ET values using the eddy covariance method for a marshland in Northeast China. They found that the Penman model overpredicted the mean measured ET for the growing season by 35%. Yoder et al. (2005) compared ET estimations from eight different equations with measured lysimeter ET. Yoder et al. (2005) found that the FAO 56 PM equation performed better followed by the original Penman equation.

2.5.2 CIMIS Penman Method

The CIMIS Penman method calculates grass reference ET (ET_0) using the Penman combination equation, as modified by Pruitt and Doorenbos (1977), with a wind function that was developed at the University of California, Davis (Ventura et al., 1999 and Temesgen et al., 2005). The CIMIS Penman method uses $a_w = 0.29$ and $b_w = 0.53$ for $R_n > 0$ and $a_w = 1.14$ and $b_w = 0.40$ for $R_n \leq 0$. These coefficients are applied hourly using Equation (14) where ET_0 is in mm/h, R_n is in MJ/m^2/h and $k_w = 0.268$ (ASCE-EWRI, 2005). Temesgen et al. (2005) showed that the CIMIS Penman method correlated well with the FAO 56 PM equation for daily time step and with the ASCE Standardized equation (ASCE EWRI, 2005) for both daily and hourly time steps for 37 different studied sites in the state of California, USA. The limitation of this method is that this method may not be applicable in different climatic conditions as the coefficients were mainly developed for the climatic condition of California.

2.5.3 Penman-Monteith Equation

Monteith (1965) introduced some crop resistance terms in the original Penman equation and the equation later became the well-known "Penman-Monteith" (PM) ET equation. This equation is physically based and its robustness has been demonstrated as it does not require local calibrations, provided there are complete input data (Temesgen et al., 2005; Allen et al., 1999). This equation does not have any wind function; rather it has aerodynamic and surface resistance terms. The wind function in the Penman equation is calculated empirically whereas the aerodynamic and surface resistance terms are calculated using physically based and semi-empirical equations, respectively. Aerodynamic resistance (r_a) is the resistance to molecular and turbulent diffusion of water vapor between leaf surfaces and the air above the canopy at a reference height (Robins, 1974). Surface resistance (r_c) is the resistance to the diffusion of water vapor within the evaporating surface (Monteith et al., 1965). The popular reference ET equations like the FAO 56 PM equation and the ASCE Standardized Reference PM ET equations are also based on the Penman Monteith equation. Equation (18) is the so called Penman-Monteith (1965) equation:

$$\lambda E = \frac{\left(\Delta(R_n - G) + \rho C_p (e_s - e_a)/r_a\right)}{\Delta + \gamma(1 + r_c/r_a)} \tag{18}$$

In the PM equation, all other parameters except r_c are relatively straightforward to calculate. A procedure has been developed to calculate r_c for grass and alfalfa surfaces. For this reason, to calculate the actual crop ET, the procedure is to first calculate the reference crop ET considering the grass or alfalfa as the reference crop surface and then multiplying the reference ET by the an appropriate crop coefficients. Direct use of the PM equation (Equation 18) to calculate actual crop ET (for any crop type_ is very rare in practice, although some researchers have tried this recently, which will be discussed in subsequent paragraphs.

2.6 Priestley Taylor Method

Priestley and Taylor (1972) developed a semi-empirical equation to calculate potential evaporation (λE or ET), which is applicable for partial equilibrium condition. Their equation is as follows:

$$\lambda E = \alpha \frac{\Delta(R_n - G)}{\Delta + \gamma} \tag{19}$$

where α is a variable that can range from 1.15 to 1.50 depending on the surface type, climate and season. For water surfaces under condition of minimal advection, Priestley and Taylor (1972) approximated the value of α as 1.26. The value of α will be different for different crops and open water bodies. Researchers are still working on finding appropriate value of α for different surfaces. Hobbins et al. (2001) found the value of α as 1.3177 for vegetation while using a calibration subset of 92 basins. This method is more suitable to find the ET rate on a large scale which is more applicable in hydrology. The disadvantage of this method is that it is not applicable in advective condition. This method is simpler to use than the PM equation as it has less parameters and variables.

2.7 Fixed Surface Resistance Approach

Allen et al. (1998) developed guidelines for computing crop water requirements or ET_c, in the FAO 56 PM paper. They recommended using the PM equation to calculate reference grass ET based on surface canopy resistance and aerodynamic resistance. They also tabulated the crop coefficient (K_c) values for the initial, mid and end stages for various crops based on previous researchers' findings. The FAO 56 PM equation was modified mainly to calculate daily crop ET; however, the authors claimed that it can also be used to calculate hourly crop ET if the needed input hourly weather data are available. The FAO 56 PM method assumes a fixed surface resistance of 70 s/m for the entire day or 24 hours. The fixed surface resistance value for 24 hours has been used for the daily or hourly FAO 56 PM equation.

In many parts of the world, including the US, alfalfa is used as another reference crop instead of clipped grass. Alfalfa can tolerate harsh weather conditions compared to the grass surface cover. Keeping this in mind, in 1999, the Irrigation Association (IA) requested the Evapotranspiration in Irrigation and Hydrology task Committee of the American Society of Civil Engineers (ASCE)-Environmental and Water Resources Institute (EWRI) to establish and define a benchmark reference evapotranspiration equation. Later, the ASCE EWRI ET task committee, in 2005, developed a reference evapotranspiration equation which is applicable for both tall (alfalfa) and short (grass) reference crops. As part of the standardization, the ASCE Penman-Monteith (ASCE-PM) equation and associated equations for calculating aerodynamic and bulk surface resistance have been combined and condensed into a single equation that is applicable to both surfaces (ASCE-EWRI, 2005). For the ASCE standardized PM equation, there is one fixed surface resistance for daytime and another fixed surface resistance for nighttime for each reference crop, hence this method is an improved version over the FAO 56 PM (Allen et al., 1998) equation.

2.7.1 FAO 56 Penman-Monteith Equation

The FAO 56 PM equation was based on the Penman-Monteith equation. The FAO 56 PM method defines the reference crop as a hypothetical crop with an assumed height of 0.12 m having a surface resistance of 70 s/m and an albedo of 0.23, closely resembling the evaporation of an extensive surface of green grass of uniform height, actively growing and adequately watered. Equation (18) can be approximated to Equation (20) after using the aerodynamic and surface resistance equations, which is the FAO 56 PM equation.

$$ET_0 = \frac{0.408\Delta\left(R_n - G\right) + \left(\dfrac{\gamma 900}{T_a + 273}\right) u_2 \left(e_s - e_a\right)}{\left(\Delta + \gamma(1 + 0.34u_2\right)} \tag{20}$$

where,

ET_0 = grass reference ET (mm/d);

R_n = net radiation at the crop surface ($MJ/m^2/d$);

G = soil heat flux density ($MJ/m^2/d$).

The equations used for aerodynamic and surface resistances while deriving FAO 56 PM equation are as follows:

$$r_a = \frac{\ln\left(\dfrac{z_m - d}{z_{om}}\right) \ln\left(\dfrac{z_h - d}{z_{oh}}\right)}{k^2 U_z} \tag{21}$$

where r_a is the aerodynamic resistance (s/m) for neutral atmospheric conditions, z_m is height of wind speed measurements (m), z_h is height of humidity measurements (m), d is zero plane displacement height (m) = $0.67h_c$, h_c is the crop height (m), $z_{om} = 0.123\ h_c$ is the roughness length governing transfer of momentum (m), $z_{oh} = 0.1\ z_{om}$ is the roughness length governing transfer of heat and water vapor (m), (Allen et al., 1998), k is von-Karman's constant (taken as 0.41), and u_z is wind speed at height z (m/s).

$$r_c = \frac{r_l}{0.5LAI} \quad (22)$$

where r_c is the canopy resistance (s m^{-1}), r_l is the daily average stomatal resistance (which is assumed as 100 s m^{-1} for clipped grass and full cover alfalfa), and LAI is leaf area index (m^2/m^2).

Lopez-Urrea et al. (2006) found that the FAO 56 PM equation performed better under semiarid climatic conditions of Albacete, Spain, as it agreed well with measured lysimeter ET compared to other versions of Penman equations and Hargreaves and Samani (1985) equation.

2.7.2 ASCE-EWRI Standardized Penman Monteith Evapotranspiration Equation

The ASCE Standardized Reference Evapotranspiration Equation (ASCE EWRI, 2005) is based on the Penman-Monteith equation, with some simplification and standardization on the aerodynamic and surface resistances. This equation is applicable for both tall (alfalfa) and short (grass) reference surfaces. A grass reference crop is defined as an extensive, uniform surface of dense, actively growing, cool-season grass with a height of 0.12 m, and not short of soil water; whereas a full cover alfalfa reference crop is defined as an extensive, uniform surface of dense, actively growing alfalfa with a height of 0.50 m, and not short of soil water (ASCE-EWRI, 2005). The equation is as follows:

$$ETsz = \frac{0.408\Delta(R_n - G) + \gamma C_n u_2 \frac{e_s - e_a}{T_a + 273}}{\Delta + \gamma(1 + C_d U_2)} \quad (23)$$

where,

ET_{sz} = standardized reference crop evapotranspiration for short crop (grass) (ET_{os}) or tall crop (alfalfa) (ET_{rs}) surfaces (mm/d for daily time steps or mm/h for hourly time steps);

R_n = calculated net radiation at the crop surface (MJ/m^2/d for daily time steps or MJ/m^2/h for hourly time steps);

G = soil heat flux density at the soil surface (MJ/m^2/d for daily time steps or MJ/m^2/h for hourly time steps).

The values for C_n for the short and tall references are 900 and 1600, respectively, for the daily time steps whereas the C_n are 37 and 66 for hourly time steps for both crops, respectively. Similarly the values for C_d for the short and tall reference crops are 0.34 and 0.38 for daily time steps whereas the C_d values for a short crop are 0.24 and 0.96 for daytime and for nighttime, respectively. For a the tall alfalfa reference crop, the C_d values are 0.25 and 1.7 for daytime and nighttime, respectively. C_n is a function of the time step and aerodynamic resistance whereas C_d is a function of the time step, surface resistance and aerodynamic resistance (ASCE-EWRI, 2005).

Irmak et al. (2005) found a good correlation between the ASCE Standardized ET_o equation results and the FAO 56 PM ET_o resultscalculated on hourly time steps. However, the FAO 56 PM method estimated 5% to 8% lower ET_o compared to the ASCE Standardized ET_o. According to Irmak et al. (2005) the results may be due to the higher surface resistance values during daytime periods in the FAO 56 PM equation. The authors also compared the daily ET_o given by the ASCE Standardized daily ET_o equation with the sum of the hourly ET_o calculated using the ASCE Standardized hourly equation. They observed that the daily ET_o values were generally higher than the sum of the hourly ET_o and they recommended to use the hourly ET_o values especially in advective condition.

2.7.3 Valiantzas Model

Valiantzas (2006) developed a set of equations to determine the ET_o rate based on simplifications made to the Penman (1963) equation. His purpose was to enable ET computation with limited meteorological data. Valiantzas (2013a) then improved these equations and claimed that his model performed equivalent in accuracy to the Penman (1963) equation. His simplified equation to calculate grass reference ET is as follows:

$$ET_o \approx 0.0393 R_s \sqrt{T_a + 9.5} - 0.19 R_s^{0.6} \varphi^{0.15} + 0.048(T_a + 20)(1 - \frac{RH}{100}) u_2^{0.7} \quad (24)$$

where, ET_o is the grass reference ET (mm/d), R_s is the measured or estimated incoming solar radiation (MJ/m^2/d), T_a is the mean daily air temperature (°C), φ is the latitude of the site (radians), RH is the relative humidity (%) and u_2 is the mean wind speed at 2 m height (m/s). He also developed an equation to calculate reference ET when the wind speed data is not available. The equation is as follows:

$$ET_o \approx 0.0393R_s\sqrt{T_a+9.5}-0.19R_s^{0.6}\varphi^{0.15}+0.078(T_a+20)(1-\frac{RH}{100}) \quad (25)$$

Valiantzas (2013b) also developed a set of equations to calculate reference ET for arid and humid regions. His equation to calculate reference ET with two different aerodynamic term weighted factors is as follows:

$$ET_o \approx 0.0393R_s\sqrt{T_a+9.5}-2.4\left(\frac{R_S}{R_a}\right)^2-0.024\left(T_a+20\right)\left(1-\frac{RH}{100}\right)+W_{aero}0.066(T_a+20)(1-\frac{RH}{100})u^{0.6} \quad (26)$$

where, R_a is the extraterrestrial radiation ($MJ/m^2/d$) and W_{aero} is an empirical weighted factor. The value of W_{aero} is as follows:

$W_{aero} = 0.78$ when RH > 65%; $W_{aero} = 1.067$ when RH ≤ 65%.

when the wind speed data is not available, the following equation applies:

$$ET_o \approx 0.0393R_s\sqrt{T_a+9.5}-2.4\left(\frac{R_S}{R_a}\right)^2+C_u(T_a+20)\left(1-\frac{RH}{100}\right) \quad (27)$$

where, $C_u = 0.054$ when RH > 65% and 0.083 when RH ≤ 65%.

Valiantzas' model might be a good substitute when some weather data are missing. However, when there are good quality data available, the use of the mechanistic Penman-Monteith equation are more appropriate than the empirical methods.

2.8 Variable Surface Resistance Approach

All of the above mentioned equations, in section 2.7, calculate ET for a reference crop surface; which is either grass or alfalfa. In order to calculate the actual crop ET, the current practice is to multiply the reference crop ET with a crop coefficient (K_c). Crop coefficients have been developed for different crop growth stages for various crops. However, Katerji and Rana (2006) have pointed out that the difference of ± 40% could be observed between the K_c values reported by Allen et al. (1998) and the experimentally obtained K_c values from different researchers. Based on findings from previous researchers (Rana et al., 1994; Steduto et al., 1996; Ventura et al., 1999; Lecina et al., 2003; Pereira, 2005; de Medeiros et al., 2006). Katerji and Rana (2006) indicated that there is up to 18% of underestimation and 13.4% of overestimation in ET_o in semi-arid regions and humid regions, respectively, due to the use of fixed r_c values. Hence, the cumulative error from reference ET calculation and the use of K_c may be significant which could be a concern for irrigation water management purposes. In order to address this problem, some researchers have started to use variable surface resistance instead of fixed surface resistance to calculate actual crop ET directly without using the crop coefficient approach. This new approach is also called the one step crop ET estimation approach as there is no need of using the crop coefficients (a two-step approach). The variable surface resistance approaches that have been developed so far are discussed below:

2.8.1 Jarvis Model

Jarvis (1976) developed a multiplicative model to calculate stomatal resistance from weather parameters including air temperature, vapor pressure deficit, leaf water potential and ambient carbon dioxide (CO_2) concentration. However, the Penman-Monteith equation requires the bulk surface resistance and hence the knowledge of stomatal resistance only may not be enough to calculate ET. The upscaling of the stomatal resistance to the canopy level is required to calculate the bulk surface resistance. Alves and Pereira (1999) objected the methodology adopted by Jarvis, as they questioned the validity of the multiplicative model and also they expressed doubt in the assumption of weather parameters acting independently.

2.8.2 Katerji and Perrier (KP) Model

Katerji and Perrier (1983) found that a linear relationship can be established between the two ratios r_c/r_a and r^*/r_a, where r^* is a climatic resistance term. They developed the following empirical relation:

$$\frac{r_c}{r_a}=a\frac{r^*}{r_a}+b \quad (28)$$

where, a and b are empirical calibration coefficients requiring experimental determination. The resistance, r^*, is defined as:

$$r^*=\frac{\Delta+\gamma}{\Delta\gamma}\frac{\rho c_p D}{R_n-G} \quad (29)$$

where D is the vapor pressure deficit (kPa), the units of R_n and G are W/m^2.

Katerji and Rana (2006) reported that the coefficients a and b have already been developed for alfalfa, rice, grass, lettuce, sweet sorghum, sunflower, grain sorghum, soybean, clementine orchard and sloping grassland. The coefficients have also been adapted for water stress conditions (Rana et al., 1997, 2001). Rana et al. (1997) claimed that the coefficients "a" and "b" have multi-local validity (i.e. they do not change with the site but only with the crop species).

The downside of this method is that there seems to be no physical meaning for these coefficients. In addition, the coefficients need to be tested for different crop species. Alves and Pereira (1999) indicated that Equation (28) is only valid for periods where the Bowen ratio varies between -0.3 and 0.3.

2.8.3 Todorovic Model

Todorovic (1999) presented a mechanistic approach to calculate surface resistance using weather variables. His summarized methodology is as follows:

$$t = \frac{\gamma}{\Delta} \frac{D}{(\Delta + \gamma)} \quad (30)$$

Firstly, t which is the difference between actual canopy temperature and canopy temperature (°C) in wet conditions is calculated using Equation (30). Then, using a quadratic Equation (31), X, which is the ratio of surface resistance (r_c) to climatological resistance (r_i), is calculated.

$$X = \frac{a + \sqrt{b^2 - 4ac}}{2a} \quad (31)$$

where,

$$a = \frac{\Delta + \gamma\Delta}{\Delta + \gamma} \gamma D \quad (32)$$

$$b = -\gamma Y t \quad (33)$$

$$c = -(\Delta + \gamma)t \quad (34)$$

The climatological resistance (r_i) can be calculated using:

$$r_i = \frac{\rho c_p D}{\gamma(R_n - G)} \quad (35)$$

In Equation (33), Y is the ratio of climatological resistance (r_i) to aerodynamic resistance (r_a). The units of all the resistances, which is the reciprocal of conductance, are in s/m. Rn and G in Equation (35) are in W/m^2.

Secondly, after finding X, r_c is calculated by multiplying X by r_i. The calculated r_c will be inserted in the Penman-Monteith Equation (18) to calculate the actual crop ET. The actual crop ET can be defined as the rate of ET that occurs under field-environmental-surface condition which may depart from the ideal "standard" conditions.

Lecina et al. (2003) evaluated the KP and Todorovic models based on measured ET using lysimeter and eddy covariance systems. Based on their finding, they recommended to use the Todorovic model to calculate the hourly ET for semiarid windy conditions. Furthermore, Lecina et al. (2003) didn't find any improvement in ET estimates using the KP model. On the other hand, Shi et al. (2008) found that the KP model agreed better with the measured eddy covariance ET values for half-hourly and daily ET by summing the half-hourly ET values. They reported that the Todorovic model overestimated ET by about 30% in their experimental site in China.

2.8.4 Li et al. Model

Li et al. (2009) found some errors in the Todorovic model in the derivation of "t". Li et al. (2009) derived "t" as:

$$t = \frac{\gamma D C}{\Delta(\Delta + \gamma)} \quad (36)$$

Li et al. proved that Todorovic's method missed the term C while deriving "t". The missing parameter C was described as shown in Equation (37):

$$C = \frac{\left(\frac{\Delta}{\gamma}\right) \cdot \left(\frac{1}{r_i}\right) + \frac{1}{r_a}}{\left(1 + \frac{\gamma}{\Delta}\right)\left(\frac{1}{r_c}\right) + \left(\frac{\gamma}{\Delta}\right)\left(\frac{1}{r_a}\right)} \quad (37)$$

In their article, Li et al. (2009) replaced C with (1 + D/D_0), where D_0 is a parameter which accounts for the response of t to vapor pressure deficit (D). They used D_0 as 1.5 kPa for their research, which they claimed was applicable for the winter wheat crop in the North China Plain. They also mentioned that the value of D_0 can vary with crops and climatic conditions.

Li et al. (2009) showed that Todorovic model severely underestimated the canopy temperature and sensible heat flux and severely overestimated the latent heat flux. On the other hand, their model gave acceptable results for latent heat flux at both half-hourly and hourly time scales. The limitation of this method is that there there seems to be no physical meaning of D_o and the value of D_o is needed to compute "t" and ultimately "r_c".

2.8.5 Shuttleworth Model

Shuttleworth (2006) introduced the concept of the crop independent blending height (50 m) to use as a reference height instead of 2 m reference height to enable the one step ET (ET_a) calculation for different crops. Shuttleworth and Wallace (2009) used the existing PM equation and then calculated r_c as a function of weather variables and K_c values documented in FAO 56 PM publication (Allen et al., 1998). They worked on a daily time step instead of hourly; hence their model is not applicable for hourly ET estimation. They concluded that the use of fixed crop coefficients (K_c) to calculate actual crop ET can be problematic. The authors mentioned that the recommended K_c values are said to be appropriate for wind speeds of 2 m/s and "humid" conditions with 45% relative humidity. Whenever the weather conditions differ then the reported values of K_c cannot provide reasonable estimates of ET. The authors also mentioned that the FAO 56 PM equation and the Priestley-Taylor equation with $\alpha = 1.26$ give identical ET values in "humid" conditions. They showed from their Equation (11) that the ambient weather changes the value of the K_c via the values of the climatological resistance and wind speed. They developed a relationship between r_c and K_c where r_c was a function of K_c, r_s^1 and r_s^2, where r_s^1 and r_s^2 could be calculated using their Equations (14) and (15) or performing interpolation from their Table 1 (See Appendix). The authors concluded that use of their approach will yield estimates of ET as good as those given by the FAO 56 PM model in humid conditions whereas it improves ET estimation for arid climates and for taller crops.

2.8.6 Irmak and Mutibwaa Model

Irmak et al. (2008) were able to upscale stomatal resistance (leaf scale) to the whole canopy surface resistance for maize using photosynthetic photon flux density (PPFD), leaf area index (LAI) for sunlit and shaded leaves, solar zenith angle, direct and diffuse solar radiation. They measured the stomatal resistance using porometers. They developed their model for corn and then successfully validated it for soybean with recalibration of some parameters (Irmak & Mutiibwa, 2008; Irmak et al., 2008; Mutiibwa & Irmak, 2010, 2012; Irmak et al., 2013). Irmak and Mutiibwa (2009) showed that the estimation of crop ET using the one step approach was better (an improvement) compared to the two step approach (i.e. using the reference crop ET multiplied by the crop coefficients). The one step approach ET was within 2 per cent of measured ET using the Bowen Ratio instrument whereas for the two step ET calculation, there was no distinct pattern of over or under estimation. On the other hand, the two step ET method underestimated actual ET (measured) especially when there was high evaporative demand according to the authors. This result suggests that the use of fixed surface resistance while calculating the reference ET is not appropriate (Irmak & Mutibwaa, 2009).

Irmak and Mutibwaa (2009) were able to modify the Jarvis model, which they referred to as Modified-Jarvis-model (NMJ) and showed that their model is an improvement to the older version, as NMJ model improved the stomatal resistance estimation by 10% in RMSD (root mean square deviation) when compared to the measured stomatal resistance using a dynamic diffusion porometer. Irmak and Mutiibwa (2010) developed a set of empirical equations for nonstressed maize crop to calculate r_c from weather variables. They used the measured ET from a Bowen Ratio instrument, then inverted the PM equation to back calculate r_c. Then, they used a linear regression technique to find the relationship between r_c and a set of weather variables. Irmak et al. (2013) also developed similar set of empirical equations for soybean crop to calculate r_c from weather variables.

Table 1 below summarizes all the methods that have been discussed in the body of this article along with equations associated with those models.

Table 1. Summary of equations of different ET estimation methods

	Methods	Equations
1.	Blaney-Criddle (1942)	$u = kf$
2.	Thornthwaite (1948)	$PET = 16(10T_a / I)^a$
3.	Hargreaves (1975) (for deg. F)	$ET_0 = 0.0075 R_s T_F$
	Hargreaves-Samani (1985), (deg. C)	$ET_0 = 0.0022 R_s (T_a + 17.8) TD^{0.5}$
4.	Christiansen (1968)	$E = K\, R_a\, C$
5.	Penman Related Equations	
5.1	*Original Penman Equation (1948)*	$ET = \dfrac{(\Delta(R_n - G) + k_w(e_s - e_a) f(u)\gamma)}{\lambda(\Delta + \gamma)}$ $f(u) = a_w + b_w u_2$
5.2	*CIMIS Penman Method*	$a_w = 0.29$, $b_w = 0.53$ for $R_n > 0$ $a_w = 1.14$, $b_w = 0.40$ for $R_n \leq 0$
5.3	*Penman Monteith (1965)*	$\lambda E = \dfrac{0.408\Delta(R_n - G) + \left(\dfrac{\gamma 900}{T_a + 273}\right) u_2 (e_s - e_a)}{(\Delta + \gamma(1 + 0.34 u_2)}$
6.	Priestley Taylor (1972)	$\lambda E = \alpha\, \Delta(Rn - G) / (\Delta + \gamma)$ where $\alpha = 1.26$ for water surfaces with minimum advection
7.	Fixed Surface Resistance Approach	
7.1	*FAO 56 PM (1998)*	$ET_0 = \dfrac{0.408\Delta(R_n - G) + \left(\dfrac{\gamma 900}{T_a + 273}\right) u_2 (e_s - e_a)}{(\Delta + \gamma(1 + 0.34 u_2)}$
7.2	*ASCE-EWRI Standardized PM Equation (2005)*	$ET_{sz} = \dfrac{0.408\Delta(R_n - G) + \gamma C_n u_2 \dfrac{(e_s - e_a)}{T_a + 273}}{(\Delta + \gamma(1 + C_d u_2)}$
7.3	*Valiantzas Model (2006, 2013)*	$ET_o \approx 0.0393 R_s \sqrt{T_a + 9.5} - 0.19 R_s^{0.6} \varphi^{0.15} + 0.048(T_a + 20)(1 - \dfrac{RH}{100}) u^{0.7}$
8.	Variable Surface Resistance Approach	
8.1	*Katerji-Perrier Model (1983)*	$\dfrac{r_c}{r_a} = a \dfrac{r^*}{r_a} + b$ $r^* = \dfrac{\Delta + \gamma}{\Delta\gamma} \dfrac{\rho c_p D}{R_n - G}$
8.2	*Todorovic Model (1999)*	$t = \dfrac{\gamma}{\Delta} \dfrac{D}{(\Delta + \gamma)}$
8.3	*Li et al. Model (2009)*	$t = \dfrac{\gamma DC}{\Delta(\Delta + \gamma)}$ $C = \dfrac{\left(\dfrac{\Delta}{\gamma}\right) \cdot \left(\dfrac{1}{r_i}\right) + \dfrac{1}{r_a}}{\left(1 + \dfrac{\gamma}{\Delta}\right)\left(\dfrac{1}{r_c}\right) + \left(\dfrac{\gamma}{\Delta}\right)\left(\dfrac{1}{r_a}\right)}$
8.4	*Shuttleworth Model (2006, 2009)*	See Appendix for details

Table 2. Advantages, limitations and application timestep of different ET estimation models

	Methods	Variables Used	Advantages	Limitations	Application Timestep
1.	Blaney-Criddle Method (1942)	T_a, T_F, p, k, K	Simplicity	ET underestimation in general	Monthly
2.	Thornthwaite Method (1948)	T_a	Simplicity	ET underestimation in advective condition	Monthly
3.	Hargreaves and Samani (1985)	R_s, T_F. T_a, R_a	Simplicity	Problems of over and under estimation of ET	Weekly
4.	Christiansen Method (1968)	K, R_a, C	More or less accurate to predict ET for monthly timestep	Not accurate to calculate ET for daily or shorter timesteps.	Monthly
5.	Penman Related Equations				
5.1	Original Penman Equation (1948)	Δ, R_n, G, e_s, ea, γ, f(u), λ	Physical equation based on the combination of surface energy balance equation and aerodynamic equation	Wind speed function is difficult to obtain. The equation was mainly developed for evaporation from free water surfaces.	Daily, hourly
5.2	CIMIS Penman Method	Δ, R_n, G, es, e_a, γ, f(u), λ	a_w and b_w coefficients used in f(u) are easy to obtain. Also applicable for hourly timesteps.	The coefficients used in this equation were developed for Californian condition, hence may not be applicable elsewhere.	Hourly
5.3	Penman Monteith equation (1965)	Δ, R_n, G, e_s, e_a, γ, ρ, c_p, r_a, r_c	Physical equation with the inclusion of r_c.	It is difficult to directly implement this equation to calculate actual crop ET, as r_c is difficult to obtain.	Daily, hourly
6.	Priestley-Taylor Method (1972)	α, Δ, γ, R_n, G	Relatively simple	ET underestimation mainly in advective condition.	Daily
7.	Fixed Surface Resistance Approach				
7.1	FAO 56 PM Equation (1998)	Δ, R_n, G, e_s, e_a, γ, Ta, u_2	Considered very accurate to calculate grass reference ET on daily basis.	May not be applicable to apply for hourly timestep.	Daily
7.2	ASCE-EWRI Standardized PM Equation (2005)	Δ, R_n, G, e_s, e_a, γ, Ta, u_2	It can calculate both grass and alfalfa reference crop ET on both hourly and daily timesteps.	Kc needs to be developed also for alfalfa reference surfaces. The use of fixed r_c for entire day may induce some errors in estimating reference ET.	Daily, hourly
7.3	Valiantzas Model (2006, 2013)	T, R_s, φ, RH, u_2	Relatively simple, can be used when some parameters like wind speed is missing.	It is semi-empirical, so may not be accurate enough as PM equation.	Daily
8.	Variable Surface Resistance Approach				
8.1	Jarvis Model (1976)		New concept to calculate stomatal resistance	It is not easy to obtain canopy resistance (r_c) from stomatal resistance.	
8.2	Katerji-Perrier Model (1983)	Δ, R_n, G, e_s, e_a, γ, T_a, u_2, a, b, r_a	Relatively simple to calculate actual crop ET in one step process.	This is empirical method. The coefficients "a" and "b" need to be tested for different species and also for different climatic conditions.	Daily, hourly
8.3	Todorovic Model (1999)	Δ, R_n, G, e_s, e_a, γ, T_a, u_2, r_a	Mechanistic equation to calculate r_c	Some flaws in the procedure as shown by Li et al. (2009).	Hourly
8.4	Li et al. Model (2009)	Δ, R_n, G, e_s, e_a, γ, T_a, u_2, r_a	Relatively simple to implement.	Only applicable for winter wheat crop in North China Plain.	Hourly
8.5	Shuttleworth Model (2006, 2009)	Δ, R_n, G, e_s, e_a, γ, Ta, u_2, r_a, K_c	Provides one step ET for daily timestep based by calculating r_c based on K_c.	Complicated to use, r_c is a function of FAO 56 Kc values, in other words, r_c depends on the accuracy of K_c.	Daily
8.6	Irmak and Mutibwaa Models (2008, 2009, 2010, 2013)	PPFD, LAI	Already implemented their models successfully for corn and soybean.	Complicated to use. Needs many variables to calculate r_c including PPFD (photosynthetic photon flux density).	Hourly

3. Discussion

Accurate estimation of crop ET is important and significant contributions from various researchers have been made until now. The full version Penman-Monteith equation is considered to be a robust method to calculate crop ET. However, it also has some limitations. The equation uses aerodynamic and surface resistance terms, where the aerodynamic resistance is relatively straight forward to calculate. However, the calculation of surface resistance is not easy. The use of the fixed surface resistance approach is a simplification of the true diurnal dynamics of this resistance, even for a reference crop under standard conditions. Hence, the modelling of the surface resistance would help in estimating crop ET with more accuracy.

Some authors (Lascano & van Bavel, 2007; Lascano, Van Bavel, & Evett, 2010; Paw, 1987, 1988, 1992; Tracy et al., 1984) point out that the exclusion of surface temperature while deriving the PM equation can induce some errors especially when the surface temperature and air temperature are significantly different. The other problem is with the linearity assumption of saturation vapor pressure and temperature curve. Lascano and Van Bavel (2007) and Lascano et al. (2010) calculated the surface temperature and ET simultaneously by iteration technique using fixed surface resistance (which they referred as the Recurvise Combination Method or RCM ET) and concluded that the RCM yielded ET rates very close to measured lysimeter ET values while the PM equation underestimated ET as much as 25%. Paw (1988, 1991) developed a fourth order, third order and second order equations to solve the energy budget equation for latent heat flux, analytically. They claimed that these methods were superior compared to the PM method; as the error associated with the linearity assumption is corrected by using higher degree polynomials.

Some researchers (Dodds et al., 1997) highlighted that the use of the PM equation in advective condition would underestimate ET as the equation is not totally capable of incorporating the horizontal flow of sensible heat flux.

Shuttleworth and Wallace (1985) developed a two layer ET model, which could incorporate evaporation from the ground surfaces as well as the transpiration from the plant canopy. Their model is thus different from the Penman and Penman related equations which are basically a big leaf one layer ET model. Their model is more useful to calculate the ET rate from sparse canopy.

Regarding the one step ET estimation, KP model lacks physical meaning and it can only be applied when the Bowen ratio is between -0.3 and 0.3 (Alves and Pereira (2000), as cited by Lecina et al. (2003)). Also the coefficients used in the equation may vary among locations (Lecina et al., 2003). The Todorovic model (1999) is a mechanistic model to calculate surface resistance. However, Li et al. (2009) showed a missing term in his equation. Li et al., in their 2009 article, didn't try to use the improved Todorovic model, but instead used the simpler empirical relation to get the missing term, which was only applicable for winter wheat crop. The Shuttleworth model (2009) is an improved model of the existing method of two step ET estimation; however, the surface resistance used in his model is a function of the FAO 56 PM crop coefficient (Allen et al., 1998), which may not be accurately transferable to different environmental, crop and soil conditions. In addition, the Shuttleworth model is only applicable for daily time steps; thus, it may not work for hourly time steps. Irmak and Mutibwaa tried to upscale the surface resistance from the measured stomatal resistance; however, the method needs several field level data acquisition including the photosynthetic photon flux density, which is not easy to obtain.

4. Future Work

By reviewing the literature on ET estimation up-to-date, from various researchers, it is evident that still there is a need to further improve the estimation of reference as well as actual crop ET. The calculation of actual crop ET with the one step approach, which does not use crop coefficients, seems challenging but promising. However, if accomplished the error on the estimation of ET could be minimized using a (hourly) variable surface resistance. Supplementary, there has been some progress in calculating the surface resistance directly; in particular publications from Katerji and Perrier, Todorovic, Li et al, Shuttleworth, and Irmak and Mutibwaa. The extension of the Shuttleworth model for hourly time steps, to calculate r_c, could be an advancement to improve the ET estimation using the one step approach. Also, the validation of Irmak and Mutibwaa's model at field level, for various crops, could be another contribution. Finally, another alternative would be to develop a new r_c model which would be robust enough to be applied to different crops under different climatic conditions.

Acknowledgements

The authors would like to acknowledge the funding agencies Colorado Agricultural Experiment Station, Colorado State University (CSU), and the USDA National Institute of Food and Agriculture (NIFA) for financially supporting this work.

References

Allen, R. G., Jensen, M. E., Wright, J. L., & Burman, R. D. (1989). Operational estimates of evapotranspiration. *Agron. J., 81*, 650-662. http://dx.doi.org/10.2134/agronj1989.00021962008100040019x

Allen, R. G., Periera, L. S., Raes, D., & Smith, M. (1998). Crop evapotranspiration. *Guidelines for computing crop water requirements* (FAO Irrigation and Drainage Paper no. 56, p. 300). Rome.

Alves, I., & Pereira, L. S. (1999). Modelling surface resistance from climatic variables? *Agric. Water Manage, 42*, 371-385. http://dx.doi.org/10.1016/S0378-3774(99)00041-4

American Society of Civil Engineers (ASCE). (1990). Evapotranspiration and Irrigation Water Requirements. *Manuals and Reports on Engineering Practice* (ASCE No. 70). New York.

ASCE-EWRI. (2005). The ASCE Standardized reference evapotranspiration equation. In R. G. Allen, I. A. Walter, R. Elliott, T. Howell, D. Itenfisu & M. Jensen (Eds.), *Environmental and Water Resources Institute (EWRI) of the American Society of Civil Engineers, ASCE, Standardization of Reference Evapotranspiration Task Committee Final Report* (p. 213). Reston, VA: American Society of Civil Engineers (ASCE). Retrieved from http://ascelibrary.org/doi/abs/10.1061/40499(2000)126

Bautista, F., Bautista, D., & Delgado-Carranza, C. (2009). Calibration of the equations of Hargreaves and Thornthwaite to estimate the potential evapotranspiration in semi-arid and subhumid tropical climates for regional applications. *Atmosphera, 22*(4), 331-348.

Blaney, H. F., & Criddle, W. D. (1962). Determining consumptive use and irrigation water requirements. *Agricultural Research Service Tech Bull 1275* (p. 59). U.S. Dept. Agr.

Christiansen, J. E. (1968). Pan evaporation and evapotranspiration from climatic data. *J. Irrig. Drain. Div.*, 243-265.

De Medeiros, G. A., Arruda, F. B., & Sakai, E. (2006). Crop coefficient for irrigated beans derived using three reference evaporation methods. *Agric. Forest Meteorol., 135*(1-4), 135-143. http://dx.doi.org/10.1016/j.agrformet.2005.11.010

Dodds, P. E., Meyer, W. S., & Barton, A. (2005). A review of methods to estimate irrigated reference crop evapotranspiration across Australia. *CRC for Irrigation Futures Technical Report No. 04/05.*

Doorenbos, J., & Pruitt, W. O. (1977). Crop water requirements. *Irrig. and Drain. Paper No. 24.* Rome, Italy: Food and Agric. Organization, United Nations.

Hargreaves, G. H. (1975). Moisture availability and crop production. *Trans. ASAE, 18*(5), 980-984. http://dx.doi.org/10.13031/2013.36722

Hargreaves, G. H. (1977). World water for agriculture. *Agency for International Development, 177.*

Hargreaves, G. H., & Samani, Z. A. (1985). Reference crop evapotranspiration from temperature. *Appl. Eng. Agric., 1*(2), 96-99. http://dx.doi.org/10.13031/2013.26773

Hargreaves, G. L., Hargreaves, G. H., & Riley, J. P. (1985). Irrigation water requirements for Senegal River Basin. *J. Irrig. Drain. Eng., 111*(3), 265-275. http://dx.doi.org/10.1061/(ASCE)0733-9437(1985)111:3(265)

Hobbins, M. T., Ramirez, J. A., & Brown, T. C. (2001). The complimantary relationship in estimation of regional evapotranspiration: An enhanced Advection-Aridity model. *Water Res. Res., 37*(5), 1389-1403. http://dx.doi.org/10.1029/2000WR900358

Irmak, S., & Mutiibwa, D. (2008). Dynamics of photosynthetic photon flux density and light extinction coefficient to assess radiant energy interactions for maize canopy. *Transcations of the ASABE, 51*(5), 1663-1673. http://dx.doi.org/10.13031/2013.25323

Irmak, S., & Mutiibwa, D. (2009). On the dynamics of evaporative losses from Penman-Monteith with fixed and variable canopy resistance during partial and complete maize canopy. *Transcations of the ASABE, 52*(4), 1139-1153. http://dx.doi.org/10.13031/2013.27791

Irmak, S., & Mutiibwa, D. (2010). On the dynamics of canopy resistance: Generalized linear estimation and relationships with primary micrometeorological variables. *Water Resour. Res., 46*, W008484. http://dx.doi.org/10.1029/2009WR008484

Irmak, S., Howell, T. A., Allen, R. G., Payero, J. O., & Martin, D. L. (2005). Standardized ASCE Penman-Monteith: Impact of Sum-of-Hourly vs. 24-hour timestep computations at reference weather station sites. *Transactions of the ASCE, 48*(3), 1063-1077. http://dx.doi.org/10.13031/2013.18517

Irmak, S., Mutiibwa, D., Irmak, A., Arkebauer, T. J., Weiss, A., Martin, D. L., & Eisenhauer, D. E. (2008). On the scaling up leaf stomatal resistance to canopy resistance using photosynthetic flux density. *Agric. And For. Meteorol., 148*, 1034-1044. http://dx.doi.org/10.1016/j.agrformet.2008.02.001

Irmak, S., Mutiibwa, D., Payero, J., Marek, T., & Porter, D. (2013). Modeling soybean canopy resistance from micrometeorological and plant variables for estimating evapotranspiration using one-step Penman-Monteith approach. *Journal of Hydrology, 507*, 1-18. http://dx.doi.org/10.1016/j.jhydrol.2013.10.008

Jarvis, P. G. (1976). The interpretations of the variations in leaf water potential and stomatal conductance found in canopies in the field. *Philosophical Transcations of the Royal Society of London. Series B. Biological Sciences, 273*(927), 593-610. http://dx.doi.org/10.1098/rstb.1976.0035

Katerji, N., & Rana, G. (2006). Modelling evapotranspiration of six irrigated crops under Mediterranean climate conditions. *Agric. Forest Meteorol., 138*, 142-155. http://dx.doi.org/10.1016/j.agrformet.2006.04.006

Kumar, K. K., Kumar, K. R., & Rakhecha, P. R. (1987). Comparison of Penman and Thornthwaite methods of estimating potential evapotranspiration for Indian conditions. *Theor. Appl. Climatol., 38*, 140-146. http://dx.doi.org/10.1007/BF00868097

Lascano, R. J., & Van Bavel, C. H. M. (2007). Explicit and recursive calculation of potential and actual evapotranspiration. *Agron. J., 99*, 585-590. http://dx.doi.org/10.2134/agronj2006.0159

Lascano, R. J., Van Bavel, C. H. M., & Evett, S. R. (2010). A field test of recursive calculation of crop evapotranspiration. *Transactions of the ASABE, 53*(4), 1117-1126. http://dx.doi.org/10.13031/2013.32601

Lecina, S., Martinez-Cob, A., Perez, P. J., Villalobos, F. G., & Baselga, J. J. (2003). Fixed versus bulk canopy resistance for reference evapotranspiration estimation using the Penman-Monteith equation under semiarid conditions. *Agric. Water Manage., 60*, 181-198. http://dx.doi.org/10.1016/S0378-3774(02)00174-9

Li, L., Yu, Q., Su, Z., & Tol, C. Van der. (2009). A Simple method using climatic variables to estimate canopy temperature, sensible and latent heat fluxes in a winter wheat field on the North China Plain. *Hydrol. Process, 23*, 665-674. http://dx.doi.org/10.1002/hyp.7166

Lopez-Urrea, R., Olalla, F. M. S., Fabeiro, C., & Moratella, A. (2006). Testing evapotranspiration equations using lysimeter observations in a semiarid climate. *Agricultural Water Mgmt., 85*, 15-26. http://dx.doi.org/10.1016/j.agwat.2006.03.014

Monteith, J. L. (1965). Evaporation and environment. In G. E. Fogg (Ed.), *Symposium of the Society for Experimental Biology: The State and Movement of Water in Living Organisms* (Vol. 19, pp. 205-234). Academic Press, Inc., NY.

Monteith, J. L., Szeich, G., & Waggoner, P. E. (1965). The measurement and control of stomatal resistance in the field. *Journal of Applied Ecology, 2*(2), 345-355. http://dx.doi.org/10.2307/2401484

Mutiibwa, D., & Irmak, S. (2010). On the scaling up soybean leaf level stomatal resistance to canopy resistance for one-step estimation of actual evapotranspiration. *Transactions of the ASABE, 54*(1), 141-154. http://dx.doi.org/10.13031/2013.36269

Mutiibwa, D., & Irmak, S. (2012). Transferability of Jarvis-type models developed and re-parameterized for maize to estimate stomatal resistance of soybean: analyses on model calibration, validation, performance, sensitivity and elasticity. *Transcations of the ASABE, 56*(2), 409-422. http://dx.doi.org/10.13031/2013.42688

Paw, U. K. T. (1987). Mathematical Analysis of the operative temperature and energy budget. *J. Therm. Biol., 12*(3), 227-233. http://dx.doi.org/10.1016/0306-4565(87)90009-X

Paw, U. K. T. (1992). A discussion of the Penman form equations and comparisons of some equations to estimate latent energy flux density. *Agric. And Forest Meteorol., 57*, 294-304. http://dx.doi.org/10.1016/0168-1923(92)90125-N

Paw, U. K. T., & Gao, W. (1988). Applications of solutions to non-linear energy budget equations. *Agric. And Forest Meteorol., 43*, 121-145. http://dx.doi.org/10.1016/0168-1923(88)90087-1

Penman, H. L. (1948). Natural evaporation from open water, bare soil and grass. *Proc. R. Soc. London, A 193*, 120-146. http://dx.doi.org/10.1098/rspa.1948.0037

Penman, H. L. (1963). Vegetation and Hydrology. *Technical Communication No. 53, Commonwealth Bureau of Soils*. Harpenden, England. http://dx.doi.org/10.1097/00010694-196311000-00014

Pereira, A. R. (2005). The Priestley Taylor parameter and the decoupling factor for estimating reference evapotranspiration. *Agric. Forest Meteorol., 125*, 305-313. http://dx.doi.org/10.1016/j.agrformet.2004.04.002

Pereira, A. R., & De Camargo, A. P. (1989). An analysis of the criticism of Thornthwaite's equation for estimating potential evapotranspiration. *Agric. and Forest Meteorol., 46*, 149-157. http://dx.doi.org/10.1016/0168-1923(89)90118-4

Priestley, C. H. B., & Taylor, R. J. (1972). On the assessment of surface heat flux and evaporation using large-scale parameters. *Mon. Weather Rev., 100*, 81-92. http://dx.doi.org/10.1175/1520-0493(1972)100%3C0081:OTAOSH%3E2.3.CO;2

Rana, G., Katerji, N., & Perniola, M. (2001). Evapotranspiration of sweet sorghum: A general model and multilocal validity in semiarid environmental conditions. *Water Res. Research, 37*(12), 3237-3246. http://dx.doi.org/10.1029/2001WR000476

Rana, G., Katerji, N., Mastorilli, M., & El-Moujabber, M. (1994). Evapotranspiration and canopy resistance of grass in a Mediterranean region. *Theor. Appl. Climatol., 50*, 61-71. http://dx.doi.org/10.1007/BF00864903

Rana, G., Katerji, N., Mastorilli, M., & El-Moujabber, M. (1997). A model for predicting actual evapotranspiration and canopy resistance of grass in a Mediterranean region. *Theor. And App. Climatology, 56*(1-2), 45-55. http://dx.doi.org/10.1007/BF00863782

Ravazzani, G., Corbari, C., Morella, S., Gianoli, P., & Mancini, M. (2012). Modified Hargreaves-Samani equation for the assessment of reference evapotranspiration in alpine river basins. *J. Irrig. Drain. Engrg., 138*(7), 592-599. http://dx.doi.org/10.1061/(ASCE)IR.1943-4774.0000453

Robins, P. C. (1974). A method of measuring the aerodynamic resistance to the transport of water vapor from forest canopies. *Journal of Applied Ecology, 11*(1), 315-325. http://dx.doi.org/10.2307/2402024

Sammis, T. W., Wang, J., & Miller, D. R. (2011). The transition of the Blaney-Criddle formula to the Penman-Monteith equation in the Western United States. *Journal of Service Climatology, 5*(1), 1-11.

Shi, T., Guan, D., Wang, A., Wu, J., Jin, C., & Ham, S. (2008). Comparison of three models to estimate evapotranspiration for a temperate mixed forest. *Hydrological Processes, 22*, 3431-3443. http://dx.doi.org/10.1002/hyp.6922

Shook, K., & Gray, D. M. (1997). Snowmelt resulting from advection. *Hydrological Processes, 11*, 1725-1736. http://dx.doi.org/10.1002/(SICI)1099-1085(19971030)11:13<1725::AID-HYP601>3.0.CO;2-P

Shuttleworth, W. J. (2006). Towards one-step estimation of crop water requirements. *Transactions of the ASABE, 49*(4), 925-935. http://dx.doi.org/10.13031/2013.21738

Shuttleworth, W. J., & Wallace, J. S. (1985). Evaporation from sparse crops – An energy combination theory. *Q. J. R. Meteorol. Soc., 111*, 839-855. http://dx.doi.org/10.1002/qj.49711146910

Shuttleworth, W. J., & Wallace, J. S. (2009). Calculating the water requirements of irrigated crops in Australia using the Matt-Shuttleworth approach. *Transactions of the ASABE, 52*(6), 1895-1906. http://dx.doi.org/10.13031/2013.29217

Steduto, P., Caliandro, A., Rubino, P., Mechlia, N. B., Masmoudi, M., Martinez-Cob, A., ... Faci, M. J. (1996). Penman-Monteith reference evapotranspiration estimates in the Mediterranean region. In C. R. Camp, E. J. Sadler & R. E. Yoder (Eds.), *Proceedings of the International conference on evapotranspiration and irrigation scheduling* (pp. 357-364). San Antonio, TX.

Temesgen, B., Eching, S., Davidoff, B., & Frame, K. (2005). Comparison of some reference evapotranspiration equations for California. *J. Irrig. Drain. Engrg., 131*, 1(73). http://dx.doi.org/10.1061/(ASCE)0733-9437(2005)131:1(73)

Thornthwaite, C. W. (1948). An approach towards a rational classification of climate. *Geogr. Rev., 38*, 55-94. http://dx.doi.org/10.2307/210739

Todorovic, M. (1999). Single-layer evapotranspiration model with variable canopy resistance. *J. Irrig. and Drain. Engrg., 125*(5), 235-245. http://dx.doi.org/10.1061/(ASCE)0733-9437(1999)125:5(235)

Tracy, C. R., Berkum, F. H. V., Tsuji, J. S., Stevenson, R. D., Nelson, J. A., Barnes, B. M., & Huey, R. B. (1984). Errors resulting from linear approximations in energy balance equations. *J. Therm. Biol., 9*(4), 261-264. http://dx.doi.org/10.1016/0306-4565(84)90006-8

Valiantzas, J. D. (2006). Simplified versions for the Penman evaporation equation using routine weather data. *Journal of Hydrology, 331*, 690-702. http://dx.doi.org/10.1016/j.jhydrol.2006.06.012

Valiantzas, J. D. (2013a). Simple ETo forms of Penman's equation without wind and/or humidity data. I: Theoretical development. *J. Irrig. Drain. Engrg., 139*(1), 1-8. http://dx.doi.org/10.1061/(ASCE)IR.1943-4774.0000502

Valiantzas, J. D. (2013b). Simplified reference evapotranspiration formula using an empirical impact factor for Penman's aerodynamic term. *J. Hydrol. Engrg., 18*(1), 108-114. http://dx.doi.org/10.1061/(ASCE)HE.1943-5584.0000590

Ventura, F., Spano, D., Duce, P., & Snyder, R. L. (1999). An evaluation of common evapotranspiration equations. *Irrig. Sci., 18*(4), 163-170. http://dx.doi.org/10.1007/s002710050058

Wai, N. M., Camerlengo, A., Wahab, A. K. A., & Harun, S. (2004). Estimation of evaporation and evapotranspiration in Malaysia using Penman and Christiansen methods. *Borneo Science, 15*, 23-26.

Wright, J. L. (1982). New evapotranspiration crop coefficients. *J. Irrig. Drain. Div., Am. Soc. Civ. Eng., 108*(2), 57-74.

Yoder, R. E., Odhiambo, L. O., & Wright, W. C. (2005). Evaluation of methods for estimating daily reference crop evapotranspiration at a site in the humid Southeast United States. *Applied Engineering in Agriculture, 21*(2), 197-202. http://dx.doi.org/10.13031/2013.18153

Appendix

Appendix 1. List of Abbreviations Used

Δ: slope of saturation vapor pressure with air temperature, kPa/°C.

γ: psychrometric constant, kPa/°C.

φ: latitude of site, radians.

ET: evapotranspiration rate, mm/h or mm/d or inches/d.

PET: potential ET rate, mm/h or mm/d or inches/d.

ET_0: reference evapotranspiration rate from a grass surface, mm/h or mm/d or inches/d.

ET_{sz}: reference evapotranspiration rate from a standardized surface, mm/h or mm/d.

ET_c: crop evapotranspiration, mm/h or mm/d.

u: monthly consumptive use (ET), inches/mon. or mm/mon.

U: seasonal consumptive use (ET), inches/season or mm/season.

K_c: crop coefficient developed by FAO 56 method.

k: empirical crop coefficient for monthly period.

K: empirical crop coefficient for irrigation season or growing period.

T_a: mean monthly/daily/hourly air temperature, °C.

T_F: mean monthly/daily/hourly air temperature, °F.

t: difference between actual canopy temperature and canopy temperature in wet conditions, °C.

u_2: wind speed at 2m height, m/s.

r_c: canopy surface resistance, s/m.

r_a: aerodynamic resistance, s/m.

r_l: daily average stomatal resistance: s/m.

r_i: climatological resistance, s/m.

r*: climatic resistance, s/m.

RH: relative humidity, %.

R_n: net radiation, $MJ/m^2/d$ or $MJ/m^2/h$.

R_s: incoming solar radiation, $MJ/m^2/d$ or $MJ/m^2/h$.

Ra: extraterrestrial radiation, $MJ/m^2/d$ or $MJ/m^2/h$.

e_a: actual vapor pressure, kPa.

e_s: saturation vapor pressure, kPa.

D: vapor pressure deficit, kPa.

p: monthly percentage of daytime hours of the year, %.

f: monthly consumptive use (ET) factor.

F: sum of monthly consumptive use (ET) factors for the period.

i: heat index.

I: sum of the 12 monthly heat index i.

S: measured sunshine hours times 100 divided by the number of possible sunshine hours.

K_{RS}: calibration coefficient.

TD: mean maximum minus mean minimum temperature, °C.

K: dimensionless constant developed empirically from data analysis.

C: dimensionless coefficient related to climatic parameters.

G: soil heat flux, $MJ/m^2/d$ or $MJ/m^2/h$.

f(u) = wind speed function.

J: Julian day of the year.

λ: latent heat of evaporation, MJ/kg.

ρ: air density, kg/m^3.

C_p: specific heat capacity of air at constant pressure, J/kg/K.

D: zero plane displacement height, m.

hc: crop height, m.

z_m: height of wind measurements, m.

z_h: height of humidity measurements, m.

z_{om}: roughness lenth governing momentum transfer, m.

z_{oh}: roughness length governing heat transfer.

k: von-Korman's constant (0.41).

U_z: wind speed at height z, m.

LAI: leaf area index, m^2/m^2.

C_n: numerator constant that changes with reference type and calculation time step, K mm s^3 M/g/d or K mm s^3M/g/h.

C_d: denominator constant that changes with reference type and calculation time step, s/m.

W_{aero}: empirical weighted factor.

Appendix 2. Shuttleworth Model

Shuttleworth (2006) developed an equation for crop coefficient (K_c) which is as follows:

$$K_c = \left[\frac{\frac{R_c^{50}}{u_2} + \frac{D_{50}}{D_2} r_{clim}}{\frac{302}{u_2} + \frac{D_{50}}{D_2} r_{clim}} \right] \left[\frac{(\Delta + \gamma)\frac{302}{u_2} + 70\gamma}{(\Delta + \gamma)\frac{R_c^{50}}{u_2} + \gamma (r_s)_c} \right] \tag{A1}$$

where,

R_c^{50} = aerodynamic coefficient for a crop of height h_c at blending height of 50 m;

U2 = wind speed at 2 m height;

D50 = vapor pressure deficit at blending height of 50 m (kPa);

D2 = vapor pressure deficit at 2 m height (kPa);

r_{clim} = climatological resistance = $\frac{\rho c_p D}{\Delta(R_n - G)}$ (s/m).

$(r_s)_c$ = surface resistance of crop (s/m)

The equation to calculate R_c^{50} in Equation (1) is given by:

$$R_c^{50} = \frac{1}{(0.41)^2} ln\left[\frac{(50-0.67h_c)}{(0.123h_c)}\right] ln\left[\frac{(50-0.67h_c)}{(0.0123h_c)}\right] \frac{ln\left[\frac{(2-0.08)}{0.0148}\right]}{ln\left[\frac{(50-0.08)}{0.0148}\right]} \quad \text{(A2)}$$

The equation to calculate r_{clim} in Equation (1) is given by:

$$r_{clim} = \frac{208}{u_2}\left[\frac{\alpha[\Delta + \gamma(1+0.337u_2)]}{\Delta+\gamma}\right] - 1 \quad \text{(A3)}$$

where, α is the Priestley-Taylor coefficient, which is equal to 1.26.

The equation to calculate $\left(\frac{D_{50}}{D_2}\right)$ in Equation (1) is given by:

$$\left(\frac{D_{50}}{D_2}\right) = \left[\frac{(\Delta+\gamma)302 + 70\gamma u_2}{(\Delta+\gamma)208 + 70\gamma u_2}\right] + \frac{1}{r_{clim}}\left\{\left[\frac{(\Delta+\gamma)302 + 70\gamma u_2}{(\Delta+\gamma)208 + 70\gamma u_2}\right]\left[\frac{208}{u_2}\right] - \frac{302}{u_2}\right\} \quad \text{(A4)}$$

The equation to calculate surface resistance $(r_s)_c$, according to Shuttleworth is given by:

$$(r_s)_c = \frac{r_s^1}{K_c^{FAO}} - r_s^2 \quad \text{(A5)}$$

where, K_c^{FAO} is the different Kc values of different crops published in FAO 56 document, r_s^1 and r_s^2 are given by:

$$r_s^1 = \left(\frac{\frac{R_c^{50}}{u_2} + \left(\frac{D_{50}}{D_2}\right)^{pref} (r_{clim})^{pref}}{\frac{302}{u_2} + \left(\frac{D_{50}}{D_2}\right)^{pref} (r_{clim})^{pref}}\right) \times \left(\frac{\left(\Delta^{pref}+\gamma\right)\frac{302}{u_2} + 70\gamma}{\gamma}\right) \quad \text{(A6)}$$

$$r_s^2 = \frac{\left(\Delta^{pref}+\gamma\right)}{\gamma}\frac{R_c^{50}}{u_2} \quad \text{(A7)}$$

where, Δ^{pref} is the value of Δ calculated at temperature T^{pref} (kPa/°C), T^{pref} being the air temperature when the value of K_c^{FAO} was derived (°C) and $\left(\frac{D_{50}}{D_2}\right)^{pref}$ is the value of $\left(\frac{D_{50}}{D_2}\right)$ at T^{pref}.

$(r_{clim})^{pref}$ can be obtained from Equation (3) as:

$$(r_{clim})^{pref} = 104\left[1.26\frac{[\Delta^{pref} + 1.67\gamma]}{\Delta^{pref}+\gamma}\right] - 1 \quad \text{(A8)}$$

Cassava (*Manihot esculenta* Crantz) Tuber Quality as Measured by Starch and Cyanide (HCN) Affected by Nitrogen, Phosphorus, and Potassium Fertilizer Rates

I. B. Cuvaca[1], N. S. Eash[1], S. Zivanovic[2], D. M. Lambert[3], F. Walker[1] & B. Rustrick[4]

[1] Department of Biosystems Engineering & Soil Science, University of Tennessee Institute of Agriculture, Knoxville, Tennessee, USA

[2] Department of Food Science and Technology, University of Tennessee Institute of Agriculture, Knoxville, Tennessee, USA

[3] Department of Agricultural & Resource Economics, University of Tennessee Institute of Agriculture, Knoxville, Tennessee, USA

[4] Eastern Produce Kenya Ltd, Nandi Hills, Kenya

Correspondence: N. S. Eash, Department of Biosystems Enginnering and Soil Science, University of Tennessee Institute of Agriculture, 2506 E. J. Chapman Drive, Knoxville, TN 37996-4531, USA.

Abstract

Cassava (*Manihot esculenta* Crantz) is an important subsistence crop for many poor rural families in Africa. Cassava contains cyanogenic glucosides (linamarin and lotaustralin) which liberate hydrogen cyanide (HCN) during tuber processing. Once liberated, HCN attaches to the processed tuber. Continuous consumption of processed tuber containing high HCN concentration coupled with low protein intake causes Konzo – a paralyzing disorder that impacts children and women of childbearing age. There are ways to reduce HCN concentration during tuber processing; however, this can also reduce the overall starch content in the cassava tuber. A study comprising twenty treatments consisting of different combinations of nitrogen (N), phosphorus (P), and potassium (K) fertilizer rates was initiated in 2013 in the coastal Dondo District of Mozambique to assess cassava tuber quality as measured by starch and HCN. Significant differences were observed in starch content (CSC) of unprocessed tubers due to combined addition of N, P and K fertilizer rates, sample size, and estimation procedure. However, no significant differences were observed in HCN concentration in tubers due to the addition of N, P and K fertilizer. The HCN concentration in cassava tuber appears to be a function of the physiology of the crop or possibly cassava variety rather than the environment or conditions under which the crop is grown.

Keywords: cassava, cassava quality, cyanide, fertilizer, rate, starch

1. Introduction

Cassava (*Manihot esculenta* Crantz) is a perennial tuber crop that originated in South America (Hillocks et al., 2002a). Cassava is a major staple and an important source of calories to more than half a billion people in Africa, Asia, and Latin America (Rosenthal & Ort, 2011; FAO, 2013). Its introduction in Africa dates back to the 1550s. Today Africa is the foremost cassava producing region with over half of the global production (FAO, 2013).

Cassava was introduced in Mozambique in the 1750s (Benesi, 2005). Today Mozambique is the fifth largest producer of cassava in Africa and second only to Angola in Southern Africa with an estimated average annual yield of ~6 tons per ha (Dias, 2012). Cassava production is mostly confined to nutrient-depleted soils and small scale resource-poor farmers (El-Sharkawy, 2004) who farm less than 4 hectares (Gwarizimba, 2009). Cassava is a major food crop and the second most important crop in Mozambique following maize (Ministério da Indústria e Comércio/Direcção Nacional do Comércio, 2007). According to Gwarizimba (2009), cassava contributes 6 percent of the country's GDP and 45 percent of the diet, approximately 628 kcal per person per day (Promar Consulting, 2011).

Despite being ranked the fifth largest producer of cassava in Africa (FAO, 2011), cassava production in

Mozambique can be severely limited due to two viral diseases: African Cassava Mosaic Disease (ACMD) and Cassava Brown Streak Disease (CBSD) (Hillocks et al., 2002b). CBSD was reported to be largely confined to the coastal regions of eastern and southern Africa. Recent proliferation of this disease (CBSD) has caused yield losses of up to 50 percent in the Northern provinces of Nampula and Zambezia, Mozambique (Promar Consulting, 2011) where cassava production is concentrated (Hillocks et al., 2002b).

Traditionally, cassava is grown in low input systems and nutrient-poor soils either in association with other crops (such as maize (*Zea mays*) and pigeon-pea (*Cajanus cajan*) (Hillocks et al., 2002a)) or as a sole crop, without any addition of fertilizer (El-Sharkawy, 2004) which leads to continuous mining of nutrients and subsequent soil degradation. Cassava is usually planted at the beginning of the rainy season but, unlike other crops, it (the leaves and tubers) can be harvested throughout the year as needed.

Cassava contains two cyanogenic glucosides (linamarin and lotaustralin) that liberate hydrogen cyanide (HCN) upon tissue disruption (Magnuson, 1997). Their presence in the plant, especially the tuber, has partly been associated with the poor conditions under which the crop is grown (Cardoso et al., 2005; Essers, 1988) with drought being one of the few parameters that has been investigated (de Bruijn, 1973; Nwosu & Onofeghara, 1991). Cassava HCN protects the plant against animal and insect predation (Saunders, 2012) but it is also an important determinant of tuber quality and use. Based solely on the content of HCN, cassava varieties are bitter (high HCN, can reach 250 mg per kg fresh matter or more) or sweet (low HCN, ~50 mg per kg fresh matter) (FAO, 1977). Bitter varieties are more drought tolerant (Essers, 1988) and mainly used for famine relief and nonfood starch based industrial products such as paper, ethanol, thermobioplastics, and pharmaceuticals (FAO, 2006). Sweet cassava varieties are mainly used for food. Despite its sweetness, sweet varieties can still be dangerous to humans if poorly processed tubers are used for human consumption (Braidotti, 2011). According to Cardoso et al. (2005) and Braidotti (2011), continuous consumption of poorly processed cassava tubers and their products coupled with low protein intake cause Konzo – a chronic dietary disease caused by high cassava HCN that impacts mainly young women and children in poor cassava-producing countries in eastern Africa. Konzo victims are unable to walk, stand and, in severe cases, unable to speak (Braidotti, 2011). Blanching, washing, and fermentation (Nambisan, 2011) are three common methods used to lower cassava HCN present in the tubers (Lambri et al., 2013). Although these methods are very well diffused among cassava producing countries, the growing number of Konzo victims in eastern Africa raises questions on their effectiveness in lowering HCN. While it is mandatory to use blanching, washing and/ or fermentation to reduce cassava HCN concentration in the tubers if it is to be used for human consumption, these methods reduce the actual cassava starch content (CSC) which is another important element in defining cassava tuber quality. Cassava starch can be retained in the water during the processing; however, losses can be minimized by harvesting the starch retained in the water used for fermentation or blanching. To avoid underestimation of the actual CSC in the tubers, a real time in-field method called the Underwater Weight Method (ISI, 1999) was developed and made available for commercial use. However, to date, there is very little literature on studies undertaken to assess the relationship between cassava starch, HCN, and fertilizer rates. It is hypothesized that nutrient-depleted soils and smallholder farmer inability to replenish soil nutrients with mineral fertilizer are also associated with high concentrations of HCN and starch present in the tubers. To investigate this hypothesis, a no-till cassava fertilizer study was conducted to determine nitrogen (N), phosphorous (P), and potassium (K) fertilizer rates for optimum cassava quality (low tuber HCN and high starch) for the coastal Dondo District of Mozambique. An underwater weight (International Starch Institute, 1999) and alkaline picrate (Sarkiyayi & Argar, 2010) methods were used to estimate actual CSC (%) and HCN levels, respectively.

2. Method

2.1 Experimental Site

The experiment was conducted at Milha-14 (19°25′54.0″S, 34°43′28.6″E), in the coastal Dondo District, central Province of Sofala, Mozambique, over the 2013/14 agricultural year. Mozambique is divided into ten provinces and an equal number of agro-ecological zones or regions (Ministry of Agriculture & Fisherie, 1996). These agro-ecological zones are identified based on climate, soil type, elevation, and farming system (Maria & Yost, 2006). Milha-14 falls within the agro-ecological region R5; which is suitable to very suitable for cassava production (MAF, 1996). R5 has low altitude ranging from 0-200 m above sea level, annual average temperature of 24 °C, rainfall index ranging from 1,000 to 1,400 mm, and soil texture ranging from sand to sandy loam. The soils at Milha-14 are relatively young, thus suggesting that they have been eroded and re-deposited by water. Field observations suggest that the soils at the site (Milha-14) are Inceptisols with a high water table at or near the surface for most of the year which prevents drainage and leads to near continuous waterlogging. The shallow depth at which reduced metal (iron and manganese) oxides occur suggests that an Aquept is the dominant suborder (Soil Survey Staff, 2010).

2.2 Experimental Design and Management

Twenty fertilizer treatments consisting of different combinations of contrasting N, P, and K rates (Table 1) were laid out in a completely randomized design (CRD) with four replicates each. Each fertilizer combination consisted of three nutrients total; two applied at fixed rates (N, P, and K fixed at 60, 60, and 150 kg/ha, respectively) and one (the one being tested) applied at varying rates. This was done to allow us to understand how changing each one of these fixed rates individually would affect our response variables if the remaining fixed rates were unchanged. A bitter cassava variety resistant to ACMD, Tapioca, was manually planted on untilled plots at a row by plant spacing of 1-m by 1-m (~10,000 plants per ha) to investigate the effect of combined application of N, P, and K fertilizer rates on CSC and HCN present in the tubers. Planting and fertilizer application were completed manually. Urea, single super phosphate, and potassium chloride were used as the N, P, and K sources, respectively. The N, P, and K were applied as basal fertilizer. Manual weeding was completed at the onset of the cropping season. Plants were harvested manually to avoid damaging the tubers, and the harvestable area (net plot) consisted of four plants harvested from two central rows (only) of a four row plot with four plants each (16 plants per plot total).

2.3 Cassava Starch Content (CSC) Estimation Procedure and Assessment

The CSC of fresh tubers was determined using a method known as the Underwater Weight Method (International Starch Institute, 1999), used in cassava commercialization to estimate the content of starch present in the tubers and its respective market price. This method is mostly used in areas lacking laboratory facilities and more sophisticated equipment for measuring starch; hence, it could be a good fit to small scale resource-poor farmers who grow cassava at Milha-14 – an area that still lacks basic infrastructure. The CSC was estimated by determining root (or tuber) specific gravity (SG). The middle portion of the tuber was selected, and SG was determined from tuber samples of approximately 3 kg or 3 kg and 5 kg depending on the below ground fresh tuber mass weight harvested from each plot by measuring the weight lost by peeled tubers submerged in water (if there was enough mass, both measurements were completed). CSC was estimated using three CSC estimation equations (below) used for estimating CSC in cassava commercialization. Estimates from each treatment were compared among themselves, and then assessed by comparison with the starch content estimates obtained from a parallel analysis run on a portion of eight samples randomly selected from each plot. The parallel analysis consisted of an enzymatic procedure known as the Megazyme Total Starch Procedure which has its basis on the AOAC Method 996.11 and AACC Method 76.13 with improvements (Megazyme International Ireland, 2011).

$$CSC(\%) = (SG - 1.00906)/0.004845 \quad \text{(Sungzikaw, 2008)} \tag{1}$$

$$CSC(\%) = (SG - 1.01506)/0.0046051 \quad \text{(ISI, 1999)} \tag{2}$$

$$CSC(\%) = (210.8 \times SG) - 213.4 \quad \text{(Khanthavong et al., 2012)} \tag{3}$$

2.4 Cassava Cyanide (HCN) Concentration Estimation Procedure

To determine cassava HCN concentration present in the tubers, the samples were peeled, cleaned, chopped into approximately 3-mm slices, air/sundried first, and dried to a constant weight in an oven (75 °C for 72 hours). Each sample was ground into flour and then homogenized with a No. 40 mesh sieve to ensure equal distribution of HCN in the sample. Subsequently, a 5 g subsample was drawn and each subsample was mixed with 50 mL distilled water and agitated for 16 hours (35 °C) in a shaking water bath. Extracts were then taken for analysis. The tuber HCN concentration was determined using a method with alkaline picrate (Sarkiyayi & Argar, 2010). Cassava HCN concentration was determined by measuring absorbance at 490 nm using a spectrophotometer (UV-2101 PC, Shimadzu).

Similar to CSC (%), results of cassava HCN concentration (mg per 100 g dry sample) analysis are reported on a dry weight basis (Table 1).

2.5 Data Analysis

Starch and cyanide data were analyzed using SAS 9.3, Cary NC. Statistical analyses were performed using ANOVA and means compared (Table 1) using LSD at $\alpha = 0.05$.

3. Results and Discussion

The results of the two quality parameters analyzed in this study consist of (1) the estimate of the starch content (CSC%) and (2) the concentration of HCN (mg per 100 g dw) present in the tubers as presented in Table 1. CSC estimates obtained using the Underwater Weight Method differed significantly due to fertilizer addition ($p < 0.05$) except when a 3 kg sample size was used ($p > 0.05$). Average estimates obtained using the 3 kg sample size revealed the highest CSC (65.9%) for both treatments 3 and 17 (0-60-90 and 60-60-90 kg/ha N-P_2O_5-K_2O, respectively) and the lowest CSC (44.2%) for treatment 10 (60-0-150 kg/ha N-P_2O_5-K_2O) whereas average estimates obtained using the 5 kg sample size revealed the highest CSC (49.8%) for treatment 13 (60-90-150 kg/ha N-P_2O_5-K_2O) and the lowest CSC (22.1%) for treatment 8 (75-60-150 kg/ha N-P_2O_5-K_2O) (Table 1).

In general, an overall decreasing trend in CSC (%) estimates was observed with increasing addition of N (treatments 5 through 9) for both 3 kg and 5 kg sample sizes (Figures 1 and 2), and increasing addition of K (Figure 1, treatments 14 through 20) for 3 kg sample size only. On the other hand, an increasing trend in CSC (%) estimates was observed with increasing addition of P (Figures 1 and 2, treatments 10 through 13) for both sample sizes (3 kg and 5 kg). Previous studies showed that the above ground plant weight is directly related to N (Gardner et al., 1985; Chintala et al., 2012a; 2012b). Ukaoma and Ogbonnaya (2013) found similar results in that N fertilization must be effectively matched with sufficient P and K else yields can be depressed significantly. Therefore, excess N or adding N alone can favor more biomass production at the expense of tuber growth as added N may promote vegetative growth while simultaneously limiting carbohydrate storage in the tuber.

Figures 1 and 2 show that CSC (%) estimates from the three estimation Equations (1), (2), and (3) adopted in the underwater weight method did not differ significantly within each sample size (3 kg and 5 kg samples), $p < 0.05$. However, significant differences were detected between CSC (%) estimates of the two sample sizes (3 kg vs 5 kg sample, Figure 3) ($p < 0.05$). These differences in CSC (%) estimates due to sample size disagree with the findings of Sungzikaw (2008) who detected insignificant differences in CSC (%) estimates due to changing the size of cassava tuber samples from 1 kg to 3 kg, and 3 kg to 5 kg. On the other hand, in determining the dry matter content (DMC) of cassava tubers using specific gravity method, Teye et al. (2011) reported differences in DMC due to differences in specific gravities obtained from different tuber portions (top, middle and bottom). This could, therefore, explain the differences we detected between estimates derived from different sample sizes since not all treatments we tested yielded enough tubers for us to draw 5 kg samples and in some cases even 3 kg samples, thus, forcing us to rely on portions of the tubers other than the middle.

Figure 4 shows a comparison between the average CSC (%) estimates from contrasting sample sizes (3 kg vs 5 kg) and methods or estimation procedures (the Underwater Weight Method vs the Megazyme Total Starch Procedure). It was observed that the Megazyme Total Starch Procedure had relatively higher CSC (%) than the Underwater Weight Method and all estimates obtained from the three estimation Equations (1), (2), and (3) regardless the sample size. This suggests that both the Underwater Weight Method and the estimation Equations (1), (2), and (3) underestimate the actual CSC (%) when compared with the Megazyme Total Starch Procedure. Furthermore, when the Underwater Weight Method is used, the actual CSC (%) may be underestimated even further if a 5kg sample is used in place of that of 3 kg. This, therefore, suggests that these estimation equations deserve further review, and that, if the Underwater Weight Method is to be used, CSC estimates close to the actual estimate can be obtained if portions of the tuber other than the middle are avoided.

Figure 5 shows cassava HCN concentration as influenced by different combinations of contrasting N, P, and K fertilizer rates, with cassava HCN concentration ranging from 46.0 mg (treatment 3: 0-60-0 kg/ha N-P_2O_5-K_2O) to 95.2 mg (treatment 4: 0-0-150 kg/ha N-P_2O_5-K_2O) per 100 g dw (Table 1). These numbers agree with the typical level of HCN reported to be present in cassava tubers (0.1 to 155 (Cardoso et al., 2005) and/or up to 200 mg per 100 g dw (European Food Safety Authority, 2009)). This, therefore, suggests that, like a normal cassava tuber, prior to its consumption, a rigorous tuber processing should be undertaken in order to bring its HCN concentration to the safe limit of 1 mg per 100 g as set by the World Health Organization (FAO/WHO, 1991).

Despite the need to lower the concentration of HCN in the tubers, it was observed that tuber HCN concentration did not differ significantly due to fertilizer addition ($p > 0.05$), which disagrees with de Bruijn (1973) who in studying the factors influencing cyanogenesis, found that cassava tuber HCN concentration increased with N, decreased with K, and did not change with P fertilization. Likewise, the Caribbean Agricultural Research and Development Institute (1992) reported higher cassava tuber HCN concentration and bitterness due to increased N addition as compared to that of K. It is likely that no significant differences in the HCN concentration present in the tubers were detected with the addition of N, P, K fertilizer due to the large variability in the data (LSD = 42.11, Table 1) which is believed to have occurred due to several biotic and abiotic factors whose impacts on the experiment could not be controlled. This includes (1) a predominantly high water table throughout the cropping season (cassava is very susceptible to waterlogging (FAO, 2013)), and (2) lack of disease-free planting material of good quality (consistent size) – a major limiting factor to high tuber yields in Africa (FAO, 2013).

Field observations suggest that these factors may have contributed to the overall variability and therefore few significant differences not only in the tuber HCN data but and the CSC response to added fertilizer (Table 1).

Due to limited availability of disease-free planting material, stem cuttings of different sizes taken from the top (green) and bottom (high lignin) portions of plants were used so that a plant density equivalent to 10,000 plants per ha could be met. As several studies observed, stem cuttings taken from portions of the plant other than the central part dehydrate very quickly, produce less vibrant shoots, and are susceptible to pests and diseases. As a result, their growth and productivity may be poor (James et al., 2000; Eke-Okoro et al., 1999; FAO, 2013) and diminish tuber quality. During the study period, rainfall was unevenly distributed with more rainfall occurring during the first two to three months after planting and again at the end of the study period. Overall, the site received a total rainfall of 1,831.6 mm which is approximately 50% above its typical average (1,000-1,400 mm). This excessive and sporadic rainfall impacted plant growth and resulted in substantial waterlogging. Coupled with a very shallow water table and a soil texture with low nutrient retention capacity (sand to sandy loam), these conditions may have resulted in nutrient (K) leaching from the soil and also favored denitrification. Additionally, in contrast to local practices, land preparation did not include tilling the plots (no-till) and building bunds prior to planting. As a result, the belowground part of the plant (tubers) remained near the water table during most of the cropping season, resulting in inefficient fertilizer/nutrient use by the crop and limited tuber growth and development. Another factor that may have contributed to the lack of CSC and tuber HCN response to added fertilizer is the fact that there were very limited options in terms of experimental sites. Site selection was based on the availability of land suitable for cultivating cassava in the study region. At harvest, it was realized that the site had been used for charcoal production in the past, a common practice in Milha-14. As several studies demonstrate, charcoal can negatively impact nutrient availability in the soil (Steiner et al., 2008; Steiner et al., 2009; Chintala et al., 2013; Chintala et al., 2014) thus resulting in low quality tubers.

Site variability, depth to water table, and inconsistency on stem cutting quality (size and thickness) all may have impacted the overall CSC and tuber HCN response to added fertilizer.

Table 1. Summary of cassava starch content (%) and cyanide (HCN) concentration in dry weight basis. Values in parentheses represent the standard error of the mean

Fertilizer Combination in kg per ha			Treatment	Cassava Starch Content (%)							HCN in mg per 100 g
				Commercial Estimation Procedure [(b)]						Megazyme Total Starch Procedure [(a)]	
				3 kg Sample [(a)]			5 kg Sample [(b)]				
N	P_2O_5	K_2O		EE I [(a)]	EE II [(a)]	EE III [(a)]	EE I [(a)]	EE II [(a)]	EE III [(a)]		
0	0	0	1	60.7 (2.3)a	62.6 (2.4)a	61.3 (2.4)a	24.9 (0.5)ab	25.6 (0.5)ab	25.2 (0.5)ab	72.5 (1.0)ab	76.3 (20.4)a
60	0	0	2	46.9 (5.8)a	48.5 (6.1)a	47.5 (5.9)a	ND	ND	ND	71.9 (1.6)b	56.8 (8.3)a
0	60	0	3	63.9 (6.6)a	65.9 (6.8)a	64.6 (6.7)a	36.9 (1.1)a	37.8 (1.2)a	37.1 (1.2)a	72.2 (0.5)ab	46 (15.5)a
0	0	150	4	54.4 (6.6)a	55.9 (6.9)a	54.9 (6.7)a	33.4 (4.5)a	34.2 (4.8)a	33.6 (4.6)a	73.3 (1.1)ab	95.2 (5.6)a
0	60	150	5	59.2 (3.2)a	60.9 (3.4)a	59.8 (3.3)a	36.8 (1.5)a	37.7 (1.6)a	37.0 (1.6)a	70.8 (1.5)b	70.7 (21.5)a
25	60	150	6	63.5 (2.4)a	65.5 (2.5)a	64.1 (2.4)a	36.6 (1.4)a	37.5 (1.4)a	49.1 (1.4)a	ND	74 (22.4)a
50	60	150	7	47.7 (5.7)a	48.8 (5.9)a	47.9 (5.8)a	42.5 (3.2)a	43.4 (3.4)a	42.7 (3.3)a	ND	71.2 (15.6)a
75	60	150	8	57.5 (3.3)a	59.2 (3.5)a	58.0 (3.4)a	22.1 (3.1)ab	22.6 (3.3)b	22.2 (3.2)ab	ND	64.7 (21.7)a
100	60	150	9	52.1 (5.8)a	53.6 (6.1)a	52.6 (5.9)a	34.3 (1.0)a	35.1 (1.1)a	34.5 (1.1)a	ND	67.9 (18.8)a
60	0	150	10	44.2 (1.3)a	45.5 (1.4)a	44.6 (1.3)a	24.1 (0.2)ab	24.7 (0.3)b	24.3 (0.2)ab	ND	63.9 (18.6)a
60	30	150	11	47.1 (1.6)a	48.6 (1.7)a	47.6 (1.6)a	36.1 (2.1)a	36.9 (2.2)a	36.3 (2.2)a	ND	72.3 (5.1)a
60	60	150	12	59.5 (2.2)a	61.3 (3.1)a	60.1 (3.0)a	ND	ND	ND	73.9 (0.6)ab	62.4 (12.7)a
60	90	150	13	61.4 (1.6)a	63.3 (1.7)a	62.0 (1.6)a	48.6 (0.7)a	49.8 (0.7)a	48.9 (0.7)a	ND	76.3 (6.4)a
60	60	0	14	57.3 (1.3)a	58.9 (1.4)a	57.8 (1.4)a	36.8 (10.9)a	37.4 (11.4)a	36.9 (11.1)a	ND	70.2 (16.1)a
60	60	30	15	63.1 (4.3)a	65.1 (4.5)a	63.8 (4.3)a	29.9 (0.6)ab	25.6 (14.8)b	25.1 (0.6)ab	ND	50.7 (19.8)a
60	60	60	16	59.6 (2.9)a	61.4 (3.0)a	60.2 (2.9)a	29.9 (0.4)ab	25.6 (0.4)b	25.2 (0.4)ab	72.7 (1.5)ab	56.8 (9.8)a
60	60	90	17	63.8 (1.5)a	65.9 (1.6)a	64.5 (1.6)a	37.8 (37.8)a	38.8 (0.3)a	38.1 (0.3)a	ND	62.7 (17.2)a
60	60	120	18	56.9 (7.5)a	58.6 (7.9)a	57.4 (7.6)a	40.0 (40)a	41.1 (3.1)a	40.4 (3.2)a	ND	53.3 (16.3)a
60	60	150	19	60.8 (0.9)a	62.7 (0.9)a	61.4 (0.9)a	36.8 (36.8)a	37.8 (0.2)a	37.1 (0.2)a	76.8 (0.0)a	57.4 (23.7)a
60	60	180	20	53.1 (5.7)a	54.6 (6.0)a	53.6 (5.8)a	33.1 (33.1)a	33.8 (4.4)a	33.7 (4.5)a	ND	62.5 (14.3)a
LSD				20.5	21.3	20.8	32.3	33.2	32.6	6.0	42.1

Note. EE = Estimation Equation: EEI: CSC = (SG – 1.00906)/0.004845 (Sungzikaw, 2008), EEII: CSC = (SG – 1.01506)/0.0046051 (ISI, 1999), EE III: CSC = 210.8*SG – 213.4 (Khanthavong et al., 2012), and ND = No Data available. Means followed by the same letter in the same column are equal at 5% level (LSD). Estimation procedures, sample sizes and EEs followed by the same letter in the same row yielded estimates that are equal at 5% level (LSD).

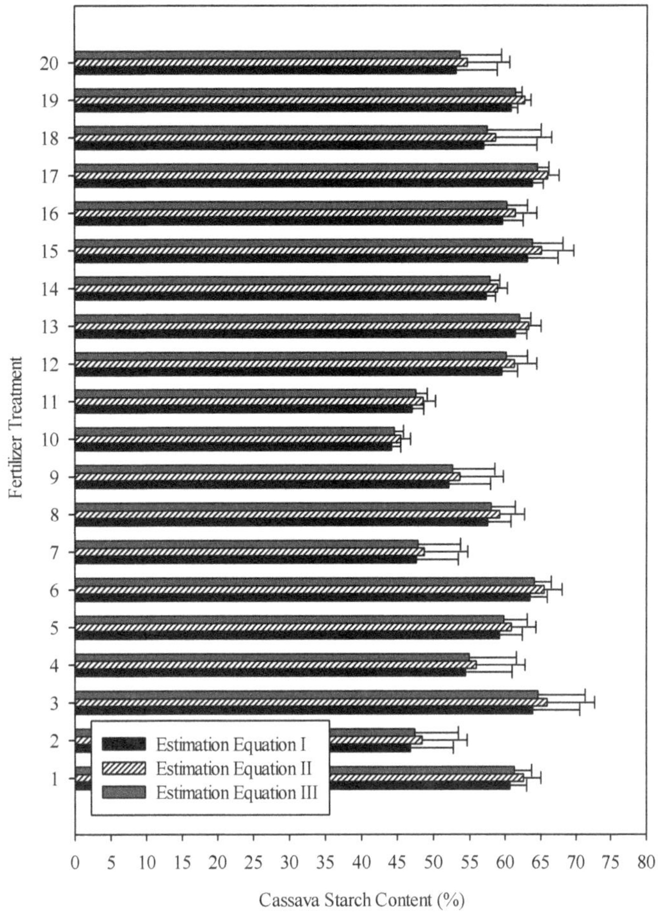

Figure 1. Cassava starch content (CSC) estimation from 3 kg tuber sample in dry weight basis. CSC estimates (X axis) from all three estimation equations are presented for each fertilizer combination (Y axis). Error bars represent the standard error of the mean. Fertilizer treatments one through twenty are different fertilizer combinations (Table 1)

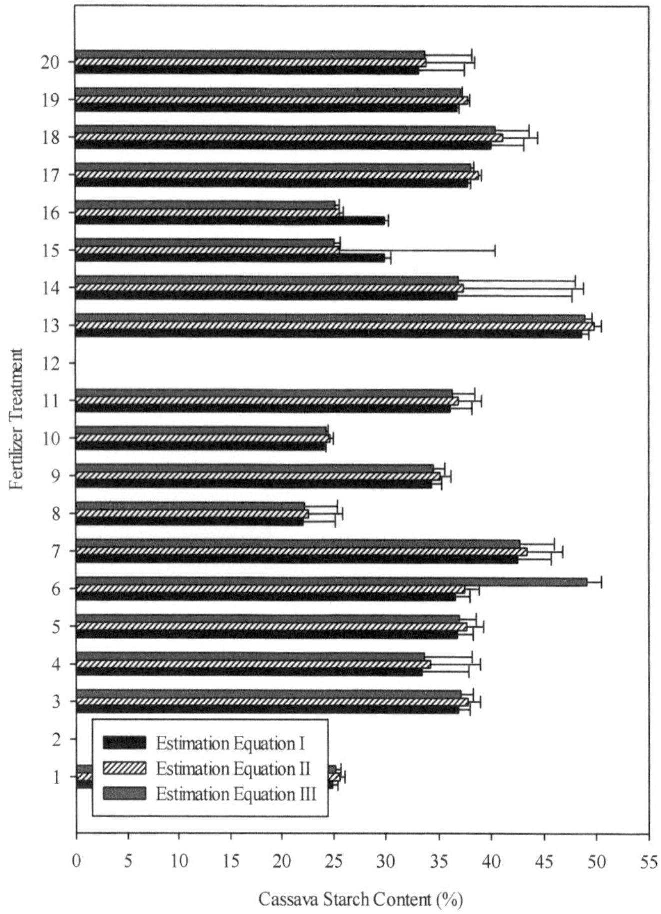

Figure 2. Cassava starch content (CSC) estimation from 5 kg tuber sample in dry weight basis. CSC estimates (X axis) from all three estimation equations are presented for each fertilizer combination (Y axis). Error bars represent the standard error of the mean. Fertilizer treatments one through twenty are different fertilizer combinations (Table 1). Treatments 2 and 12 did not yield enough tubers to estimate CSC from a 5 kg sample

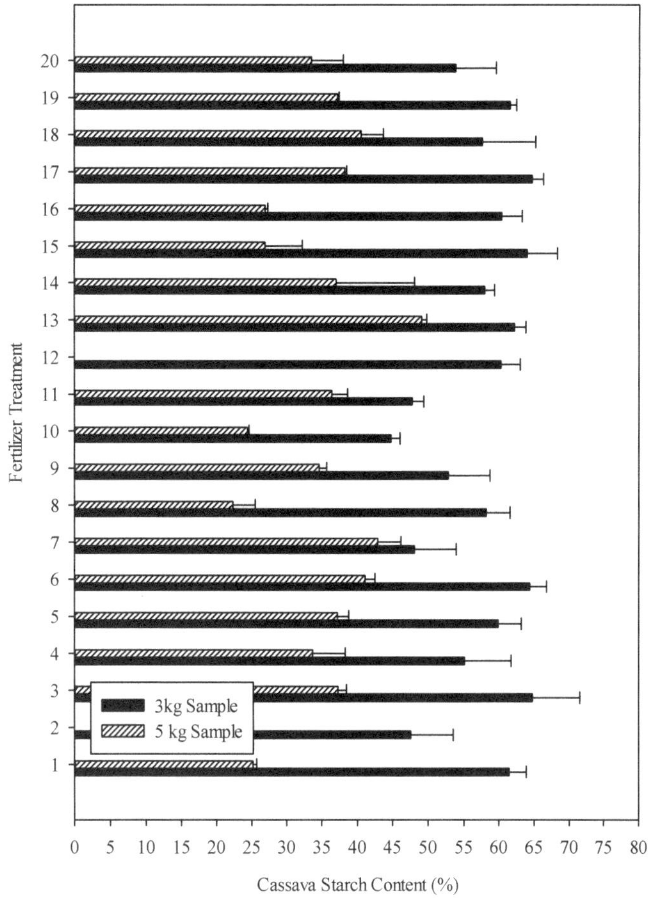

Figure 3. Comparison of cassava starch content (CSC) estimations from contrasting tuber sample sizes in dry weight basis. Average CSC estimates (X axis) from all three estimation equations and two sample sizes (3 kg and kg) are presented for each fertilizer combination (Y axis). Error bars represent the standard error of the mean. Fertilizer treatments one through twenty are different fertilizer combinations (Table 1). Treatments 2 and 12 did not yield enough tubers to estimate CSC from a 5 kg sample

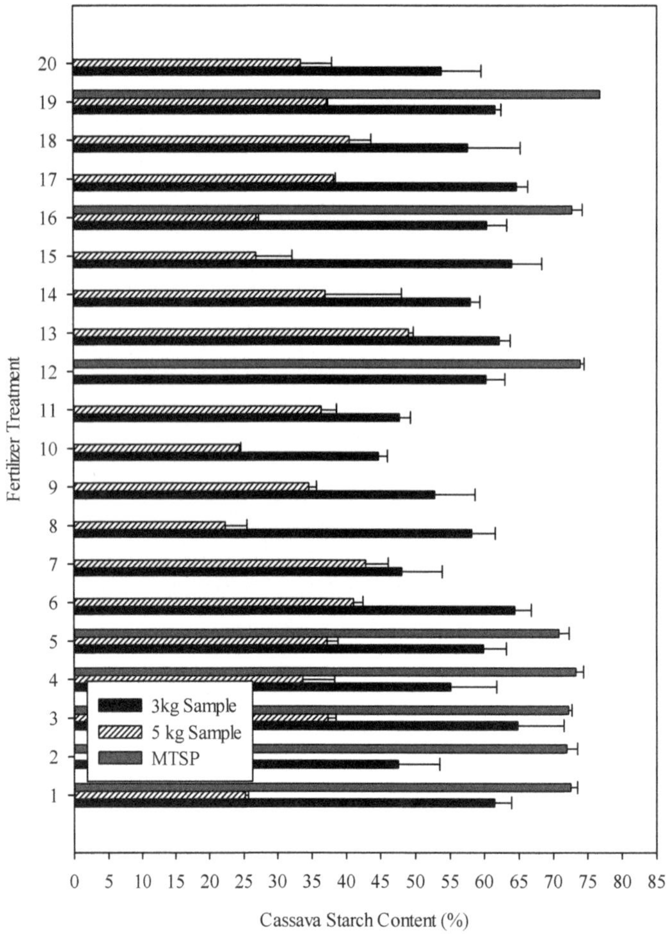

Figure 4. Comparison of cassava starch content (CSC) estimations from contrasting tuber sample sizes and procedures in dry weight basis. Average CSC estimates (X axis) from all three estimation equations and estimation procedures are presented for each fertilizer combination (Y axis). Error bars represent the standard error of the mean. Fertilizer treatments one through twenty are different fertilizer combinations (Table 1). Treatments 2 and 12 did not yield enough tubers to estimate CSC from a 5 kg sample. MTSP stands for Megazyme Total Starch Procedure. The MTSP was run on a total of eight samples randomly selected from each plot

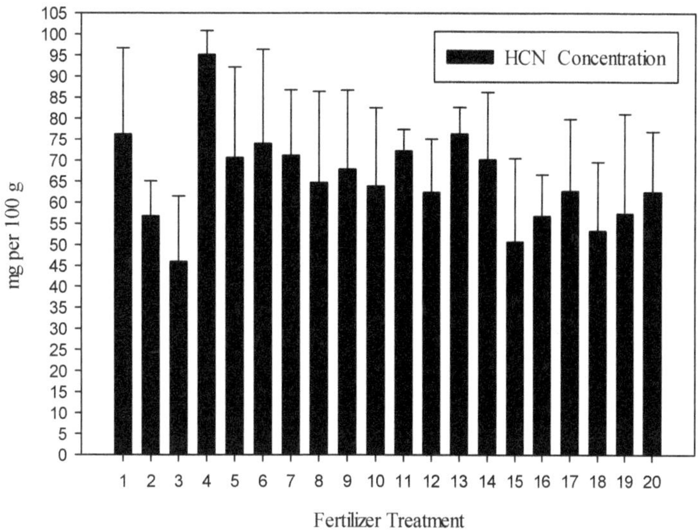

Figure 5. Cassava tuber cyanide (HCN) concentration in dry weight basis. Error bars represent the standard error of the mean. Fertilizer treatments one through twenty are different fertilizer combinations (Table 1)

4. Conclusions

Our findings suggest that cassava starch content (%) is influenced by N, P and K fertilizer rate combination, but that these differences may be confounded by sampling and content determination procedures. CSC estimated using Equations (1), (2), and (3) (used in the Underwater Weight Method) did not differ significantly; however, all three estimation Equations (1), (2), and (3) used in the Underwater Weight Method underestimated actual CSC compared with the results obtained using the Megazyme Total Starch Procedure. On the other hand, it was observed that all estimation Equations (1), (2), and (3) provide CSC estimates that approximate those obtained using the Megazyme Total Starch Procedure if a sample size of 3 kg is used, thus suggesting that famers lacking resources should use a 3 kg sample size for estimating the market price of their starch. It was also observed that cassava tuber cyanide (HCN) is unaffected by N, P and K fertilizer rate combination. This suggests that it is likely that cassava HCN is more sensitive to the type of cultivar (bitter or sweet) and crop physiology rather than the nutrient-poor conditions under which the crop is grown. Thus, adoption of low HCN cassava varieties and a more efficient use of existing methods to lowering cassava tuber HCN concentration such as blanching and fermentation should be considered for minimizing the risks of human exposure to cassava HCN and potential poisoning. It is also clear from this analysis that the three estimation Equations (1), (2), and (3) used for estimating actual CSC when the Underwater Weight Method was used deserve further attention. The response curves and overall statistical analysis suggest that high starch production at Milha-14 can be achieved with a combined fertilizer addition of 0-60-0 kg/ha N-P_2O_5-K_2O without affecting the concentration of HCN present in the tubers.

References

Benesi, I. R. M. (2005). *Characterisation of Malawian cassava germplasm for diversity, starch extraction and its native and modified properties* (Ph.D. thesis). University of the Free State. Retrieved June 3, 2014, from http://etd.uovs.ac.za/ETD-db/theses/available/etd-07042006-090839/unrestricted/BENESIIRM.pdf

Braidotti, G. (2011). *Cassava, cyanide and Konzo disease.* Focus on NGOs. Retrieved December 19, 2013, from http://aciar.gov.au/files/node/13748/cassava_cyanide_and_konzo_disease__75809.pdf

Cardoso, A. P., Mirione, E., Ernesto, M., Massaza, F., Cliff, J., Hare, M. R., & Bradbury, J. H. (2005). Processing of cassava roots to remove cyanogens. *Journal of Food Composition and Analysis, 18*(2005), 451-460. http://dx.doi.org/10.1016/j.jfca.2004.04.002

Caribbean Agricultural Research and Development Institute (CARDI). (1992). *Commercial Cassava Production. Factsheet.* Frances Chandler. Order no.: CP-F/ 16-89.

Chintala, R., McDonald, L. M., & Bryan, W. B. (2012a). Effect of soil water and nutrients on productivity of

Kentucky bluegrass systems in acidic soils. *Journal of Plant Nutrition, 35*, 288-303. http://dx.doi.org/10.1080/01904167.2012.636131

Chintala, R., McDonald, L. M., & Bryan, W. B. (2012b). Optimization of water potential and nutrient levels for Kentucky bluegrass-white clover mixture on acidic soils. *Biotechnology, Agronomy, Society and Environment, 16*, 167-177.

Chintala, R., Mollinedo, J., Schumacher, T. E., Malo, D. D., Papiernik, S., Clay, D. E., ... Gulbrandson, D. W. (2013). Nitrate sorption and desorption by biochars produced from microwave pyrolysis. *Microporous and Mesoporous Materials, 179*, 250-257. http://dx.doi.org/10.1016/j.micromeso.2013.05.023

Chintala, R., Schumacher, T. E., McDonald, L. M., Clay, D. E., Malo, D. D., Clay, S. A., ... Julson, J. L. (2014). Phosphorus sorption and availability in biochars and soil biochar mixtures. *CLEAN-Soil Air Water, 42*(5), 626-634. http://dx.doi.org/10.1002/clen.201300089

De Bruijn, G. H. (1973). The cyanogenic character of cassava (*Manihot esculenta*). *Chronic cassava toxicity* (pp. 43-48). Proceedings of an interdisciplinary workshop. London, England: Mt. Develop. Res. Centre Monogr. IDRC-OLOe.

Dias, P. (2012). *Analysis of incentives and disincentives for cassava in Mozambique*. Technical notes series, MAFAP, FAO, Rome. Retrieved June 19, 2014, from http://www.fao.org/fileadmin/templates/mafap/documents/technical_notes/MOZAMBIQUE/MOZAMBIQUE_Technical_Note_CASSAVA_EN_Oct2012.pdf

Eke-Okoro, O. N., Okereke, O. U., & Okele, J. E. (1999). Effects of weather change and planting set on growth and productivity of cassava in southeastern Nigeria. *African Journal of Root and Tuber Crops, 3*(2), 34-38.

El-Sharkawy, M. A. (2004). Cassava biology and physiology. *Plant Molecular Biology, 56*, 481-501. http://dx.doi.org/10.1007/s11103-005-2270-7

Essers, S. (1988). Bitter cassava as a drought resistant crop: A case in Mozambique. *ILEIA Newsletter, 4*(4). Retrieved August 19, 2014, from http://www.agriculturesnetwork.org/magazines/global/enhancing-dryland-agriculture/bitter-cassava-as-a-drought-resistant-crop-a-case

European Food Safety Authority (EFSA). (2009). EFSA SCIENTIFIC COOPERATION (ESCO) REPORT: EFSA Compendium of botanicals that have been reported to contain toxic, addictive, psychotropic or other substances of concern. *EFSA Journal, 7*(9), 281. Retrieved December 4, 2014, from http://www.efsa.europa.eu/en/scdocs/doc/280rax1.pdf

FAO. (1977). Cassava processing. *FAO Corporate Document Repository* (p. 17). ISBN 92-5-100171-5. Retrieved December 3, 2014, from http://www.fao.org/docrep/X5032E/x5032E01.htm

FAO. (2006). *Starch market adds value to cassava*. Retrieved December 2, 2014, from http://www.fao.org/ag/magazine/pdf/0610-1.pdf

FAO. (2011). *Food Outlook: Global Market Analysis*. Retrieved from http://www.fao.org/docrep/014/al981e/al981e00.pdf

FAO. (2013). *Save and Grow: Cassava*. A Guide to Sustainable Production Intensification, Rome. E-ISBN 978-92-5-107642-2. Retrieved from http://www.fao.org/docrep/018/i3278e/i3278e.pdf

FAO/WHO. (1991). Joint FAO/WHO Food Standards Programme. *Codex Alimentarius Comission XII* (Supplement 4). FAO, Rome, Italy.

Gardner, F. P., Pearce, R. B., & Mitchell, R. L. (1985). *Physiology of Crop Plants*. Iowa State University Press.

Gwarizimba, V. (2009). *Cotton and cassava seed systems: Malawi, Mozambique and Zambia* (p. 7, p. 11, p. 28). FAO. All Agricultural Commodities Programme, Zimbabwe.

Hillocks, R. J., Thresh J. M., & Bellotti, A. C. (2002a). Cassava: Biology, Production and Utilization. *The origins and taxonomy of cassava* (Chapter 1, pp. 1-11). CABI Publishing. New York, USA. ISBN 0851995241. http://dx.doi.org/10.1079/9780851995243.0000

Hillocks, R. J., Thresh, J. M., Tomas, J., Botao, M., Macia, R., & Zavier, R. (2002b). Cassava brown streak disease in northern Mozambique. *International Journal of Pest Management, 48*(3), 178-181. http://dx.doi.org/10.1080/09670870110087376

International Starch Institute (ISI). (1999). ISI 13-2e Determination of starch in tubers by underwater weight. Science Park Aarhus, Denmark.

James, B., Yaninek, J., Tumanteh, A., Maroya, N., Dixon, A., Salawu, R., & Kwarteng, J. (2000). *Starting a*

Cassava Farm. IPM Field Guide for Extension Agents, Nigeria. ISBN 978-131-173-8. Retrieved June 4, 2014, from http://www.infonet-biovision.org/res/res/files/1855.Starting.pdf

Khanthavong, P., Phattarakul, N., Jamjod, S., Aye, T. M., & Rerkasem, B. (2012). Effect of Stake Priming with Complete Nutrient Solution on Cassava Root and Starch Yield. *CMU. J. Nat. Sci. Special Issue on Agricultural and Natural Resources, 11*(1), 75.

Lambri, M., Fumi, M. D., Roda, A., & De Faveri, D. M. (2013). Improved processing methods to reduce the total cyanide content of cassava roots from Burundi. *Afr. J. Biotechnol, 12*(19), 2685-2691.

Magnuson, B. (1997). *Cyanogenic Glycosides*. Department of Food Science and Toxicology, University of Idaho. Retrieved from http://extoxnet.orst.edu/faqs/natural/cya.htm

Maria, R. M., & Yost, R. (2006). A survey of soil fertility status of four agro-ecological zones of Mozambique. *Soil Science, 171*(11), 902-914. http://dx.doi.org/10.1097/01.ss.0000228058.38581.de

Megazyme International Ireland (MII). (2011). *Megazyme Total Starch Assay Procedure (Amyloglucodsidase/α-Amylase Method)*. AOAC Method 996.11 and AACC Method 76.13 (and improvements). K-TSTA 07/11.

Ministério da Indústria e Comércio/Direcção Nacional do Comércio (MIC/DNC). (2007). *Sub-Sector Strategic Study on Cassava: Cassava Development Strategy for Mozambique (2008-2012)* (Vol. 1).

Ministry of Agriculture & Fisherie (MAF). (1996). *Agroecological Zones and Production Systems*. Working Document 2/b, Program of Investment in Agricultural Extension, Process of the Formulation of Proagri. Maputo, Mozambique; Ministry of Agriculture and Fisheries.

Nambisan, B. (2011). Strategies for elimination of cyanogens from cassava for reducing toxicity and improving food safety. *Food Chem. Toxicol., 49*, 690-693. http://dx.doi.org/10.1016/j.fct.2010.10.035

Nwosu, L. A., & Onofeghara, F. A. (1991). *A comparison of cyanide accumulation, leaf retention and Linamarase activity of cassava varieties during water stress* (pp. 187-193). Tropical Root Crops in a Developing Economy: Proceeding of the ninth Symposium of the International Society for Tropical Root Crops held in Accra, Ghana. October 20-26 1991.

Promar Consulting. (2011). Subsistence Agriculture Study. *The Cassava Industries in Mozambique and Tanzania: Production, processing, distribution and consumption of cassava and its related policy challenges* (pp. 79-81). Retrieved from http://www.apip-apec.com/ja/good-practices/files/The_Cassava_Industries_in_Mozambique_and_Tanzania-2011.pdf

Rosenthal, D. M., & Ort, D. R. (2011). Examining Cassava's Potential to Enhance Food Security Under Climate Change. *Tropical Plant Biology, 4*(3-4), 145-250. http://dx.doi.org/10.1007/s12042-011-9086-1

Sarkiyayi, S., & Agar, T. M. (2010). Comparative analysis on the nutritional and anti-nutritional contents of the sweet and bitter cassava varieties. *Adv. J. Food Sci. Technol., 2*, 328-334.

Saunders, D. A. (2012). *When plants bite back: A broadly applicable method for the determination of cyanogenic glycosides as hydrogen cyanide in plant-based foodstuffs*. Chemistry in New Zealand. Retrieved December 4, 2012, from http://nzic.org.nz/CiNZ/articles/2012/CiNZ%20OCT%202012_Saunders.pdf

Soil Survey Staff. (2010). *Keys to Soil Taxonomy* (11th ed.). USDA – Natural Resources Conservation Service, Washington DC 338.

Steiner, C., Garcia, M., & Zech, W. (2009). Effects of Charcoal as Slow Release Nutrient Carrier on N-P-K Dynamics and Soil Microbial Population: Pot Experiments with Ferrasol Substrate. In W. I. Woods et al. (Eds.), *Amazonian Dark Earths: Wim Sombroek's cisio* (pp. 325-338). Springer Science & Business Media B.V. http://dx.doi.org/10.1007/978-1-4020-9031-8_17

Steiner, C., Glaser, B., Teixeira, W. G., Lehmann, J., Blum, W. E. H., & Zech, W. (2008). Nitrogen retention and plant uptake on a highly weathered central Amazonian Ferralsol amended with compost and charcoal. *J. Plant Nutr. Soil Sci., 171*, 893-899. http://dx.doi.org/10.1002/jpln.200625199

Sungzikaw, S. (2008). *Measurements of starch content in cassava*. Workshop on metrology in food safety, agricultural products and product safety, Hangzhou, China.

Teye, E., Asare, A. P., Amoah, R. S., & Tetteh, J. P. (2014). Determination of the dry matter content of cassava (*Manihot esculenta,* Crantz) tubers using specific gravity method. *ARPN Journal of Agriculture and Biological Science*. ISSN 1990-6145. Retrieved December 4, 2014, from http://www.arpnjournals.com/jabs/research_papers/rp_2011/jabs_1111_332.pdf

Ukaoma, A., & Ogbonnaya, C. (2013). Effect of inorganic mineral nutrition on tuber yield of cassava (*Manihot esculenta Crantz*) on marginal ultisol of South Eastern Nigeria. *Acad. J. Agric. Res., 1*(9), 172-179.

Radish Planting Date and Nitrogen Rate for Cover Crop Production and the Impact on Corn Yields in Upstate Missouri

Leah Sandler[1], Kelly A. Nelson[2] & Christopher J. Dudenhoeffer[1]

[1] Lee Greenley Jr. Memorial Research Center, University of Missouri, Novelty, MO, USA

[2] Division of Plant Sciences, University of Missouri, Novelty, MO, USA

Correspondence: Kelly A. Nelson, University of Missouri, Greenley Research Center, Novelty, MO 63460, USA. E-mail: nelsonke@missouri.edu

Abstract

Cover crops can increase yield of a marketable crop, offer greater yield stability, reduce fertilizer inputs, suppress weeds, and interrupt disease or pest cycles. Producers increasingly look to radish (*Raphanus sativus* L.) as a cover crop, but integrating it into a cropping system with corn (*Zea mays* L.) and soybean [*Glycine max* (L.) Merr.] has been difficult. Radish following wheat (*Triticum aestivum* L.) may provide more options for farmers. Few studies have tested the effects of nitrogen (N) rates on radish production for a dual purpose cover/forage crop or whether additional N is needed because radish is often used to trap N. This study evaluated radish planting date (1 Aug., 15 Aug., 29 Aug., and 11 Sep. in 2012 and 2013) and N fertilizer amount (0, 17, 33, and 67 kg N ha^{-1}) on radish heights, chlorophyll content, and biomass production as well as their impact on corn yields the next year (2013 and 2014). Across planting dates and years, non-fertilized controls had shorter radishes than any N rate, while radish chlorophyll content increased as N rates increased. Chlorophyll and biomass content generally did not significantly increase above 33 kg N ha^{-1}. Radish foliage dry biomass production was 820 to 1670 kg ha^{-1} greater at the first planting date (1 Aug.) in 2012 and 2013 than the other planting dates. Winter annual weed suppression was greatest at the first planting date in 2012. No differences in winter annual weed suppression were observed in 2013 compared to the other planting dates. Radish planting date did not affect corn yield the following year (2013 and 2014). This study demonstrates the benefits of planting radishes in late summer prior to 1 Sep. and fertilizing with 33 kg N ha^{-1} for optimum radish biomass production following wheat in upstate Missouri.

Keywords: cover crop, corn, forage, nitrogen, no-till, radish, seeding date

1. Introduction

Cover crops are plants grown in a rotation when main crops are absent (Mohammadi, 2013). Cover crops offer many benefits, the most direct positive of which is increased yield of a marketable crop. They also increase yield stability, reduce fertilizer inputs, suppress weeds, break disease or pest cycles, increase soil and water quality, and increase nutrient cycling efficiency (Thiessen Martens et al., 2001; Snapp et al., 2005; Hoffbeck et al., 2008). Cover crops also aid in soil conservation, and reduce soil erosion, nitrate leaching, and chemical runoff (Snapp et al., 2005). Such crops increase water infiltration, retain soil moisture, improve soil tilth, and increase nutrient cycling efficiency (Teasdale, 1996). Finally, planting a cover crop replaces unmanageable weed populations such as common lambsquarters (*Chenopodium album* L.), redroot pigweed (*Amaranthus retrofelxus* L.), and some annual grasses with a manageable crop or mulch population (Teasdale, 1996). Producers considering using cover crops should determine which benefits they seek, as well as determining the seeding timing and type. For winter cover crops, a cash crop does not have to be sacrificed, and planting normally occurs in the fall after harvest of an annual cash crop (Snapp et al., 2005).

Radish's (*Raphanus sativus* L.) myriad of benefits, include improving soil aeration, suppressing weeds, capturing nutrients, and possibly increasing yields of grain crops in the crop rotaiton. Sojka et al. (1991) noted that, in addition to suppressing weeds, radish's typical cover-crop benefits include reducing soil erosion, increasing soil moisture infiltration and adding organic matter to soil. In Maryland, forage radishes planted before 1 Sep. produced 3.9 to 6.6 Mg ha^{-1} shoot dry matter and 1.3 to 3.2 Mg ha^{-1} tuber dry matter (Lawley et al., 2011). Although corn (*Zea mays* L.) tested after radish and other cover crops showed no yield difference, corn populations decreased in the several site years following annual ryegrass (*Lolium perenne* L.). Radishes have a small C:N ratio.

Lawley (2011) found that cover crops that did not overwinter, such as radish, left less residue. This simplifies spring seeding, provides warmer soils, and allows for more timely planting of subsequent crops compared to cover crops that overwinter, such as ryegrass (Lawley et al., 2011).

Radishes' planting date influences crop production (Pell at al., 1993; Pandey et al., 2009; Alam et al., 2010; Ebrahimi et al., 2013), including the vegetative and reproductive growth periods as well as the balance between them, which affects yield (Ebrahimi et al., 2013). An appropriate sowing date helps reduce damage from cold, heat, pests, diseases, and weeds. This favorable timing also coincides with climatic factors affecting production, such as coincidence of flowering with suitable temperature, which aids in plant establishment (Ebrahimi et al., 2013). Ideal radish sowing times vary depending on environment and variety. Growers often manipulate sowing times for better growth and yields (Alam et al., 2010). Timing of planting may be significant due to available light. Light treatments affect radish growth and yield (Schmitt et al., 1986; Pell et al., 1993). In a pot experiment, radish growth rates and mass decreased with decreasing light (Schmitt et al., 1986), and photoperiod influenced biomass partitioning (Pell et al., 1993). In addition to the amount of light available, the season may impact radish production (Pell at al., 1993). In growth chambers, radishes exhibited a more adverse response to soil moisture deficits in the spring than fall. Soil moisture reduced CO_2 fixation and thus subsequent biomass accumulation (Pell et al., 1993). The authors hypothesize that spring conditions led to a large relative growth rate and faster water loss by the shoot, thus favoring a negative impact of reduced soil moisture which was more evident in the spring. It is important that timing of crop cycles matches rainfall cycles (Caviglia et al., 2004).

Establishing radish after wheat (*Triticum aestivum* L.) may foster both growth and cover crop benefits. The timing of establishing a second crop early can be important because, after 1 Dec., the rate of decline in yield with delayed sowing is about 1.3% per day for sole-crops and 0.5% per day for double-crops (Caviglia et al., 2011). Sanford et al. (1973) observed that no-till double-crop systems provided the least delay in establishing a second crop. Water available to plants affected winter wheat grain production in a study of tillage and N rates on wheat production (Halverson et al., 1999). No-till and minimal tillage yielded higher than conventional tillage, with grain yields of 2022 kg ha^{-1}, 1968 kg ha^{-1} and 1801 kg ha^{-1}, respectively (Halverson et al., 1999).

In Missouri, standard practice includes adding N fertilizer to non-leguminous crops such as wheat and corn, regardless of the N form. However, few studies have evaluated how N rates affect radish production (Sanchez et al., 1991; Hochmuth et al., 2009) or whether N fertilizer is needed because radish traps N. Nitrogen fertilizer increased seed yield of radish by 1.4 q ha^{-1}, but yields increased to 2.3 q ha^{-1} when 30 kg ha^{-1} of phosphorus was added (Sharma, 2000). In Pakistan, adding N fertilizer significantly increased radish biomass and yield compared to non-fertilized controls (Asghar et al., 2006). However, a Florida study found no response with N fertilizer and so did not recommend it for radish production (Hochmuth et al., 2009). Other research determined that radishes used soil mineral N, and that N fertilization did not increase radish yields in that system (Sanchez et al., 1991). The objective of this research was to evaluate radish planting date and N fertilizer rate on radish height, chlorophyll content, and biomass production as well as the impact on corn yield the following year.

2. Materials and Methods

Field research took place at the University of Missouri Lee Greenley Jr. Memorial Research Center near Novelty (40°01′N, 92°11′W) in 2012 and 2013 while the rotational impact on corn was evaluated in 2013 and 2014, respectively. In the experiment's split-plot design, the main plot was planting date, and the sub-plot was N fertilizer rate. Four replications were planted in 3 by 10.5 m plots on a Kilwinning silt loam (fine, montmorillonitic, mesic, Vertic Ochraqualfs). In a field that had previously been winter wheat, 'Tillage radish' (Cover Crop Solutions LLC, Lititz, PA) were no-till drilled (Great Plains Ag, Salina, KS) in 19 cm rows at 10 kg seed ha^{-1} approximately every two weeks. In 2012, the seeding dates were 1 Aug., 17 Aug., 30 Aug., and 12 Sep. In 2013, the dates were 1 Aug., 14 Aug., 28 Aug., and 11 Sep. Planting dates were averaged as (1 Aug., 15 Aug., 29 Aug., and 11 Sep.) for figures and tables throughout the manuscript. Nitrogen in the form of ammonium nitrate (34-0-0) at 0, 17, 33 and 67 kg N ha^{-1} was applied by hand spreader across all radish planting dates on 5 Oct. 2012 and 12 Sep. 2013 . Radish foliage heights were recorded at 0, 6, 13, 16, 23, 34, and 40 days after N fertilizer treatment (DAT). Clethodim ((E)-2-[1-[[(3-chloro-2-propenyl)oxy]imino]propyl]5-[2-(ethylthio)propyl]-3-hydroxy-2-cyclohexen-1-one) at 0.07 kg ai ha^{-1} and crop oil concentrate at 2.4 L ha^{-1} were applied on 29 Oct. 2012 and 12 Sep. 2013 to reduce interference with wheat. On 12 Nov. 2012 and 22 Oct. 2013, chlorophyll index measurements of the uppermost fully expanded radish leaf were determined using a SPAD meter (Minolta SPAD-502, Konica Minolta Optics, Inc.). Radish top biomass was collected from a 0.3 by 0.3 m quadrat on 13 Nov. 2012 and 12 Nov. 2013. Samples were dried and weights recorded.

Common chickweed [*Stellaria media* (L.) Vill.] and henbit (*Lamium amplexicauli* L.) suppression was visually rated (0=no control to 100=complete control) prior to a burndown herbicide application in the spring the following year prior to planting corn (2013 and 2014). Glyphosate (*N*-(phosphonomethyl)glycine) at 0.84 kg ae ha^{-1} plus saflufenacil (N'-[2-chloro-4-fluoro-5-(3-methyl-2,6-dioxo-4-(trifluoromethyl)-3,6-dihydro-1(2H)-pyrimidinyl)benzoyl]-N-isopropyl-N-methylsulfamide) at 0.025 kg ai ha^{-1} plus diammonium sulfate (DAS) at 20 g L^{-1} plus methylated seed oil at 1% vol./vol. plus was applied on 26 April 2013. A premix of acetochlor (2-chloro-2'-methyl-6'-ethyl-N-ethoxymethylacetanilide) plus atrazine (2-chloro-4-(ethylamino)-6-(isopropylamino)-s-triazine) at 3.4 kg ai ha^{-1} plus glyphosate at 0.84 kg ae ha^{-1} plus DAS at 20 g L^{-1} was applied on 6 May 2014.

Corn (DKC 62-97) was no-till planted (John Deere 7000, Moline, IL) at 81,500 seeds ha^{-1} in 76 cm rows on 8 May 2013 and 11 April 2014. Anhydrous ammonia was applied (John Deere 2510H, Moline, IL) preplant at 170 N ha^{-1} on 26 April 2013 and at 160 kg N ha^{-1} on 8 April 2014. On 27 June 2013, plots were maintained weed-free with a premixture of acetochlor plus clopyralid (3,6-dichloro-2-pyridine carboxylic acid, monoethanolamine salt) plus flumetsulam (N-(2,6-difluorophenyl)-5-methyl-1,2,4-triazolo-[1,5a]-pyrimidine-2-sulfonamide) plus at 1.2 kg ai ha^{-1} plus glyphosate at 0.84 kg ae ha^{-1} plus non-ionic surfactant at 0.25% vol./vol. on 21 May followed by glyphosate at 0.84 kg ae ha^{-1} plus mesotrione (2-[4-(methylsulfonyl)-2-nitrobenzoyl]-1,3-cyclohexanedione) at 0.105 kg ai ha^{-1} plus DAS at 20 g L^{-1}. On 26 June 2014, there was a single postemergence application of glyphosate at at 0.84 kg ae ha^{-1} plus topramezone ([3-(4,5-dihydro-3-isoxazolyl)-2-methyl-4-(methylsulfonyl) phenyl] (5-hydroxy-1-methyl-1H-pyrazol-4-yl) methanone) at 0.012 kg ai ha^{-1} plus crop oil concentrate at 1% vol./vol.. Corn plant density was recorded before harvest. Chlorophyll index was determined using a SPAD meter (Minolta SPAD-502, Konica Minolta Optics, Inc.) at R1 (Ritchie et al., 1993). Grain yield, moisture, and test weight were determined at harvest using a Wintersteiger Delta (Salt Lake City, UT) equipped with a Harvest Master BDS800 (Juniper Systems Inc., Logan, UT) system.

All data were subjected to ANOVA (SAS, 2010). Radish heights and chlorophyll content were reported separately for year, planting date, and N rate due to significant interactions. Radish biomass production was separated by year and planting date, but analyzed across N rates due to a lack of significant 2- and 3-way interactions. Winter annual weed suppression and corn main effects were presented in the absence of significant 2- and 3-way interactions, except for chlorphyll content measurements, which had a significant N x radish seeding date interaction. An interaction between radish seeding date and year was presented for common chickweed control, henbit control, and yield. Means were separated using Fisher's Protected LSD (*P=0.1*).

3. Results and Discussion

3.1 Environmental Conditions

Annual precipitation for 2012 and 2013 was below the 10-year average (Figure 1). In 2012, from planting of radishes in August to a killing freeze in late Nov., rainfall was 339 mm (Figure 1a), compared to the 10-year average of 481 mm. Although total precipitation for 2012 was 722 mm, only 215 mm of rain fell from May through August, which was 210 mm below the 10-year average (data not presented). Thus, radishes were planted following drought conditions. On the last day of August, 60 mm of precipitation occurred from storms following Hurricane Isaac (Figure 1a). In 2013, although total precipitation was 1003 mm (data not presented), only 140 mm of rain fell from June through August, and no rain fell in August (Figure 1b). Precipitation in September and most of October was only 123 mm, which is 58 mm below the 10-year average. The last two days of October saw 77 mm of rainfall. And so conditions for radishes in 2013 were drier than 2012.

The summer of 2012's abnormally high temperatures averaged 23.7 °C from May through August, with 30.7 °C as the average high (Figure 2a). During the August-October radish growth period, the average temperature was 17.4 °C. In contrast, 2013 was relatively cool, with an average average high summertime temperature of 27.3 °C and an average temperature of 21.2 °C (Figure 2b). The July-October average temperature was 19.9 °C. Due to below-average rainfall and high average temperatures, particularly in 2012, the results of this research during 2012 were achieved under drought conditions and in 2013 under "flash drought" conditions.

3.2 Radish Height

Planting date and N amount affected radish heights (Table 1). Across planting dates, radish heights reached a maximum on 26 Oct. 2012 (23 DAT) and 22 Oct. 2013 (40 DAT) (Table 1 and Figure 3). In general, earlier planting dates had taller radishes than later plantings for all measurement dates across N rates (Table 1). In 2012, the first planting date was taller than the rest, with the exception of 34 DAT for 17 and 67 kg N ha^{-1}, when the first and second planting dates were taller than the other three planting dates. In 2013, the first planting date had greater heights than the other planting dates, with the exception of 13 and 40 DAT at 67 kg N ha^{-1} when the first two planting dates were taller than the last two (Table 1).

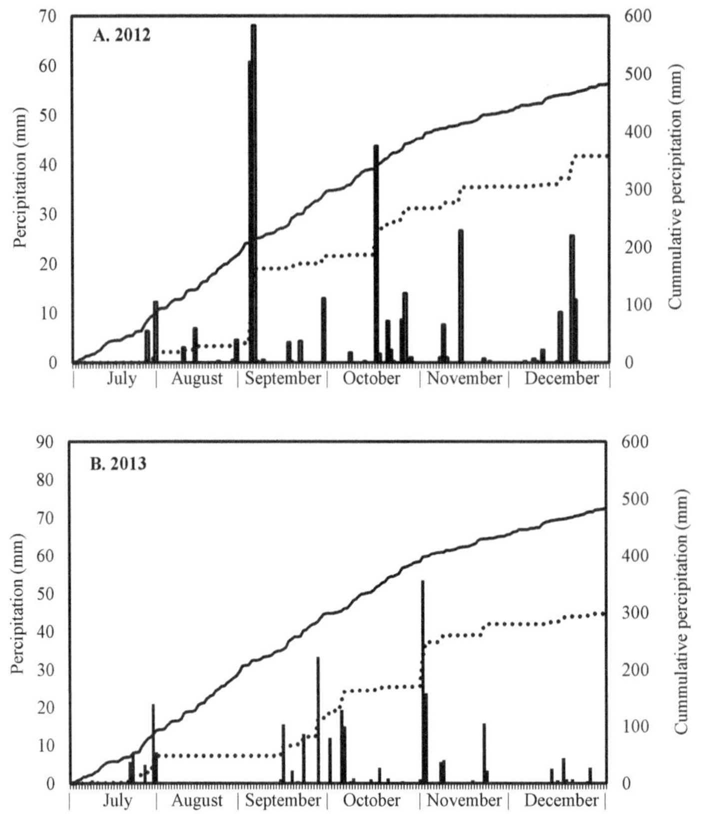

Figure 1. Daily (bar) and cumulative precipitation data for individual years (dash line) and 10-year average (solid line) for experiments in 2012 (A) and 2013 (B)

Nitrogen rates also impacted radish height. Across planting date and year, non-fertilized controls generally had shorter radishes than any N amount, indicating the need for additional N to maximize radish production for livestock grazing. In 2012, the N amount with the greatest radish heights across planting date was 33 kg N ha^{-1} (Figure 1). The 67 kg N ha^{-1} amount did not increase heights (Figure 1). However, 2013's results were different. With the exception of 13 DAT at the first planting date; 6, 13, and 40 DAT at the second planting; 13 and 40 DAT at the third planting; and 6 DAT at the fourth planting, radish heights showed no differences due to N rates, which was probably due to limited rainfall during this period (Table 1).

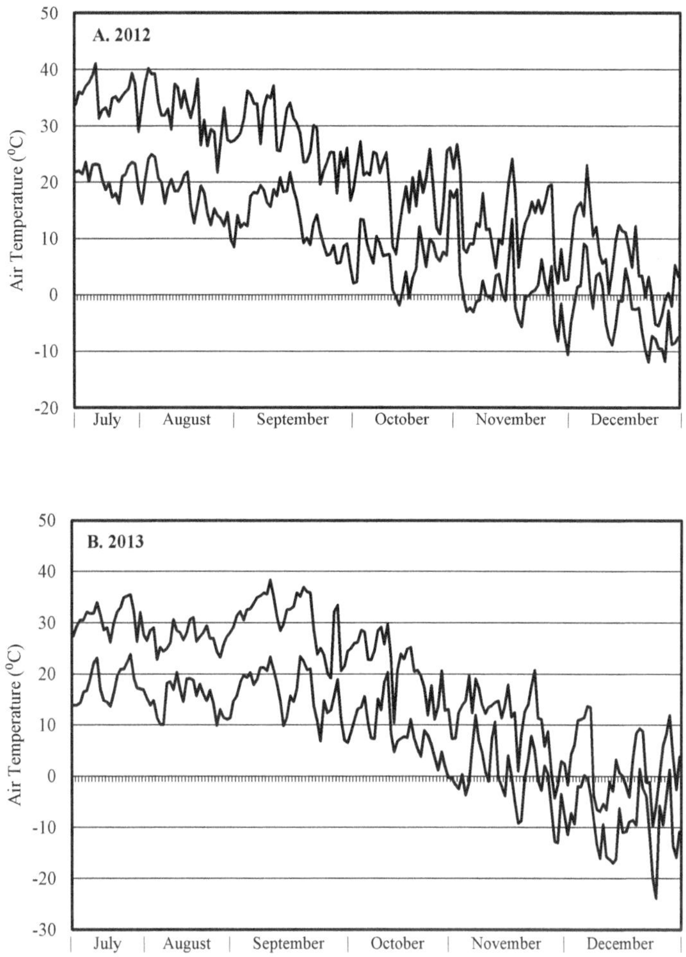

Figure 2. Daily maximum and minimum air temperature data for experiments in 2012 (A) and 2013 (B)

The few prior studies of the influence of planting date on radish heights, support this study's results that earlier dates foster greater growth. In other studies, radishes planted earlier had greater yields than those planted later (Pandey et al., 2009; Alam et al., 2010; Ebrahimi et al., 2013). However, available light affected crop growth rates (Liu et al., 2013; Zhang et al., 2014). According to a study in California, a 16-hour photoperiod was optimal for flowering of radish (Suge & Rappaport, 1968). It is probable that radishes planted earliest had the best light as well as temperature conditions for optimal growth and height (Figure 3). Radishes reached maximum height during the third week of Oct. (Figure 3, Table 1). Finally, research completed in India reported that radish height increased with N rate by 8.6 to 22 cm compared to the non-fertilized control (Sharma, 2000). The authors attributed this to possible increased cell division and elongation (Sharma, 2000). Similar patterns occurred with increased radish height as N rates increased. Although this study did not measure residual N, 33 kg N ha^{-1} is recommended for optimal radish heights.

Table 1. Radish heights at all four planting dates 6, 13, 16, 23, 34, and 40 days after fertilizer treatment (DAT). Data were separated by N rates (kg ha^{-1}) and planting dates in 2012 and 2013

DAT	Radish seeding date†	2012					2013				
		0‡	17	33	67	LSD (*P*=0.1)	0	17	33	67	LSD (*P*=0.1)
		cm									
6	1 Aug.	31.1	33	34.9	32.4	3.8	27.9	27.3	30.5	26	6.1
	15 Aug.	25.4	25.4	25.4	25.4	NS	15.2	16.5	19.7	17.8	1.8
	29 Aug.	20.3	20.3	20.3	20.3	NS	6.4	8.9	7.6	7.6	0.8
	11 Sep.	10.2	10.2	10.2	10.2	NS	2.5	2.5	3.2	2.5	4.1
	LSD (*P*=0.1)	1.8	1.7	0.7	1.4		3.3	3.3	6.1	4.5	
13	1 Aug.	32.4	36.2	38.1	36.2	2.8	38.7	38.1	34.9	36.2	10.4
	15 Aug.	27.3	28.6	29.8	28.6	2.8	21	22.9	26	34.3	2.5
	29 Aug.	26	27.9	27.9	27.3	2.3	9.5	14	10.8	11.4	1.0
	11 Sep.	12.7	12.7	12.7	12.7	NS	7	7.6	8.3	7.6	4.6
	LSD (*P*=0.1)	1.7	2.2	1.4	3.3		6.7	4.9	7.5	5.6	
16	1 Aug.	35.6	40.7	42.5	40.6	3.3	43.2	44.5	44.5	42.5	7.1
	15 Aug.	29.2	32.4	34.9	33.7	3.6	36.9	66.7	36.2	39.4	3.8
	29 Aug.	27.3	30.5	29.8	28.6	2.8	20.3	21.6	21	22.2	1.3
	11 Sep.	17.1	17.1	16.5	17.1	1.3	14	13.3	14	14	6.4
	LSD (*P*=0.1)	3.6	4.1	2.7	2.9		7.5	4.8	7.2	4.8	
23	1 Aug.	34.9	41.9	47.6	47	4.6	49.5	45.7	52.7	49.5	7.6
	15 Aug.	26.8	36.9	41.3	40	3.3	38.7	37.5	42.5	45.1	6.6
	29 Aug.	27.3	35.6	34.9	38.1	5.3	31.8	39.8	27.9	33.7	2.8
	11 Sep.	19.7	21	22.2	22.2	2.5	22.2	22.2	22.9	23.5	11.9
	LSD (*P*=0.1)	3.7	4.8	4	3.1		13.3	6.3	8.3	5.7	
34	1 Aug.	36.2	34.9	41.3	40	6.1	50.8	49.5	49.5	48.9	7.6
	15 Aug.	29.2	34.9	37.5	36.2	4.6	44.5	41.3	46.4	44.5	4.8
	29 Aug.	26	33	31.7	31.1	3.6	34.9	38.1	37.5	40	4.1
	11 Sep.	18.4	18.4	18.4	17.8	0.8	31.1	29.8	31.8	31.1	8.1
	LSD (*P*=0.1)	4.2	3.1	3.6	4.3		8	7.6	9	6	
40	1 Aug.	34.3	36.8	40.6	38.7	4.1	55.2	52.1	52.7	55.9	11.2
	15 Aug.	27.9	30.5	33	34.3	5.8	45.7	47.6	50.2	54.6	8.6
	29 Aug.	26.7	29.2	31.1	30.5	3.8	43.8	40	43.2	44.5	4.1
	11 Sep.	19.1	19.1	19.1	19.1	NS	35.6	31.8	33	39.4	7.6
	LSD (*P*=0.1)	5.3	5.1	2.8	3.7		9.4	8.2	7.3	5.8	

Note. †Radish planting dates were 1 Aug., 17 Aug., 30 Aug., and 12 Sep. 2012, while in 2013 radish planting dates were 1 Aug., 14 Aug., 28 Aug., and 11 Sep; ‡N application amount (0, 17, 33, and 66 kg N ha^{-1}).

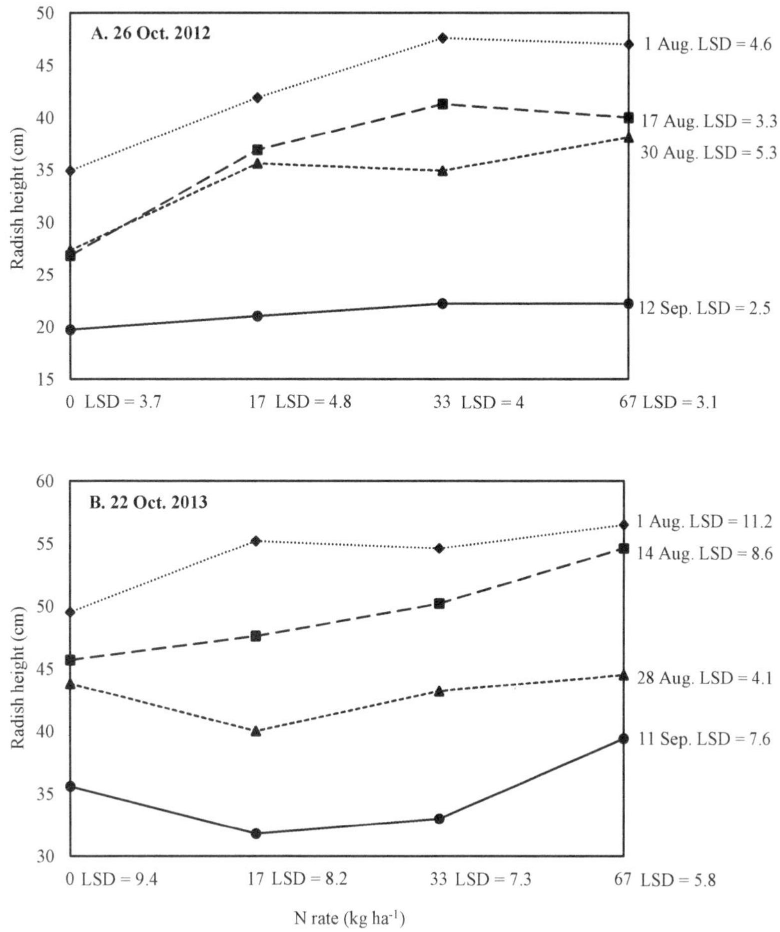

Figure 3. Radish heights for different planting dates on (A) 26 Oct. 2012 (23 days after treatment) and (B) 22 Oct. 2013 (40 days after treatment). Comparisions within a planting date are valid (LSD ($P = 0.1$) values along the right vertical axis). Comparisons within a fertilizer amount are valid (LSD ($P = 0.1$) values along the horizontal axis). Radish planting dates were labeled along the right side of the figure

3.3 Radish Chlorophyll Index

Radish chlorophyll content was variable and was impacted by year, planting date, and N rate, with a year x planting date x N rate interaction ($P = 0.09$). In 2012, the first planting date (17 Aug.) had the greatest chlorophyll content all planting dates regardless of N application rate (Figure 4). However, in 2013, the second and third planting dates (14 and 28 Aug.) had significantly greater chlorophyll content than the other planting dates.

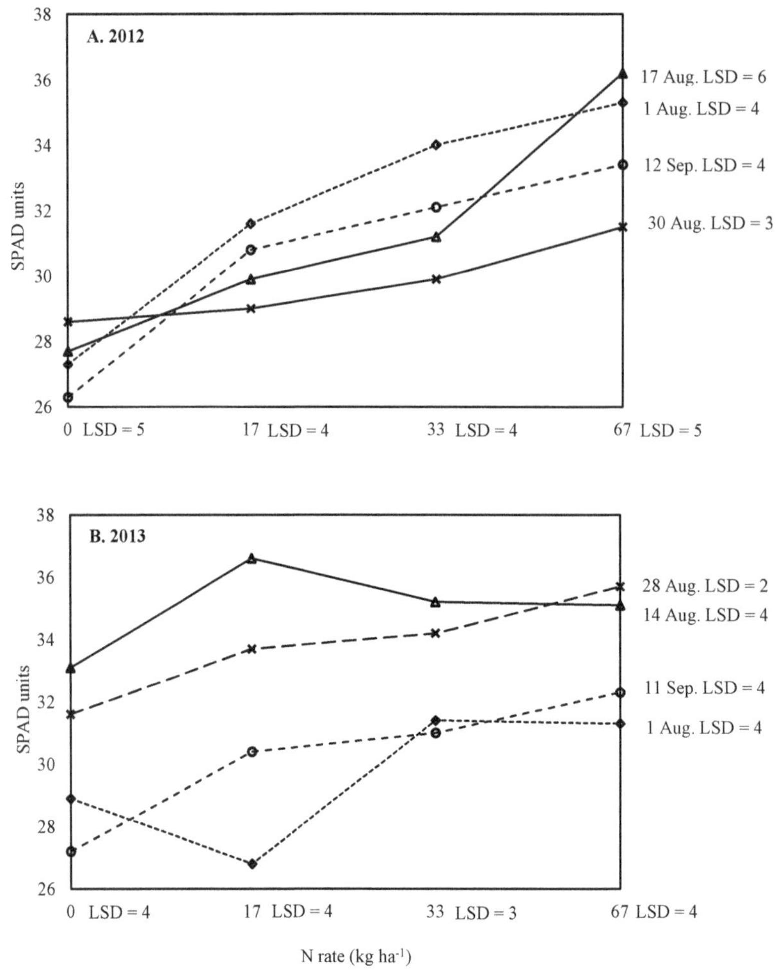

Figure 4. Chlorophyll index (SPAD units) for radish leaves. Data were separated by planting date and N fertilizer rate in (0, 17, 33, and 67 kg N ha^{-1}) in 2012 and 2013. Comparisions within a planting date are valid (LSD ($P = 0.1$) values along the right vertical axis). Comparisons within a fertilizer amount are valid (LSD ($P = 0.1$) values along the horizontal axis). Radish planting dates were labeled along the right side of the figure

Nitrogen rates also affected radish chlorophyll content. Across planting dates in 2012, increased rates of N generally led to increased chlorophyll content (Figure 4). When compared to the non-treated control, 67 kg N ha^{-1} increased chlorophyll content by 2.9 to 8.5 SPAD units except for the 12 Sep. planting date. Interestingly, chlorophyll content was not significantly greater above 33 kg ha^{-1}. In 2013, chlorophyll content did not follow N rates as closely, probably due to drier conditions which could affect the availability of N to the crop. At the last two planting dates (28 Aug. and 11 Sep.) the greatest chlorophyll content recorded was at the greatest N level (67 kg ha^{-1}); however, the previous two planting dates reached their largest chlorophyll levels at 17 kg N ha^{-1} (Aug. 14) and 33 kg N ha^{-1} (1 Aug.). The chlorophyll index indicated that 33 kg N ha^{-1} was sufficient for radishes.

3.4 Radish Biomass

Planting date significantly impacted radish foliage biomass yield (Figure 5). Across N rates, a year x planting date interaction ($P = 0.0003$) indicated that planting date affected radish growth. Across years, the first planting date (1 Aug.) produced the greatest biomass yields, with 820 to 1670 kg ha^{-1} more biomass compared with the later planting dates (Figure 5). Across all planting dates, 2013 saw greater radish production than 2012. It is possible that 2013's cooler conditions provided better growing conditions for radishes regardless of planting date. The optimal planting date for the greatest biomass yields was prior to 1 Sep.

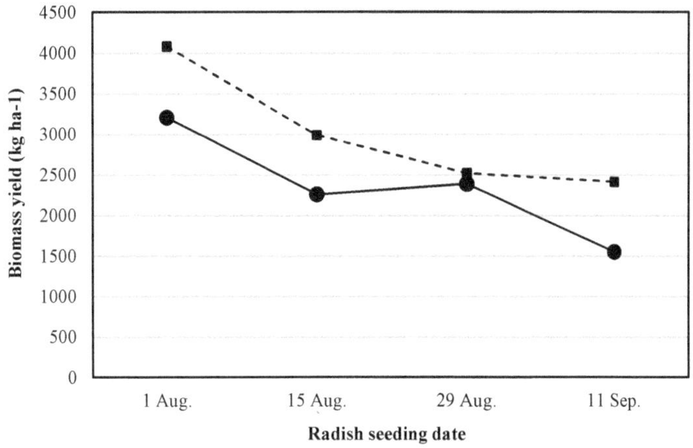

Figure 5. Radish foliage dry biomass of different planting dates in 2012 (circles and solid line, LSD = 340) and 2013 (squares and dash line, LSD = 635). Data were averaged over N amounts

Earlier studies evaluating how planting date affects radish production report similar findings to those shown here. In Bangladesh, sowing time significantly affected growth and yield, with the 1 Nov. planting date producing the highest root yield (81.8 Mg ha^{-1}) and 1 Dec. producing the lowest (68.7 Mg ha^{-1}) (Alam et al., 2010). In addition, 1 Nov. produced the greatest root length per plant (25.6 cm $root^{-1}$), and 1 Dec. the shortest (23.5cm $root^{-1}$). The 1 Nov. planting date produced the maximum number of leaves (16.3 $plant^{-1}$) and leaf length (59.6 cm $leaf^{-1}$) (Alam et al., 2010). All parameters showed a decreasing trend as sowing date was delayed. Pandey's study in India (2009) showed similar results when radishes were planted every ten days beginning on 9 Nov. The second planting date (19 Nov.) produced the maximum yield (28.4 Mg ha^{-1}), root length (24.45 cm $root^{-1}$) and root weight (0.097 kg $root^{-1}$). The last planting date (9 Dec.) produced the least yield (23.2 Mg ha^{-1}), root length (22.31 cm $root^{-1}$) and root weight (0.008 kg $root^{-1}$). The authors concluded that a delay in planting significantly reduced radish yield (Pandey et al., 2009). Ebrahimi's (2013) study in Iran planted black radishes (*Raphanus sativus* var. *niger*) on 10, 20, and 30 Sep. Yields started at 21.9 Mg ha^{-1} for the 10 Sept. planting and decreased to 9.8 Mg ha^{-1} for the 30 Sept. planting. Later plantings had shorter days and decreased light interception, as well as fewer degree days required for crop growth (Ebrahimi et al., 2013).

In this study, radish biomass was similar at 33 and 67 kg N ha^{-1} when averaged over planting dates (data not presented). Thus, 33 kg N ha^{-1} was sufficient for optimal forage production during a limited autumn growth period. Although research concludes that N fertilizer improves crop yields (Vos et al., 1998; Kristensen & Thorup-Kristensen, 2004; Dean & Weil, 2009; Constantin et al., 2010; Sapkota et al., 2012) few studies evaluate how N fertilizer affects radish in particular. In Florida, N fertilization did not impact radish yield in 7 of 8 experiments (Sanchez et al., 1991). The total fertilizer-derived N from marketable radishes averaged only 2 to 8 kg N ha^{-1}, prompting the authors to determine that the radishes used soil mineral N, and that N fertilization did not increase radish yields in that system (Sanchez et al., 1991). The recommended N fertilizer amount in Pakistan was 60 kg N ha^{-1}, which increased fresh radish biomass by 73% and yield by 14%, with an increase of 0.19 kg $plant^{-1}$ and 32 Mg ha^{-1} when compared to a non-fertilized control (Asghar et al., 2006). Based on our research, we recommend planting before 1 Sep. with at least 33 kg N ha^{-1}.

3.5 Winter Annual Weed and Corn Response the Following Year

The only effect of fall N rate on winter annual weed or corn response was its interaction with seeding date on chlorophyll index readings. An early radish planting date (1 Aug.) suppressed common chickweed and henbit greater than later planting dates in 2013 (Table 2). However, in 2014 no differences among planting dates were observed. Research has shown that Brassica species includeing radish have significant weed suppression (Altieri et al., 2011; Lawley et al., 2011, 2012; Malik et al., 2008; Norsworth, 2003; Stivers-Young, 1998; Uremis et al., 2009). Norsworthy (2003) showed that germination and radical growth of all weed species [sickle pod (*Senna obtusifolia* L.), prickly sida (*Sida spinosa* L.), and yellow nutsedge (*Cyperus esculentus* L.)] evaluated were reduced by extract of forage radish in comparison to non-treated controls. Greater weed suppression coincided with greater radish biomass production in 2013 (Table 2) which corresponded with Teasdale (1996) who reviewed cover crop research and reported that control of weeds increased with greater amounts of cover crop residue biomass. In western New York, Brassica cover crops including oilseed radish residues significantly suppressed weeds with earlier planted crops (25 Aug.) and had particularly greater winter annual weed [common chickweed (*Stellaria media* L.), henbit (*Lamium amplexicaule* L.), and malva (*Malva moschata* L.)] suppression than later planting dates (Stivers-Young, 1998). Corn was planted earlier in 2014 along with an autumn that was dry and the winter was extremely cold (data not presented). Taken together, these factors may explain weed suppression differences between years, which delayed the overall burndown herbicide application to an early postemergence treatment in 2013. Lawley et al. (2012) observed early spring suppression of common chickweed and henbit; however, radish showed almost complete weed control in early spring while control declined throughout the growing season as summer annual weeds started emerging. Forage radish provided complete weed suppression in early fall through winter and into early March when weed cover was only 0-3%; however, weed suppression from radishes did not extend to the next growing season and a postemergence herbicide was needed to be applied to avoid a yield reduction in corn due to weed interference (Lawley et al., 2011). Thus it is possible that winter annual weed suppression was greater in 2014 due to an earlier burdown application prior to allowing growth of weeds later in the season.

Radish seeding date did not affect corn plant density. According to the chlorophyll index at R1, corn plants following early-planted radish with 17 and 33 kg N ha^{-1} were generally greener than the non-treated control, while corn from the last planting date generally was greener when 0 or 66 kg N ha^{-1} was applied. Differences in corn chlorophyll content varied based on radish planting date and N applied for radish production. Corn grain moisture (188 g kg^{-1}) and test weight (42.3 g cm^{-3}) were greatest following early-planted radish; however, no significant difference in grain yields occurred in moderate- (2013) and high-yield (2014) environments at preplant N rates typical for this region. This was similar to other research evaluating corn's rotational response to cover crops with radish (Lawley et al., 2011). However, there is some research that has shown a yield increase in subsequent crops such as sweet corn following radish (Malik et al., 2008) which was related to increased weed suppression. Radish taproots may also reduce soil compaction and provide root channels for the rotational crop. Channels produced by cover crop roots in the autumn and winter, when soils are relatively moist, may facilitate the penetration of compacted soils by subsequent crop roots in summer when soils are relatively hard and dry (Chen & Weil, 2011; Creswell & Kirkegaard, 1995). Kirkegaard et al. (1993) hypothesized that tap-rooted Brassica crops produce channels in the dense subsoil, which were utilized by the subsequent crop to access water and nutrients as well as increase yield, and they reported that in seasons with adequate rainfall, the yield advantage of wheat grown after the Brassica crop was in the range of 15-25% greater compared with a wheat mono-crop which was unlike our research. William and Weil (2004) suggested that roots of summer crops grew following channels created by preceding cover crops which provided lower resistance channels for soybean roots to search for water in the subsoil late in the season during drought thus increasing soybean yields to 200 kg ha^{-1} and the impacts of preceding cover crops were greatest during severe drought and highly compacted soil conditions. Evaluating radish, rapeseed and winter rye (*Secale cereale* L.) as cover crops followed by corn, radish and rapeseed were the most effective in reducing the effects of soil compaction on corn with corn having more roots at a 45 cm depth in the forage radish treatments (Chen & Weil, 2011). It is possible that we did not observe increases in subsequent corn yields due to sufficient rain fall during the corn groth periods in 2013 and 2014 (data not presented) and thus possible impacts from the radish were not observed.

Table 2. Winter annual weed control prior to burndown herbicide application and corn response to radish planting date the previous year. Data were averaged over N application amount and year in the absence of a significant interaction.

Radish seeding date	Chickweed		Henbit		Plant density	SPAD				Moisture	Grain density	Yield	
	2013	2014	2013	2014		0†	17	33	66			2013	2014
	----------------- % -----------------				No. ha^{-1}					g kg^{-1}	g cm^{-3}	Mg ha^{-1}	
1 Aug.	79	100	81	100	64,390	53.3	55.0	55.3	53.7	188	42.3	9.73	12.24
15 Aug.	51	100	72	100	68,450	52.9	53.4	53.7	54.1	185	41.4	9.54	12.12
29 Aug.	66	100	51	100	62,890	53.1	53.2	55.1	52.3	187	41.3	9.17	12.06
11 Sep.	31	100	29	100	67,930	55.6	53.8	52.2	54.9	182	41.1	9.10	12.93
LSD (*P*=*0.1*)	--------9-------		--------9--------		NS	---------------2.2--------------				5	1.2	-------1.00-----	

Note. †N application amount (0, 17, 33, and 66 kg N ha^{-1}).

4. Conclusion

This research took place during drought conditions in 2012 and 2013. In 2012, limited precipitation in summer preceded the planting of radishes, and 2013 saw a flash drought with no precipitation in August and little rainfall September through October. Planting date and N rate affected radish heights. In general, earlier planting dates produced taller radishes than later ones for all measurement dates across N rates. Across planting date and year, non-fertilized controls had shorter radishes than any N rate. The greatest radish heights were produced at 33 kg N ha^{-1}, while radish chlorophyll content did not significantly increase after 33 kg N ha^{-1}. Radish foliage dry biomass production was greatest at the first planting date (1 Aug.) across years. When compared to the other planting dates, foliage dry biomass was 820 to 1670 kg ha^{-1} greater with the early planting date. In 2013, winter annual weed suppression was greatest at the first planting date (1 Aug.), but in 2014 no differences in winter annual weed suppression occurred compared to the other planting dates. Radish planting date had no effect on corn yield the following year (2013 and 2014). This research demonstrates that for optimum biomass production, radishes should be planted in late summer prior to 1 Sep. and fertilized with 33 kg N ha^{-1}, and that radish planting date likely will not increase corn grain yields in upstate Missouri.

References

Alam, M. K., Farooque, A. M., Nuruzzaman, M., & Uddin, A. F. M. J. (2010). Effect of sowing time on growth and yield of three radish (*Raphanus sativus* L.) varieties. *Bangladesh Res. Pub. J., 3*, 998-1006.

Altieri, M. A., Lana, M. A., Bettencourt, H. V., Kieling, A. S., Comin, J. J., & Lovato, P. E. (2011). Enhancing crop productivity via weed suppression in organic no-till cropping systems in Santa Catarina, Brazil. *J. of Sustainable Agric., 35*, 855-869. http://dx.doi.org/10.1080/10440046.2011.588998

Asghar, H. N., Ishaq, M., Zahir, Z. A., Khalid, M., & Arshad, M. (2006). Response of radish to integrated use of nitrogen fertilizer and recycled organic waste. *Pak. J. Bot., 38*, 691-700.

Caviglia, O. P., Sadras, V. O., & Andrade, F. H. (2004). Intensification of agriculture in the south-eastern Pampas I. Capture and efficiency in the use of water and radiation in double-cropped wheat–soybean. *Field Crops Res., 87*, 117-129. http://dx.doi.org/10.1016/j.fcr.2003.10.002

Caviglia, O. P., Sadras, V. O., & Andrade, F. H. (2011). Yield and quality of wheat and soybean in sole- and double-cropping. *Agron. J., 103*, 1081-1089. http://dx.doi.org/10.2134/agronj2011.0019

Chen, G., & Weil, R. R. (2011). Root growth and yield of maize as affected by soil compaction and cover crops. *Soil & Tillage Res., 117*, 17-27. http://dx.doi.org/10.1016/jstill.2011.08.001

Constantin, J., Mary, B., Laurent, F., Aubrion, G., Fontaine, A., Kerveillant, P., & Beaudoin, N. (2010). Effects of catch crops, no till and reduced nitrogen fertilization on nitrogen leaching and balance in three long-term experiments. *Agric. Ecosys. Enviro., 135*, 268-278. http://dx.doi.org/10.1016/j.agee.2009.10.005

Dean, J. E., & Weil, R. R. (2009). Brassica cover crops for nitrogen retention in the Mid-Atlantic Coastal Plain. *J. Environ. Qual., 38*, 520-528. http://dx.doi.org/10.2134/jeq2008.0066

Ebrahimi, M., Hassandokht, M. R., & Payvast, G. H. (2013). Effect of sowing dates on some quantitative and qualitative traits of three landraces of black radish (*Raphanus sativus* var. *niger*). *Advances in Environ. Bio., 7*, 136-140.

Halverson, A. D., Black, A. L., Krupinsky, J. M., & Merrill, S. D. (1999). Dryland winter wheat response to tillage and nitrogen within an annual cropping system. *Agron. J., 91*, 702-707. http://dx.doi.org/10.2134/agronj1999.914702

Hochmuth, G., Hanlon, E., Snyder, G., Nagata, R., & Schueneman, T. (2009). *Fertilization of sweet corn, celery, romaine, escarole, endive, and radish on organic soils in Florida*. Florida Coop. Ext. Serv. BUL313.

Hoffbeck, C., Ruhland, J., Miller, J., & Millar, J. (2008). Benefits of cover crops in no-till wheat stubble. *Field Facts: Pioneer Agron. Sci., 8*, 1-4.

Kirkegaard, J. A., Angus, J. F., Gardner, P. A., & Creswell, H. P. (1993). Benefits of Brassica break crops in the southeast wheatbelt. Proc. 7th Australian Agronomy Conference Adelaide. *Aust. Soc. of Agron.*, 282-285.

Kristensen, H. L., & Thorup-Kristensen, K. (2004). Root growth and nitrate uptake of three different catch crops in deep soil layers. *Soil Sci. Soc. Am. J., 68*, 529-537. http://dx.doi.org/10.2136/sssaj2004.0529

Lawley, Y. E., Teasdale, J. R., & Weil, R. R. (2012). The mechanism for weed suppression by a forage radish cover crop. *Agron. J., 104*, 205-214. http://dx.doi.org/10.2134/agronj2011.0128

Lawley, Y. E., Weil, R. R., & Teasdale, J. R. (2011). Forage radish cover crop suppresses winter annual weeds in fall before corn planting. *Agron. J., 103*, 137-144. http://dx.doi.org/10.2134/agronj2010.0187

Liu, B., Liu, X., Li, Y., & Herbert, S. J. (2013). Effects of enhanced UV-B radiation on seed growth characteristics and yield components in soybean. *Field Crop Res., 154*, 158-163. http://dx.doi.org/10.1016/j.fcr.2013.08.006

Malik, M. S., Norsworthy, J. K., Culpepper, S., Riley, M. B., & Bridges Jr., W. (2008). Use of wild radish (Raphanus raphanistrum) and rye cover crops for weed suppression in sweet corn. *Weed Sci., 56*, 588-595. http://dx.doi.org/10.1614/WS-08-002.1

Mohammadi, G. R. (2013). Alternative weed control methods: A review, weed and pest control – Conventional and new challenges. In S. Soloneski (Ed.), *InTech*. http://dx.doi.org/10.5772/54164

Norsworthy, J. K. (2003). Allelopathic potential of wild radish (*Raphanus raphanistrum*). *Weed Tech., 17*, 307-313. http://dx.doi.org/10.1614/0890-037X(2003)017[0307:APOWRR]2.0.CO;2

Pandey, B. P., Basnet, K. B., Bhatta, M. R., Sah, S. K., Thapa, R. B., & Kandel, T. P. (2013). Effects of row spacing a direction of sowing on yield and yield attributing characters of wheat cultivated in Western Chitwan, *Nepal. Agric. Sci., 4*, 309-316.

Pell, E. J., Sinn, J. P., Eckardt, N., Vinten Johansen, C., Winner, W. E., & Mooney, H. A. (1993). Response of radish to multiple stress II. Influence of season and genotype on plant response to ozone and spoil moisture deficit. *New Phytol., 123*, 153-163. http://dx.doi.org/10.1111/j.1469-8137.1993.tb04541.x

Ritchie, S. W., Hanway, J. J., & Benson, G. O. (1993). How a corn plant develops (p. 21). *Spec. Rep. 48*. Iowa State Coop. Ext. Serv., Ames.

Sapkota, T., Askegaard, M., Laegdsmand, M., & Olesen, J. E. (2012). Effects of catch crop type and root depth on nitrogen leaching and yield offspring barley. *Field Crops Res., 125*, 129-138. http://dx.doi.org/10.1016/j.fcr.2011.09.009

Sanchez, C. A., Ozaki, H. Y., Schiler, K., & Lockart, M. (1991). Nitrogen fertilization of radishes on Histosols: Response and ^{15}N recovery. *HortSci., 26*, 865-867.

Sanford, J. O., Myhre, D. L., & Merwine, N. C. (1973). Double cropping systmes involving no-tillage and conventional tillage. *Agron. J., 65*, 978-982. http://dx.doi.org/10.2134/agronj1973.00021962006500060040x

SAS Institute. (2010). *SAS user's guide*. SAS Inst., Cary, NC.

Schmitt, J., Ehrhardt, D. W., & Cheo, M. (1986). Light-dependent dominance and suppression in experimental radish populations. *Ecology, 67*, 1502-1507. http://dx.doi.org/10.2307/1939081

Sharma, A. K. (2000). Effect of nitrogen and phosphorus on seed yield in radish (*Raphanus sativus* L.). *Agric. Sci, Digest, 20*, 46-49.

Snapp, S. S., Swinton, S. M., Labarta, R., Mutch, D., Black, J. R., Leep, R., ... O'Neil, K. (2005). Evaluating cover crops for benefits, costs, and performance within cropping system niches. *Agron. J., 97*, 322-332.

Sojka, R. E., Karlen, D. L., & Busscher, W. J. (1991). A conservation tillage update from the Costal Plain Soil and Water Conservation Center of South Carolina: A review of previous research. *Soil and Tillage Res., 21*, 361-376. http://dx.doi.org/10.1016/0167-1987(91)90031-R

Stivers-Young, L. (1998). Growth, nitrogen accumulation, and weed suppression by fall cover crops following early harvest of vegetables. *HortSci., 33*, 60-63.

Suge, H., & Rappaport, L. (1968). Role of gibberellins in stem elongation and flowering in radish. *Plant Physiol., 43*, 1208-1214. http://dx.doi.org/10.1104/pp.43.8.1208

Teasdale, J. R. (1996). Contribution of cover crops to weed management in sustainable agricultural systems. *J. Prod. Agric., 9*, 475-479. http://dx.doi.org/10.2134/jpa1996.0475

Thiessen Martens, J. R., Hoeppner, J. W., & Entz, M. H. (2001). Legume cover crops with winter cereals in Southern Manitoba: Establishment, productivity, and microclimate effects. *Agron. J., 93*, 1086-1096. http://dx.doi.org/10.2134/agronj2001.9351086x

Uremis, I., Arslan, M., Uludag, A., & Sangun, M. K. (2009). Allelopathic potentials of residues of 6 brassica species on johnsongrass [*Sorghum halepense* (L.) Pers.]. *African J. of Biotech., 8*, 3497-3501.

Vos, J., van der Putten, P. E. L., Hussien, M. H., van Dam, A. N., & Leffelaar, P. A. (1998). Field observations on nitrogen catch crops II. Root length and root length distribution in relation to species and nitrogen supply. *Plant and Soil, 201*, 149-155. http://dx.doi.org/10.1023/A:1004367530320

William, S. M., & Weil, R. R. (2004). Crop cover root channels may alleviate soil compaction effects on soybean crop. *Soil Sci. Soc. Am. J., 68*, 1403-1309. http://dx.doi.org/10.2136/sssaj2004.1403

Zhang, L., Allen Jr., L. H., Vaughan, M. M., Hauser, B. A., & Boote, K. J. (2014). Solar ultraviolet radiation exclusion increases soybean internode lengths and plant height. *Agri. and Forest Meteorol., 184*, 170-178.

21

An Analysis of Institutional Factors Influencing Vegetable Production amongst Small-Scale Farmers in Six Vegetable Projects of the Nkonkobe Local Municipality

Phatela M. Raleting[1] & Ajuruchukwu Obi[1]

[1] Department of Agricultural Economics & Extension, University of Fort Hare, Alice, South Africa

Correspondence: Phatela M. Raleting, Department of Agricultural Economics & Extension, University of Fort Hare, Alice Private Bag X1314, Alice, South Africa. E-mail: mraleting@yahoo.com

Abstract

The specific roles of institutions in mediating production and marketing within the smallholder sector have not been fully investigated and understood especially in the parts of South Africa designated independent homelands prior to the end of Apartheid. This paper investigated institutional factors influencing vegetable production in six small-scale vegetable projects in Alice town in the Nkonkobe Municipality of Eastern Cape Province of South Africa. Amidst worsening poverty in the wider society it was the intention to know how vegetable production can contribute to enhancing food security and if it is in a position to do so. Seeking some insights on effectiveness of the agrarian reforms on smallholder farmers in South Africa, the objectives of the study were to identify and explore institutional factors that influence vegetable production. The data were drawn from all the 62 farmers in the projects investigated. Descriptive analysis and binary logistic regression were employed to analyze the data and explain the patterns of interactions among the identified institutional factors influencing vegetable production. The study results revealed that some institutional factors need to be addressed to enhance vegetable production. The binary logistic results show that formal rules and informal norms are important in vegetable production. The most significant institutional variables revealed by the analysis were attributes of the formation and organizational structure of the projects, land tenure, extension service, collective action in production and marketing. The findings suggest that institutional changes in respect to aforementioned variables and other complementary institutions such as contract farming and credit access can significantly contribute to increased, efficient and sustainable vegetable production.

Keywords: institutional factors, vegetable production, livelihoods, poverty, food security

1. Introduction

In the literature it has been affirmed that institutional factors (both formal and informal norms) influence and shape the way individuals interact to undertake economic activity (Pejovich, 1999; Williamson, 2001; Lekovic, 2011). This would normally influence the costs of production and/or exchange (Coase, 1992; Lal, 1999). Besides, according to North (1993), the development strategy of any country is embodied within the way societies view and interpret the world around them. Since formal rules are designed and implemented by human beings, an analysis of institutional factors is necessary to gain understanding of existing institutional arrangements that would contribute to improved livelihoods and hence economic development (Adebeyi & Obasa, 2004). Brown (2000) also supported this view by postulating that development is not based on how much one has but on how much one does with what is owned. It is, therefore, not surprising why positive policy reforms have been a success in some countries but a failure in others. Hence, the assertion that sound policy reform is a function of the interdependency of formal rules and informal norms (Hou & Smith, 2009) comes quite handy to policy makers or development practitioners as they grope their way through the difficult and uncertain policy terrain right from the policy formulation to the implementation (Greenberg, 2010).

Besides, studies analyzing the role of institutions have of late been receiving special attention in natural resource management (Mokhahlane, 2009; Kirkland, Hunter, & Twine, 2005), marketing management and agricultural production costs (Coase, 1992; North, 1993; Lal, 1999; Jari & Fraser, 2009). An integrated, holistic approach that seeks to address agricultural challenges should therefore be viewed as that which includes institutional factor analysis on specific agricultural enterprises with special reference to both production and marketing outcomes. Rightly observed, when smallholder farmers are faced with high transaction costs they resort to producing mainly for home consumption (World Food Programme, 2009), although Randela (2005) has noted that the size of a household influences decision making on what and how much to produce. In addition to the interdependent nature of production and marketing institutions claimed by Coase (1992), a number of 'new', emerging institutional factors call for special attention.

In South Africa it has been claimed that developing markets for smallholder farmers can significantly attract them into mainstream formal marketing, with positive implications for production (Jacobs, 2008; Jari & Fraser, 2009; Chikazunga & Paradza, 2012; Ferris et al., 2014). A linkage of the working mechanism of production and market institutions is no doubt necessary in this regard. As North (1993) and IFPRI [International Food Policy and Research Institute] (2006) assert, economists still know very little about the dynamics of institutional changes. In support of this, Hou and Smith (2009) cautioned that the effects of informal norms (as a bridge between formal rules and governmental policy outcome) should not easily be undermined for positive policy reform. Evidently, the dynamics of informal norms are not easy to understand as they take time to change and the very same formal rules themselves (Hou & Smith., 2009; Van Schalkwyk et al., 2012) are implemented via informal norms based on their interpretation by policy makers or government officials.

For South Africa where a number of policy shifts have occurred since 1994 aimed at uplifting the previously deprived Black farmers, these considerations assume exceptional significance. Such policy shifts included with them some programmes such as market deregulation, Land Redistribution for Agricultural Development and its successor schemes and programmes, the programme of black economic empowerment in agriculture known as Agri-BEE, the Comprehensive Agricultural Support Programme (CASP) and Micro-Agricultural Finance of South Africa (MAFISA), to name just a few. In one way or the other, they were all aimed at increasing production, productivity and market access for smallholder farmers. Why progress has remained slow and at inadequate levels is sufficiently confounding that urgent response is mandatory. At the same time, policy makers have to grapple with new issues such as the impact of climatic change, forces of international trade due to globalization some of which have triggered off the rising food prices that came to a head in 2008 and still continuing, despite a slight respite a few years ago. There is no denying that South Africa's smallholder farmers are more than ever faced with high risks and uncertainty both in production and marketing.

Besides, the influence of 'interaction" of proposed policy reforms and evolving informal rules and norms have most probably retarded progress in smallholder development initiatives. Perhaps the pace at which the formal rules change is not properly aligned with the rate of change in the informal norms: a misalignment which has the tendency to disturb the institutional interdependency. The rest of this paper describes the project area, and presents the framework for the analysis before turning to the methodology and procedures followed in conducting the study. The paper then presents the results of the analysis with respect to the conception, management arrangements, operational modalities, asset ownership, and the nature of the relationship with governmental structures. Conclusions are drawn from these and the paper winds up with a set of recommendations.

1.1 Description of the Study Area

The study was carried out in the Lovedale1 & Lovedale2 Vegetable Cooperatives, Nompumelelo Farmers Association, Lower Gqumahashe vegetable Cooperative, Somxada Food Security Association and the Sigabuliamatyathenga Vegetable Cooperative of Alice town in the Nkonkobe Local Municipality. The Nkonkobe Local Municipality falls under Amatole District Municipality of the Eastern Cape Province. Alice, a small town where UFH is situated, lies between latitude 32°56′ South and longitude 26°50′ East in southern South Africa in Eastern Cape Province. Alice town, where the six projects investigated are situated, is about 120 kilometers northwest of East London. Approximately 74% of the people living within Nkonkobe Local Municipality area are highly affected by poverty (Nkonkobe Municipality, 2009). The majority of the population, approximately 61%, resides in rural areas, while as many as 20% of the population resides in farms and scattered settlements (Nkonkobe Municipality, 2009). Alice has a population 65 472 individuals. The climate varies from hot in summer to extreme cold in winter with heavy frost and snowfall along the hilly areas. Average annual rainfall is 640mm, and most rain falls during the summer months from October to March, with frost and sometimes snow in winter. Mean maximum monthly temperatures range from 4 °C in July to 38 °C in February. Most of the roads linking the rural settlements are generally in poor conditions. The vegetation type is mostly Thorn Bush veld dominated by acacia Karroo species.

1.2 Framework for the Study

The study entailed both qualitative and quantitative approaches. The objective was to seek an explanation outside conventional production economics for the sluggish performance of the vegetable production of smallholder farmers that perpetuates their poverty and food insecurity. The main reason to look outside conventional economics research principles and concepts is that a large body of research already exists which have failed to come up with adequate explanations for the paradox of the unrelenting poverty despite the clear and unmistakable emphasis that has been placed on rural development, community empowerment, poverty alleviation and rural employment. There clearly must be issues that have escaped the attention of conventional and neo-classical economics which are influential enough to sustain the current disadvantages.

The new institutional economics has been helpful in filling the conceptual gap in situations such as this. Some of the distinguishing characteristics of this approach include the range of variables that are modeled that are often calibrated on the basis of qualitative attributes. This paper therefore assumes a model of economic performance of vegetable farms that is heavily influenced by such qualitative attributes as management capabilities through committee involvement in production and marketing decisions, the membership of farming groups, extension contacts, among others. This perspective incorporates the transactions cost idea of Coase (1992) in which the cost of doing business is seen as a binding constraint. This is no true than in the smallholder contexts that characterize the former independent homelands where poverty is rife and even the most basic services can be inaccessible to the small farmer.

2. Methods

Data were collected using structured questionnaire. Farmers in six vegetable projects were interviewed in respect to institutional factors influencing vegetable production in the study area. In order to ensure that all questions were attended to by the respondents, the personal interview approach was used.

2.1 Sampling Techniques and Designs

The purposive sampling design which was used is a non-probability sampling method whereby the researcher purposively interviews individuals meeting the criteria of interest with the hope that they will provide an indepth information required for the research project (Leedy & Ormrod, 2005). To remove the element of bias, particularly given the small size of the total farming population in that category, all participating farmers in each project were interviewed (i.e. 5 from Lovdale1, 6 from Lovedale2, 5 from Sigabuliamatythenga, 13 from Somxada, 14 from Nompumelelo and 19 from Gqumahashe).

2.2 Data Analysis

Data were coded in Microsoft Excel and then exported into SPSS (version 19) for analysis. In addition to the descriptive analysis to profile the smallholders and their socio-economic characteristics and production and marketing environments, the binary logistic regression operation was performed to test how institutional factors influence farmers' main reasons of producing vegetables.

Binary logistic regression is useful when a dependent variable is dichotomous (Chan, 2005). The model, according to Liao (1994), has only two categories in the endogenous variable resulting from a number of categorical predictor variables. The binary regression model has been widely used in a number of fields including social sciences when investigating dichotomous response (Long, 1997). In the study area, it is assumed that a smallholder farmer is faced with two choices subject to institutional arrangement. A smallholder farmer strives to maximize his/her utility by either producing vegetables mainly for sale (considered to be fully commercialized) or for both sale and consumption (considered partial commercialized) subject to the prevailing institutional arrangement.

Using the binary logit model, Darroch and Clover (2005) also determined factors affecting the survival, growth and success of small-medium and agribusiness. Mohammed and Ortman (2005) had also used the binary regression model when investigating whether or not commercial dairy farmers were insuring their livestock. The model employed in this study tests the probability that a smallholder farmer is influenced by institutional factors to either produce vegetable for both sale and consumption or solely for sale. Thus, the binary response is set up which defines Y= 1 for situations where a farmer produced vegetables for both consumption and sale and Y= 0 where a farmer produced solely for sale. The model specification can then be written as shown in equation 1.

$$In(Y1) = \beta 0 + \beta 1x1 + \beta 2x2 ... \beta nxn \quad (1)$$

$$In(Y0) = \beta 0 + \beta 1x1 + \beta 2x2 ... \beta nxn \quad (2)$$

As shown in Equations (1) and (2), there are two probabilities with $Y0$ denoting the lower response category

(Vegetable production mainly for sale) and Y_1 representing the higher response category being when a smallholder is producing for both sale and consumption. Both equations present the outcome of the logit transformation of odd ratios which can compactly be represented as:

$$In(Y0/Y1) = \beta 0 + \beta 1x1 + \beta 2x2 ... \beta nxn \tag{3}$$

Equation (3) allows for its estimation as a linear model for which the following definitions apply:

$\beta 0$ = the constant or intercept term of the regression model; $\beta 1$, $\beta 2$, βn = the regression coefficients of the individual predictor variables; $x1 + x2 ... xn$ = Explanatory variables.

Table 1 depicts the variables included in the model and the expected sign of such variables as they influence the dependent variable.

Table 1. Summary of modeled explanatory variables

Variable name	Variable label	Coding	expected sign
Management committee meeting frequency	MCMF	0 if more frequent, 1 otherwise	+/-
Management committee strengths	MCS	0 if many, 1 otherwise	+
Management committees weakness	MCW	0 if no, 1 otherwise	+/-
Idea of association formation	IAF	0 if participants, 1 otherwise	+/-
Land ownership	LO	0 if participants, 1 otherwise	+
Working strategy in the field	WSF	0 if collectively, 1 otherwise	+
marketing strategy	WSM	0 if collectively, 1 otherwise	+/-
Presence of extension service	PES	0 if yes, 1 otherwise	+/-
conditions of roads	CR	0 good, 1 otherwise	+
Relationship to the municipality	RM	0 if good, 1 otherwise	-

When the above variables are fitted in the model in Equation (1), a typical regression model would be as shown in Equation 4:

$$In(Y0/Y1) = \beta 0 + \beta 1MCMF + \beta 2MCS + \beta 3MCW + \beta 4IAF + \beta 5LO + \beta 6WSF + \beta 7WSM + \beta 8PES + \beta 9CR + \beta 10RM \tag{4}$$

Where:

$Y0$ = Vegetable production mainly for sale; $Y1$ = vegetable production for both consumption and sale; $\beta 0$ = Constant; $\beta 0 ... \beta 10$= Co-variants.

3. Results and Discussion

The results are presented separately for the descriptive analysis and the inferential analysis.

3.1 Results of the Descriptive Analysis

Characteristics of sampled household such as gender, age, marital status and educational levels attained were analyzed. The rationale for the inclusion of the demographic features is that demographic characteristics of households are essential when analyzing economic data because such factors would influence the households' economic behavior. Demographic features of sampled household are most likely to influence their production levels and as such are relevant. Table 2 exhibits the demographic characteristics of the sampled households in the study area.

Table 2. Distribution by demographic characteristics of the sampled households in the study area (n = 62)

Variable	Siga	Lov1	Lov 2	Somxa	Nompu	Gquma	All projects
	Respondents (%)	Respondents (%)	Respondents (%)	Respondents (%)	Respondents (%)	Respondents (%)	
Age							
0-35	**40**	0	0	15	14	11	13
36-45	20	33	40	54	22	5	**26**
46-55	40	33	60	15	0	21	19
56-65	0	17	0	16	7	15	13
66-75	0	17	0	0	**46**	37	23
>75	0	0	0	0	14	11	6
	100	100	100	100	100	100	100
Gender							
Male	100	100	100	31	64	79	**71**
Female	0	0	0	**69**	36	21	29
	100	100	100	100	100	100	100
Marital status							
Married	40	67	60	36	64	**79**	**61**
Single	**60**	33	40	36	29	11	29
Widowed	0	0	0	23	7	5	8
Divorced	0	0	0	0	0	5	2
	100	100	100	100	100	100	100
Educational level							
Primary	0	50	0	39	43	26	31
Secondary	20	16	0	8	43	32	24
High school	60	34	**100**	53	7	37	**40**
Tertiary	20	0	0	0	7	5	5
	100	100	100	100	100	100	100
Occupation							
Farmer	80	100	100	100	100	95	**90**
Teacher	0	0	0	0	0	0	5
Security	0	0	0	0	0	0	3
Student	20	0	0	0	0	0	2
	100	100	100	100	100	100	100

Source: Field Survey (2011).

The six projects investigated were Siga, Lov1, Lov2, Somxa, Nompu and Gquma. Comparing the age distribution across all projects, Nompu and Gquma are dominated by much older farmers in the 66-75 years who comprise 43% and 37% of all the farmers, respectively. The results also show that it was mainly in Nompu and Gquma where the age distributions by respondents have been evenly distributed, implying that not only old aged persons engage in vegetable production. Interestingly, male farmers represented a higher percentage (71%) of the projects participants. This finding may most probably be explained by a high retrenchment rate of men from the mining sector in South Africa. The majority of sampled farmers were married except for Siga where 60% of farmers were single. All in all, 61% of the project participants were married as shown in Table 2.

The results presented in Table 2 further show that most project participants (40%) had a high school education. According to Table 2, farming plays a central role in farmers' livelihoods as 90% of all the farmers surveyed were fully employed in farming. This finding can best be explained by the fact that in Nkonkobe Municipality, unemployment rate is high (Nkonkobe Municipality, 2009). The Organizational structure of the projects was also investigated. By organizational structure, the study investigated whether or not the projects members had the management committees that oversaw the projects' operation and took decisions about resource allocation and distribution of tasks and responsibilities. The descriptive analysis on organizational structure of the projects revealed some interesting results on strengths and weakness from the point of view of the projects participants. Table 3 presents results of the values of the strengths and weakness of the management committees of the enumerated projects.

Table 3. Distribution by strengths and weakness of management committees in six projects

Variable	Siga %	Lov 1 %	Lov 2 %	Somxa %	Nompu %	Gquma %
Motivation to members	40	0	40	15	57	5
Motivation and Disputes resolutions among members	60	100	60	39	36	68
Good communication flow	0	0	0	8	0	0
No strengths	0	0	0	39	7	26
	100	100	100	100	100	100
Variable	**Siga %**	**Lov1 %**	**Lov2 %**	**Somxa %**	**Nompu %**	**Gquma %**
Poor communication flow	0	0	0	0	0	53
Lack of negotiation power for formal markets access	80	33	80	31	7	37
Lack of commitment	0	0	0	15	14	0
Failure to assist members	0	17	0	31	29	10
No weakness	20	50	20	23	50	0
	100	100	100	100	100	100

Source: Field Survey (2011).

The findings show that the management committees motivated members in all six projects. In general, motivation and dispute resolution amongst the project members seem to have been the common strengths reported in all six projects This finding seems to support the evidence from the literature that local communities have their own complex informal mechanisms for resolving their conflicts (IFPRI, 2006). Following on the strengths of the management, the survey farmers were also asked to pinpoint the weakness (if any) of their management committees. Table 3 presents the distribution of the respondents by the weaknesses of their respective management committees.

Based on respondents' perception of the weakness of their management committees, the research findings show that lack of negotiation power to lobby formal markets seems to be common across all projects. The evidence from the literature, that institutional factors have tended to constrain smallholder's access to formal markets, is supported by this finding (Jacobs, 2008; Magingxa, Zerihun, & van Schalkwyk, 2009; Jari & Fraser, 2009). More weaknesses than strengths were identified by farmers in each project.

In an attempt to predict the sustainability of each project, the survey households were asked where the idea of forming the cooperatives came from. In Siga, 100% of farmers reported that the idea came from members themselves. In Lov1 and Lov2, 100% of respondents mentioned that University of Fort Hare pioneered the formation of the association. ARDRI (Agriculture and Rural Development Research Institute at the University of Fort Hare) was reported (by 100% of the respondents) as the body that had broached the idea of cooperative formation in Gquma. In Nompu and Somxa, the survey households provided different responses on who brought the idea of project formation. The issues of who pioneered farmer's association projects seem to be critical as revealed by the binary logistic results and as such may open the scope for further research to determine the productivity and sustainability of such projects.

Given the fact that Food and Agricultural Organization (FAO) (2009) argues that ownership of farm land would affect the sustainability of the land use, the survey farmers were asked how they accessed the land on which they farm. Generally, the findings revealed that most of the land used for vegetable production is not owned by the project participants. Figure 1 illustrates different ways through which land was accessed in all projects.

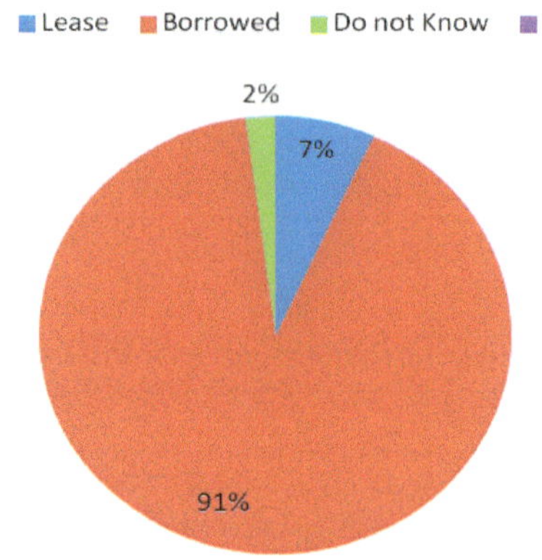

Figure 1. Land acquisition strategy in six vegetable projects (n = 62)

Source: Field Survey (2011).

With regard to land ownership and or land access, the research findings, indicate that the land reform programme, in particular the tenure systems, has not yet fully addressed the problems of the previously deprived black smallholder farmers in South Africa.

3.2 Empirical Results of the Binary Logistic Regression

From the study, it is hypothesized that smallholder farmers are influenced by institutional factors in vegetable production. This assumption implies that a smallholder farmer aims at maximizing his or her utility by either producing vegetable mainly for sale and consumption or mainly for sales subject to institutional constraints in Nkonkobe Local Municipality. Nine institutional variables and one technical variable (CR) were factored in a binary logistic regression to test their influence on farmer's main reason for producing vegetables.

The variables modeled include: management committee meeting frequency (MCMF), Management committee strength (MCS), management committee weakness (MCW), idea of association formation (IAF), land ownership (LO), working strategy on production in the field (WSF), working strategy in marketing (WSM), presence of an extension service (PES), Condition of roads (CR) and relationship to the Municipality (RM) by the survey farmers. Seven of these variables were significant at 5 percent alpha level in influencing the farmers 'main reason for vegetable production (see Table 4). The higher the value of the odds ratio for each variable in Table 4, the higher probability that such a variable would change to either direction (i.e. vegetable production mainly for sale or for both consumption and sale) if adjusted otherwise.

Table 4. Results of the logistic regression analysis

Parameter	Coef.	S.E.	Wald	P-value	Exp (B)
Constant	-10.290	5.184	7.102	0.47	0.000
Management Committees Meeting frequency	-5.752	2.158	1.397	0.008**	0.003
Manangement Committee Strengths	0.612	0.517	2.604	0.237	1.843
Management Committee Weakness	0.37	0.333	3.860	0.107	1.710
Idea of association formation	0.971	0.494	4.707	0.049**	2.640
Land Ownership	0.889	0.410	9.838	0.030**	2.432
Working strategy in the field	6.222	1.984	5.249	0.002**	503.858
Working Strategy Marketing	-2.591	1.131	3.888	0.022**	0.075
Presence of Extension Services	2.855	1.448	1.539	0.049**	17.373
Condition of Roads	-1.484	1.196	4.027	0.215	0.227
Relationship with the Local Municipality	1.310	0.653	3.940	0.045**	3.707

Note. Significance tests were carried out at the 5% level.

Source: Computed from Field Survey Data (2011).

$$In(\text{Mainly for sale/For sale \& consumption}) =$$
$$10.290 - 5.752(MCMF) + 0.612(MCS) + 0.537(MCW) + 0.971(IAF) + 0.889(LO) + 6.222(MSF) - 2.591(WSM) \quad (5)$$
$$+2.855(PES) - 1.484(CR) + 1.310RM$$

3.3 Frequency of Management Committee Meetings

Frequency of the management committees meeting seems to have a strong influence on farmers' main reason of producing vegetables. The logistic results presented in Table 4 reveal that an increase in the frequency of the management meeting can significantly influence farmers who work in projects or associations and shift their main aim of vegetable production towards the lower base response variable (i.e. mainly for sale). This demonstrates the urgency to disseminate information on cultural practices and also capacitate farmers to acquire skills and know-how. It is however important to note that despite the significance of this variable in influencing the farmers' decision in vegetable production, the complementary nature of other market institutions is crucial.

3.4 Idea of Association Formation

The results indicated that there is a weak relationship between the main reason farmers produce vegetables and the initiator of the project. From Table 4, the logistic results show that if the ideas of project formation came from the project participants themselves, they are more likely toalign their production decisions towards the high base response variable (i.e. vegetable production for both sales and consumption).

This finding shows that in situations where farmers are engaged in project identification and formulation, they tend to be flexible in planning the production decisions subject to resources, technical and institutional constraints given the fact that two options are possible in a high base response category. This finding may best explain why imposed or top-down projects have had little successes in improving livelihoods. This is likely the case when participants are supposed to work within the expectations of project pioneer(s). As with other variables, it must be acknowledged that a farmer's main reason to produce vegetables could also be influenced by other factors such as market access and family size (Randela, 2005; WFP, 2009), further highlighting the need to adopt a broad analytical perspective that combines conventional neo-classical tools and institutional economics.

3.5 Land Ownership

Land ownership seems to have a strong influence on farmer's main reason of vegetable production. At 5 percent significance level, the results in Table 4 show that if farmers do not have title to the land they farm, they are likely to shift their production decision towards a high response variable being for both home consumption and sale. In this respect, even if the essential infrastructure required for commercialization were to be put in place, it could have still been somehow hard for smallholder farmers to be fully commercialized owing to the lack of land ownership. When farmers lack title to the land used for farming, they resort to making short term production decisions since they could feel that they are not stable on such land (FAO, 2009). The results in relation to land ownership in this regard, suggest that whether or not a farmer has access to the market, he could be fully commercialized when he holds title to the land used.

3.6 Working Strategy in the Field

Farmers working collectively can pool together some farm resources thereby securing more produce for the market (Chikazunga, 2011). The prior expectation was that positive relationship between vegetable production for sale and the collectively operating projects would exist. However, the results presented in Table 4 revealed that famers who work collectively are influenced to produce vegetables for both sales as well as consumption. Given the high significance level of this variable (p = 0.002), the result suggests that when farmers work collectively in the field there is greater likelihood that they would produce vegetable for both sale and consumption, pointing up the crucial food security implications. It is more likely that this occurs as people working together have different aspirations which will most likely reinforce each other (Brown, 2000).

3.7 Marketing Strategy

It is argued that collective marketing improves market access by smallholder farmers (Jacobs, 2008; Jari & Fraser, 2009). The -2.591 coefficient of this variable in the logistic model represents the likelihood that collective marketing would influence smallholder farmers to shift their main reason for producing vegetables towards sales. The significance of this variable (p = 0.02) in influencing farmers main reason for producing vegetables clearly justifies the strengths of the variable and show that farmers marketing their produce individually are less likely to produce mainly for sale. Hence, this finding supports that of IFPRI (2006) postulating that collective action helps farmers access markets which they individually could not access. The results however, indicate that those farmers who market their produce individually would most probably shift also produce for consumption because the degree of market access is minimized by the higher transport costs and lower social capital they experience (Chikazunga, 2007; Jari & Fraser, 2009).

3.8 Presence of an Extension Service

Based on the results of the logistic regression presented in Table 4, the relationship seems to exist between farmers' main reason for vegetable production and the presence of an extension service that delivers extension messages to farmers. The delivery of an extension service seems to influence farmers' main reason for the production of vegetables mainly for sale and consumption given the coefficient of 2.855. The implication of this coefficient in the logistic regression is mainly reflected in the strengths of this variable which influences the consumption as well as the sale of vegetables either way. In order words, the absence of an extension service would influence farmers to produce vegetable mainly for sale. It is most likely that the current extension services provided to farmers are mostly oriented to subsistence and food security assurance, with farmers selling only the surplus. The fact that FAO (2007) has discovered that the extension workers rarely if ever plan the work along marketing challenges faced by farmers may best explain why it seems from the results that the absence of extension service would influence farmers produce for both sale and consumption. Nevertheless, it cannot be concluded with some certainty that the absence of extension services would always influence farmers to produce vegetables for both sale and consumption. In practice, what this result implies is that the current extension service provided to farmers needs to be reviewed in such a way that it orientates the smallholder farmers more strongly to marketing. Hence, it is more likely that most of the smallholder farmers would be fully commercialized (i.e. producing vegetables mainly for sales) given an enabling market environment, *ceteris paribus*.

3.9 Reasons for the Unique Relationship with the Municipality

The results presented in Table 4 on reasons for the kind of relationship (normally based on perceptions as informal norms) of the project participants and the local Municipality influenced their main vegetable production decisions in the study area. The logistic results show that the current relationship of the project participants and the local municipality influenced them to produce vegetables for sale and consumption at ($p = 0.045$). Given its significance in the model, this variable (based on perceptions) if adjusted to the other direction could influence farmers to produce mainly for sale. This is especially true in view of the fact that Hou and Smith (2009) cautioned that the influence of informal norms, as a bridge between formal rules and policy outcome, should not be undermined. It must, however, be noted that changing of norms and perception would require strategic approaches, such as provision of incentives or penalties, given the fact that they take time to change.

4. Conclusion

In the Nkonkobe Local Municipality nine institutional factors were identified and analyzed on the basis of their influence on small-scale vegetable producers' decision. Of all such factors identified, the frequency of the management committees meeting was significant in shaping the farmers' decisions in vegetable production. As the management committee convenes meetings frequently, there are increased chances that project member would produce mainly for sale. An issue of who brought up the idea of project formation was also significant to the extent that it influenced farmers to produce vegetables for sale and consumption. Given the fact that there was confusion as to who pioneered association formation in most projects, it is most likely that commercialization policies (i.e. as hypothesized to represent mainly for sale) are likely to be a success if farmers initiate their own projects. On the other hand, land ownership in the study area seemed to be a subject of concern as the majority of farmers (see Figure 1) were farming on leased land. The inferential statistics suggests that in the absence of secure land tenure, it is most unlikely that farmers would be fully commercialized (i.e. production mainly for sale). Similarly, the results of the study revealed that different types of collective action must be noted in policy formulation and/or recommendations. This is especially the case in view of the fact that farmers who worked collectively in production were influenced to produce for both sale and consumption but those who marketed the produce collectively were influenced to make production decisions that are inclined more towards sales.

Consequently an argument that collective action in marketing can help make farmers access more rewarding markets seems plausible (IFPRI, 2006; Jari & Fraser, 2009). Presence of an extension service to the farmers does not necessarily mean farmers would be influenced to produce mainly for sale. In the projects investigated, most farmers received an extension service but such a service influenced them to produce for both sale and consumption. This could most probably be explained by the fact that extension workers seldom address marketing opportunities and challenges to farmers (WFP, 2009). The relationships with the local municipality influenced farmers' production and marketing. Although such a relationship between the municipality and farmers is largely influenced by differing perceptions and political affiliations, the research finding suggests that if the very same perceptions and political positions change in the opposite direction, it is most likely that farmers would produce vegetables mainly for sale or be fully commercialized. Therefore, strategies proposed in linking smallholder farmers to profitable markets should take into account the role the municipality could play in this regard. Perhaps, this is most likely the case given that the Municipality has responsibility for the provision of basic market access requisites such as roads and market information among others.

Given the fact that institutional factors influence production and marketing costs (Coase, 1992; North, 1993; Lal, 1999), it is natural to expect that they would influence farmers' decisions to either produce for sale, consumption or for both. This is especially true as discovered by WFP (2009) that when smallholder farmers fail to access more profitable markets, they normally resort to produce mainly for home consumption.

Nevertheless, Monde (2003) and Randela (2005) had cautioned that own agricultural production for home consumption should not be easily undermined as it contributes to improved household food security. The fact that most of the surveyed farmers did not have secure tenure rights and that the majority did not receive consistent visits from the extension service could best explain why some land was left idle (i.e in Somxa and Gquma for instance). Besides, sustainable and efficient land resources use is compromised when individuals lack secure tenure rights on the land use (Makosholo, 2005; FAO, 2009). This raises a serious concern as about 90% of the survey farmers were using leased land across all six projects.

The problem statement for this investigation was that the institutional factors influence the way individuals interact in vegetable production and hence economic activity. The problem that needed to be examined was whether the existing or imposed formal rules on project participants do influence them to make production decisions that make them derive more benefits on sustainable and efficient basis in vegetable production.

A number of institutional aspects identified influenced decision making in vegetable production by the survey farmers and it was observed that a limitation of one or more of the identified institutional factors is likely to outweigh the positive impact of the other institutional factors. Owing to the fact that most of the surveyed farmers confirmed that they lacked ownership of land, clearly signals retarded development and inappropriate land use practices. Factored in a logistic model, land ownership was significant such that the results showed that in the face of lack of ownership, farmers are unlikely to be fully commercialized. Provision of title deeds to farmers is necessary for increased vegetable production and long-term productivity of farm lands. Similarly, the fact that the farmers who had relatively large land or plot sizes such as those in Somxa and Gquma (i.e. where some land was left idle during 2010/2011 growing season) explains that institutional support is needed and that exploiting the interdependency of institutional factors as well as the availability and suitability of the resources is crucial to smallholder development.

Contract farming (though did not exist amongst all survey farmers) is crucial as a tool for enhancing smallholder farmers' market access based on the evidence from the literature. It encourages commercialization of agriculture and promotes the development of entrepreneurial skills among smallholder farmers (Kirsten & Satorius, 2002).

Similarly, access to and use of credit by the survey farmers was investigated. None of the farmers in the study area had used credit. Given the evidence from the literature (Hoff & Stiglitz, 1990; Swinnem & Gow, 1999; Seibel, 2000; FAO, 2009) and the observations during field survey in 2011, the use of credit by smallholder farmers was low. In this regard it is recommended that access to credit and its use (i.e. where farmers do not have alternative means of cash) could help farmers address problems of production inputs such as seeds, fertilizers, including irrigations facilities, storage facilities and transport (Salami, Kamara, & Brixiova , 2010; de Klerk et al., 2013).

However, access to credit should be made on interest rate that smallholder farmers can afford while at the same time earning some income. It is in this sense that one notes that market access correlates with credit access because if farmers do not have the market to sell the produce so as to repay the loan, then the financial institutions are reluctant to lend. The same applies to farmers who do not have collateral. The banks or financial institutions would be reluctant to issue loans (FAO, 2009). Even those smallholder farmers who have collateral must be linked to markets (Jacobs, 2008). Such links would lead tomutual benefits for both the financial institutions and clients or farmers.

Finally, the analysis and discussion of the results have conclusively shown that when farmers are constrained by some institutional factors, the positive outcome of the existing institutional factors on smallholder farmers may be unrealizable. However, for increased vegetable production in the Nkonkobe Local Municipality, a more involved immediate reconsideration of weak institutional factors identified is crucial. The availability and suitability of the physical resource attributes such as land and infrastructure are crucial in complementing the effectiveness of institutional factors. Given the foregoing discussion, it is concluded that some institutional factors need some modifications accordingly so as to complement the effectiveness of the proposed agricultural policies in the study area. There is a need for institutional change more especially when development practitioners and farmers claim the commercialization of agriculture in the Nkonkobe Municipality.

Acknowledgements

The authors wish to express gratitude to the survey farmers, the Water Research Commission, N. Monde, N. Lebitsa, and N. Madolo, for their respective roles in the original study on which this paper is based.

References

Adebiyi, M. A., & Obasa, B. B. (2004). *Institutional framework, interest rate policy and financing of Nigeria manufacturing sub-sector.* Retrieved April 10, 2011, from http://www.tips.org.za/files/institutional_framework_interes_rate_policy_obasa_adbiyi.pdf

Brown, L. R. (2000). *In a search for a system model for decision making behaviour of first Generation black commercial farmers in Border-keiskamahooek Region of Eastern Cape Province* (Unpublished PhD Thesis). University of Fort Hare.

Chan, Y. H. (2005). Multinomial logistic regression analysis. *Singapore Medical Journal, 46*(6), 259-269. Retrieved April 2, 2011, from http://www.sma.org.sg/smj/4606/4606bs1.pdf

Chikazunga, D. (2011). *Collective action and smallholder farmers' participation in agricultural markets.* Retrieved October 14, 2011, from http://www.findavenue.co.za/AEASA/Presentations/Chikazunga%202.pdf

Chikazunga, D., & Paradza, G. (2012). Smallholder farming: A panacea for employment creation and enterprise development in South Africa? Lessons from the Pro-Poor Value Chain Governance Project in Limpopo Province. *Strategies to overcome poverty and inequality: Towards Carnegie III.*

Coase, R. H. (1992). The Institutional structure of Production. *The American Economic Review, 82*(4), 713-719.

Darroch, M. A. G., & Clover, T. A. (2005). Factors Affecting the Survival, Growth and Success of Small, Medium and Micro Agribusiness in KwaZulu-Natal, South Africa. *International food and agribusiness management association world foodand agribusiness symposium* (p. 27). Chicago, USA.

de Klerk, M., Fraser, F., & Fullerton, K. (2013). *The Status of Agricultural and Rural Finance in South Africa.* FinMark Trust.

FAO. (2007). *Approaches to linking producers to markets: A review of experiences to date.* Food and Agricultural Organisational of theUnited Nations, Rome, Italy.

FAO. (2009). *Rethinking public policy in agriculture: Lessons from distant and recent history.* Cambridge, London.

Ferris, S., Robbins, P., Best, R., Seville, D., Buxton, A., Shriver, J., & Wei, E. (2014). *Linking Smallholder Farmers to Markets and the Implications for Extension and Advisory Services.* MEAS Discussion Paper Series on Good Practices and Best Fit Approaches in Extension and Advisory Service Provision, Washington, USAID.

Flom, P. L. (2011). *Multinomial and ordinal logistic regression using PROC LOGISTIC.* Retrieved from April 9, 2011, from http://www.nesug.org/proceedings/nesug05/an/an2.pdf

Greenberg, S. (2010). Status report on Land and Agricultural policy in South Africa. *Research report 40.*

Hoff, K., & Stiglitz, J. (1990). Imperfect Information and Rural Credit Markets – Puzzles and Policy Perspectives. *World Bank Economic Review, 4*(3), 235-250. http://dx.doi.org/10.1093/wber/4.3.235

Hou, Y., & Smith, D. L. (2009). Informal Norms as a Bridge between Formal Rules and Outcomes of Government Financial Operations: Evidence from State Balanced Budget Requirements. *Journal of Public Administration Research and Theory, 20*(3), 655-678. Retrieved from July 16, 2010, from http://jpart.oxfordjournals.org/content/20/3/655.short

IFPRI. (2006). *Collective action and property rights for sustainable development: Understanding Collective Action.* Retrieved from April 1, 2011, from http://www.ifpri.org/sites/default/files/pubs/2020/focus/focus11/focus11 _02.pdf

Jacobs, P. (2008). *Market Development and Smallholder Farmers-A Selective Literature Survey.* Retrieved from April 14, 2010, from http://www.Scincedirect.com/science

Jari, B., & Fraser, G. (2009). An analysis of institutional and technical constraints influencing agricultural marketing amongst smallholder farmers in the Kat river valley, Eastern Cape, South Africa. *African Journal of Agricultural Research, 11*(4), 1129-1137.

Kirkland, T., Hunter, M., & Twine, W. (2005). *"The bush is no more": Insights on Institutional Change and Natural Resource Availability in Rural South Africa.* Working paper. University of Colorado at Boulder.

Kirsten, J., & Satorius, K. (2002). Linking agribusiness and small-scale farmers in developing countries: Is there a new role for contract farming? *Development Southern Africa, 19*(4). http://dx.doi.org/10.1080/0376835022000019428

Lal, D. (1999). Culture, democracy and development: Role of formal and informal Institutions and development. Retrieved August 23, 2010, from http://www.imf.org/External/pubs/FT/seminar/1999/reforms/lal.htm

Leedy, P. D., & Ormrod, J. E. (2005). *Practical research. Planning and design.* Pearson Merrill prentice hall-Ohio. USA.

Lekovic, V. (2011). Interaction of formal and informal institutions – impact on economic success. *Facta Universitatis – Economics & Organizations, 8*(4), 357-370.

Liao, F. T. (1994). *Interpreting Probability Models.* Sage Publications Inc, New Delhi.

Long, J. S. (1997). *Regression Models for Categorical and Limited Dependant Variables: Advanced Quantitative Techniques in the Social Sciences* (Vol. 7). Sage Publications Inc, Thousand Oaks.

Magingxa, L., Zerihun, G., Alemu, & van Schalkwyk, H. D. (2009). Factors influencing access to produce markets for smallholder irrigators in South Africa. *Development Southern Africa, 26*(1), 47-58. http://dx.doi.org/10.1080/03768350802640081

Makosholo, M. (2005). *Comparative advantage of Long-term crops in Lesotho* (MSc Thesis). University of Free State, South Africa.

Mohammed, M. A., & Ortmann, G. F. (2005). Factors influencing Adoption of Livestock Insurance by Commercial Dairy Farmers in Three Zobatat of Eritrea. *Agrekon, 44*(2), 172-186. http://dx.doi.org/10.1080/03031853.2005.9523708

Monde, N. (2003). *Household Food Security in Rural areas of Central Eastern Cape* (PhD Thesis). University of Fort Hare

Mokhahlane, M. (2009). Institutional factors affecting the use of communal rangeland in the Eastern Cape Province of South Africa. MSc Thesis, University of Fort Hare, South Africa.

Motteux, N. (2001). The Development and Co-ordination of Catchment Fora through the empowerment of Rural Communities, Catchment Research Group. *WRC Report K5/1014.* Water Res. Commission, Pretoria.

Nel, E., & Davies, J. (1999). *Farming against the odds: An examination of the challenges facing farming and rural development in the Eastern Cape province of South Africa* (p. 22). Department of Geography, Rhodes University, Grahamstown.

Nkonkobe Municipality. (2009). Nkonkobe Integrated Development Programme. *Draft IDP Review 2011/12, The Honourable Mayor.* Retrieved April 22, 2015, from http://mfma.treasury.gov.za/Documents/01.%20Integrated%20Development%20Plans/2011-12/02.%20Local%20Municipalities/EC127%20Nkonkobe/EC127%20Nkonkobe%20%20IDP%20-%201112.pdf

North, D. (1993). *Institutions, institutional change and economic performance.* Cambridge: Cambridge University Press.

Pejovich, S. (1999). The Effects of the Interaction of Formal and Informal Institutions on Social Stability and Economic Development. *Journal of Markets & Morality, 2*(2), 164-181.

Randela, R. (2005). Integration of emerging cotton farmers into the commercial agricultural economy (Unpublished PhD thesis). University of the Free State, Bloemfontein.

Salami, A., Kamara, A. B., & Brixiova, Z. (2010). *Smallholder Agriculture in East Africa: Trends, constraints and opportunities.* The working paper. Retrieved May 6, 2011, from http://www.afdb.org/fileadmin/uploads/afdb/Documents/Publications/WORKING%20105%20%20PDF%20d.pdf

Schultz, T. (1964). *Transformation of Traditional Agriculture.* New Haven: Yale University Press.

Seibel, H. D. (2000). Agricultural Development Banks: Close them or reform them. *Finance and Development, 37*(2), 1-4. Retrieved August 12, 2011, from http://www.imf.org/external/pubs/ft/fandd/2000/06/pdf/seibel.pdf

Swinnen, J. F. M., & Gow, H. R. (1999). Agricultural credit problems and policies during the transition to a market economy in Central Eastern Europe. *Food Policy, 24*(1), 21-47. http://dx.doi.org/10.1016/S0306-9192(98)00067-0

Van Schalkwyk, H. D., Groenewald, J. A., Fraser, G. C. G., Obi, A., & van Tilburg, A. (eds) (2012). Unlocking Markets to Smallholders – Lessons from South Africa. *Series No. 10.* Wageningen, Wageningen Academic Publishers and Mansholt Publications. http://dx.doi.org/10.3920/978-90-8686-168-2

WFP. (2009). Fighting hunger world wide: 3 ways to help smallholder farmers help themselves. *World Food Programme.* Retrieved August 26, 2010, from http://www.wfp.org/stories/3-ways-help-smallholder-farmers-help-themselves

Williamson, O. E. (2000). The New Institutional Economics: Taking Stock, Looking Ahead. *Journal of Economic Literature, 38*(3), 595-613. http://dx.doi.org/10.1257/jel.38.3.595

22

Milk Consumption and Chronic Disease Risk – The Strategy or Challenge to Avoid and Eliminate "Unwanted" Compounds and Contaminants

Gordana M. Kocic[1], Tatjana Jevtovic-Stoimenov[1], Dusan Sokolovic[1] & Hristina Kocic[2]

[1] University Nis, Medical Faculty, Nis, Serbia

[2] University Maribor, Medical Faculty, Maribor, Slovenia

Correspondence: Gordana Kocic, University Nis, Medical Faculty, Nis, Serbia.

Abstract

Milk consumption is a subject of many opposing opinions, regarding its epigenetic influence on genomic and proteomic cell events, what may change inflammatory, metabolic, oxidative and neurocognitive function. For some of milk ingredients, such as saturated lipids, lactose and some casein forms, the pathogenetic mechanisms were documented. Certain products, including some enzymes, low-weight partially digested peptides and non-protein nitrogen compounds, are still in the serious observations. To avoid „unwanted" compounds is complex, because of different conditions of their appearance, what in some extent determines if they can be avoided in pre-dairy manufacturing or eliminated in a dairy manufacturing and post-dairy processing and packing. Pre-dairy manufacturing should represent new approaches toward genetic selections of dairy breeds regarding milk biochemical composition (β-caseins). A number of contaminants, from pesticides to drugs, toxins, hormones, heavy metals, pharmaceuticals, phthalates, carcinogens, synthetic chemicals, or veterinary drugs may appear in a pre-dairy process through the plant material (forage) or artificially via (uncontrolled) cattle treatment. A proper growing and planting conditions, antifungal chemical treatment or dry collection of feedstuffs, or destruction of mycotoxins by biodegradation process may decrease their content. Unwanted compounds of milk, which may be eliminated during dairy manufacturing and packing, may be xanthine oxidase, uric acid, saturated fats and lactose. Beside low fat milk production, the quality improvement in milk fatty acids composition can be achieved by changing the animal feeding type. Regarding lactose or its components, glucose and galactose, it should be clearly distinguished the sugar-free milk from milk containing digested lactose, on the pack nutrition label. The main contaminants which appear, because of dairy products packing in plastic bottles and cups, are phthalates. Good agricultural practices (GAPs) concerning dairy industry should be directed to integrate different preventive strategies and to satisfy personalized nutrition concept.

Keywords: milk, xanthine oxidase, uric acid, β-casein type A1, lactose, phtalates

1. Introduction

In recent years the nutrigenomics tried to search for dietary factors which may contribute to development of various (the most common) chronic diseases, such as atherosclerosis, cardiovascular events, diabetes, chronic inflammatory conditions, autoimmune disorders, allergies and malignancies, by making the integration of genomic science and nutrition (Simopoulos, 2010). Regarding above mentioned diseases, milk consumption has become a subject of different, often opposing opinions, regarding its effect on inflammatory, metabolic, oxidative and neuro-cognitive function (Mead, 2007). Among the certain things are that milk and dairy products represent the best providers of calcium, other minerals, vitamins and essential amino acids (Figure 1).

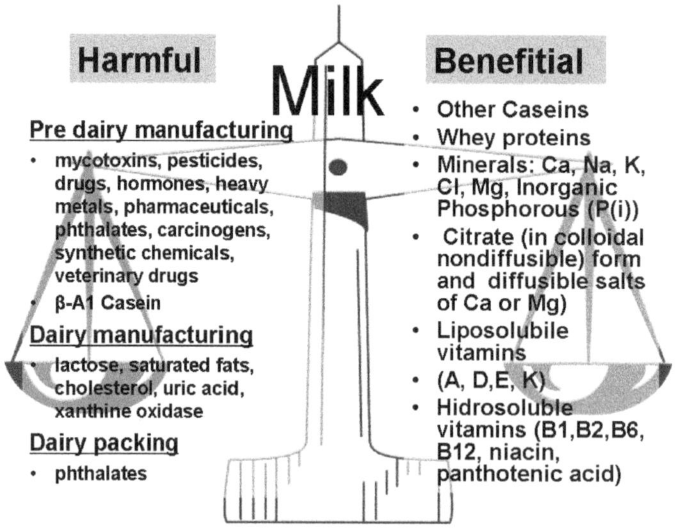

Figure 1. Beneficial and possible harmful milk ingredients and contaminants

The ingredients that may contribute to atherogenic potential of milk are different. For some of them, such as saturated lipids, lactose and some casein forms, the mechanisms of atherogenic effect were documented (Laugesen et al., 2003; Segall, 1994, 2002; Warensjo et al., 2010; Markey et al., 2014). Other ingredients or products, like some enzymes, low-weight partially digested peptides and non-protein nitrogen compounds, are still in the serious observations, as it was published by Awasthi et al. (2012), and Kocic et al. (2014). To avoid "unwanted" compounds is complex, because of different conditions of their appearance, what in some extent determines if they may be avoided in pre-dairy manufacturing or eliminated in a dairy manufacturing and packing.

2. Unwanted Compounds Which Could Be Avoided during Pre-Dairy Manufacturing

Pre-dairy manufacturing should represent new approaches toward genetic selections of dairy breeds not only regarding higher milk production, but regarding milk biochemical composition. Beside this, an adequate nutrition program should be directed to avoid any unwanted toxic substances.

2.1 β-Casein Type A1

Caseins are the main protein components in cows' milk (consisting about 80% of milk proteins) representing the major source of essential amino acids. Casein is a complex protein, belonging to phosphoproteins and three major groups (αS1, αS2, β, and κ casein) have been identified. They are present in specific milk micellar structures (Elliot et al., 1999). Experimental β-casein type A feeding may lead to increased level of homocysteine, which is a documented risk factor for atherosclerosis and to increased oxo-LDL (Laugesen & Elliot, 2003). Significant number of publications pointed out a causal link between β-casein type A1 and development of Type 1 diabetes, since the enhanced levels of antibodies to β-casein in children with Type 1 diabetes were obtained. Partial digestion of β-casein type A1 by digestive tract proteolytic enzymes can produce casomorphins, a family of small peptides with potent immune and opioid activity. Up to now, about 30 opioid peptides were isolated from milk, such as types of casomorphins 4, 5, 6, 7, 8, 9, 11, 13, 21, all of them isolated from bovine β-casein. Only two types of opioid bioactive peptides, known as exorphins, were isolated from α-casein (Kostyra et al., 2004). The types 7 and 8 were isolated from human milk as well. Their opioid activity increases with their hydrophobicity, due to higher hydrophobic interaction and affinity to opioid receptors, similar to morphine affinity. The appearance of Type 1 diabetes may be immunologically linked to the homology between the amino acid sequence of β-casomorphin7 (BCM7) and the pancreatic glucose transporting molecule GLUT2. The symptoms of autism and schizophrenia, other auto-immune conditions and milk intolerance are also linked to β-casein-A1. Regarding the A types of caseins, the difference between bovine A1 and A2 β-casein is only one amino acid of 209, since A2 β-casein has a proline at position 67, while A1 β-casein has a histidine. Most probably this point mutation occurred several thousand years ago. The histidine at position 67 in A1 β-casein accelerated partial casein digestion, with the subsequent release of the opioid peptide BCM7, which can easily cross the blood brain barrier. It might probably lead to neuropsychiatric behavior patterns, observed in autism and schizophrenia (Meisel & FitzGerald, 2000). The opinion that genetically β-casein A1 phenotypes are responsible for the "wrong milk" points out practical genomics to careful phenotyping for A1/A2 status and subsequent A2 cattle selection.

2.2 Pre-Dairy Chemical Contaminants, Drugs and Toxins

A number of contaminants, from pesticides to drugs, toxins, hormones, heavy metals, pharmaceuticals, phthalates, carcinogens, synthetic chemicals, or veterinary drugs may appear in a pre-dairy process through the plant material (forage) or artificially via (uncontrolled) cattle treatment by the accidental feeding and further transmission of contaminants or its possible metabolite into the milk of dairy cattle.

The plant toxin goitrin (5-viniloksazolidin-2-thione) exerts toxic effects on the thyroid gland, considered to be strumogenic. Toxins belonging the group of pyrrolizidines are present in Senecio plants jacobea (Bachmann et al., 1985).

The mycotoxins, especially aflatoxins (AF), are experimentally documented to act as the strong hepatotoxicants, which may induce liver and kidney cancer, teratogenicity and mutagenicity. The hazard ratio increases with the fact that the pasteurization and sterilization conditions are not able to destroy these toxins, which were documented to be heat resistant. Up to now several AF chemical forms and metabolites were documented to exist, such as B1,B2,G1,G2 and M. Metabolism of AF in liver is very complex and may occur in different ways, such as hydroxylation, epoxydation, hydratation or demethylation. Each of these steps mainly depends on catalytic action of liver corresponding microsomal enzymes, among which of essential importance belongs to cytochrome P450. Aflatoxicol appears as a secondary metabolite of Aflatoxin B1 (AFB1) after its reduction by the digestive tract flora. The appearance of aflatoxin M_1 in milk of cattle, sheep and goats corresponds to the ingested aflatoxin B_1, representing a biomarker of AF exposure, inspite of the fact that only 1.5% of AFB_1 can be excreted as the metabolite AFM_1 (Kiermeier, 1973). Other AF metabolites like AFM2 and AFM4 are present in a very small, almost negligible amount. Metabolites of AF are highly reactive compounds, they can bind covalently to DNA, making DNA adducts, what may induce DNA mutation. The additional damaging effect may appear from oxidative damage of cells, because of reduced glutathione depletion. Excess of reactive oxygen species may also led to programmed cell death (apoptosis). Epoxidation (8, 9, AFB1 epoxide) is of greatest importance for AF carcinogenic or delayed mutagenic potential, because of its electrophylic attack to nitrogen or oxygen of purine bases (AFB1-N7 guanine adduct). At the same time, epoxide level directly correlates with reactive oxygen species (ROS) level. From carcinogenic point of view, the adduct formation may be of importance by inducing mutation of p53 suppressor gene (Elaton & Gallanger, 2004). Main sources of mycotoxins are different feedstuffs, like forages, concentrates and preserved feeds, such as silage, hay, and straw (O'Brien, 2005). Aflatoxins may occur especially in energy-rich feedstuffs concentrates, such as the grains, corn, soybean, oil plants like sunflower seeds. Mycotoxins like fumonisins and zearalenone may be found in maize derived products, while trichothecenes and ergot alkaloids may be found in cereal grains (Nawaz et al., 1997). Ergovaline may be found in pasture grasses (Cheeke, 1995). After absorption they can be excreted into milk. Good agricultural practices (GAPs) is directed mainly toward different preventive strategies, such as proper growing and planting conditions, antifungal chemical treatment or dry collection of feedstuffs. As the new promising strategies are the approaches regarding possible destruction of mycotoxins by biodegradation process, using different microbes, like probiotic bacteria, enzymes, metals salts and surfactants (Salminen et al., 2010; Das et al., 2014).

Artificially present pesticides or heavy metals may appear in soils and plants as a result of use of agricultural chemicals, industrial pollutant or car exhaust. Heavy metals, like Pb, Cd, Fe, Hg, Cu, represent persistent environmental hazard, causing a multi-organ failure, followed by many metabolic abnormalities, but followed also with the economic losses due to disease treatment costs (Bischoff et al., 2014). The pollution with Pb may occur by industrial processes emissions, like the traffic petrol, the smoke, but also because of the use of different paints. As a typical cumulative poison, it can be accumulated in bones and bone marrow, in the kidney and liver tissue, leading to the inhibition of hem synthesis and development of chronic anemia, encephalopathy, reproductive function failure. The carcinogenic and delayed mutagenic effects were also documented. The contamination of milk may be due to feeding of the cows with fodder collected near the big roads. Similar toxic effects can be diagnosed by Cd poisoning, with deposition and consequent damage of parenchimatous organs, like kidneys, liver, and lungs.

Pesticides represent the most persistent pollutants in the environment, ingested through water or forages, causing contamination of milk. Among pesticides, most often are the lypophilic chemicals, exerting very high affinity to bind to lipid structures they can be easily accumulated in different tissue lipoprotein structures. Such types are chlorinated pesticides, organophosphates, herbicides, fungicides and anti-helmintic chemicals. They can cause a number of systemic toxic effects, such as anemia, reproductive failure, parenchimatous organs damage (liver and kidney) or brain damage. Some lactic acid bacteria may be useful in their degradation (Villarroel et al., 2013; Witczak et al., 2013; Zhou et al., 2015).

Antibiotics and sulphonamides can be present in cow milk as veterinary agents, after treatment of mastitis. Phthalates, phenolic compounds, detergents and desinfectants and their metabolites may usually get into water as a result of use of pharmaceuticals, agricultural chemicals, households chemicals or from industrial wastes (Khaniki, 2007; Awasthi et al., 2012).

An increasing focus on prevention of the pollution by radioactive compounds should be emphasized as well, because of possible local nuclear accidental catastrophes or other possible ways of contamination during war conditions. Among radioactive elements, the concentrations of radioactive cesium (134 and 137) Cs, iodine and uranium may be detected in dairy products from contaminated regions (Miyazaki et al., 2013; Belles et al., 2013).

All of the above mentioned unwanted compounds may be detected during milk collection and can be avoided before the dairy manufacturing process by high quality control.

3. Unwanted Compounds Which Could Be Eliminated during Dairy Manufacturing and Packing

3.1 Milk Homogenization, Xanthine Oxidase Liberation and Urate Level

Commercial liquid milk is commonly homogenized. Fresh, unprocessed milk contains different purine and pyrimidine compounds mainly as adenine and guanine ribonucleotides or free bases (hypoxanthine and xanthine). Purine ribonucleotides which are usually present in milk, can be degraded to uric acid by milk fat globule xanthine oxidase, during the dairy process of milk homogenization (Harrison, 2002; Harrison, 2006; Michalski, 2007). Milk fat globules are secreted after synthesis in the endoplasmic reticulum of the mammary gland. Since xanthine oxidase is located in the milk fat globule in its inactive form, procedures of homogenization of milk induces its liberation from fatty globules, what is also followed by a subsequent activation. Xanthine oxidase may also be activated during milk intestinal digestion (Harrison, 2002). In the above reactions, a strong free radical, superoxide anion is liberated. The activity of xanthine oxidase in breast milk seems to be beneficial in the newborn period, as a major bactericidal element for the prevention of gut infection (Stevens et al., 2000). But in a later age, the activity of xanthine oxidase may be harmful because of its prooxidative function. A hypothesis about possible atherogenic effect of xanthine oxidase proposed that XO after its gastrointestinal absorption may across into the bloodstream, where it can damage endothelium by binding, immune reactivity and *in situ* free radical production (Harrison, 2002). Since enzymes, as protein structures, can be inactivated (denaturated) by using high temperature, the UHT sterilization would be the most suitable way of its inactivation.

Regarding the presence of uric acid in milk, its possible harmfull effects were documented in primary hyperuricemia (gout, urate nephrolithiasis, Lech-Nychan syndrome), secondary hyperuricemia (preeclampsia, cardiovascular diseases, hypertension, diabetes mellitus, stroke, dementia, obesity, alcohol abuse, tumor lysis syndrome, AIDS, psoriasis, high sport demands, age over 65 and "stress manager disease") (Sculley et al., 1992; Dincer et al., 2002; Iseki et al., 2004; Schretlen et al., 2007; Tatli et al., 2008; Gonzalez et al., 2008). High uric acid in blood may increase several times cardiovascular risk or stroke and can enhance oxidation of LDL lipoproteins (Feig et al., 2008). Hyperuricemia may be linked to disorders in cognitive functioning and in dementia (Schretlen et al., 2007). In order to satisfy the demand of healthy dairy product for population under hyperuricemic risk, we produced milk almost free of purine and uric acid. Our experimental studies documented positive impact of depurinized milk on liver regenerative signaling transduction pathway, anti-apoptotic signaling, myocardial reparation, antioxidative potential and bone-marrow stem cell potential (Kocic et al., 2014a, 2014b). Urate crystals can induce immuno-inflammatory effect acting as the damage-associated mollecular patterns (DAMP), after binding to the surface of antigen-presenting cells, what induces strong IL-1β production (Foell et al., 2007; Cheng et al., 2008). Uric acid can induce mechanical endothelial damage, vascular remodeling, smooth muscle cell proliferation and endothelial dysfunction, can increase platelet adhesiveness and aggregability, and can stimulate neutrophils adhesion. Urate crystals were detected inside of the atherosclerotic plaques (Feig et al., 2008). Urates can aggravate eclamptic pregnancy, by inducing inflammation, endothelial dysfunction, hypertension, they may damage placenta, affecting in this way intrauterine growth (Bainbridge & Roberts, 2008). Since the linear correlation between the uric acid level and the appearance of symptoms of cardiovascular or renal dysfunction was not documented, it would be recommended to avoid uric acid or their source (purines) of food especially in older age (Dincer et al., 2002). In order to satisfy the demand of healthy dairy product for population prone to uric acid unwanted effects, we produced "depurinized" milk almost free of purine nucleotides and uric acid (Kocic et al., 2014a, 2014b).

3.2 Milk Fats

The milk as a source of the milk fat in an amount of 3-4%, can significantly contribute to the general nutritional input of lipids. Milk fat is composed of neutral lipids (primarily triglycerides) and polar lipids, structured as emulsified milk fat globules (MFG), finely dispersed as very small (5-10 microns in diameter) drops in the water phase. The fatty acids involved in composing milk triglycerides are very different in carbon size (from 4 to 26 carbons), mainly saturated (about two thirds), branched and linear, regarding the structure of carbon chain. About 12% of the fatty acids are short-chain, such as butyric, caproic, caprylic, capric acids, conjugated linolenic acid, sphingolipids and fat soluble vitamins (A, D, E and K). Regarding the milk fat involvement in atherogenic blood lipid profile with high LDL lipoprotein content, associated with cardiovascular risk, the main contribution belongs to saturated fatty acids, the level of which exceeds 60% in milk. The most hazardous of them are C12, C14 and C16 chain saturated milk fatty acids. Others, like short-chain fatty acids, polyunsaturated fatty acids or branched fatty acids have been documented to possess bioactive properties, having influence on gene expression, exerting hormone-like, anticancerogenic and antimicrobial activity and may decrease cardiovascular risk through the decrease of LDL cholesterol level (Parodi, 2004).

The effect of milk fats on human health may be considered to be favorable to human health, compared to other animal fat sources, but unfavorable, compared to low or fat-free milk. That is why the dairy industry has in recent years worked intensively on manufacturing the low-fat, fat-modified or fat-free milk, what involved filtering out the fat from the milk or supplementation of polyunsaturated fatty acids (Markey et al., 2014; Ferreiro et al., 2015). Above mentioned articles may suggest that low-SFA or fat-modified milk should be preferred, regarding cardio-metabolic benefit, when compared to commercially available whole milk dairy products. In the newest meta analysis a total of 22 studies were analyzed regarding the role of low-fat dairy products in preventing cardiovascular diseases. Above mentioned meta analysis supported this finding (Qin et al., 2015). The low-fat milk can be protective against development of type 2 Diabetes Mellitus through moderate insulin sensitizing effect, as recent meta analysis has been documented (Abedini et al., 2015). Beside low fat milk recommendation, the quality improvement in milk fatty acids composition can be achieved by changing the animal feeding type (organically managed cows) by supplementation of polyunsaturated fatty acids, short-changed fatty acids, antioxidants, vitamins A, E, and trace element Selenium (Martemucci & D'Alessandro, 2013; Ferreiro et al., 2015).

3.3 Lactose

Lactose represents a normal constituent and the most important sugar in milk. The Scientific Committee recommendations for nutrition and food standards agencies admitted lactose as an atherogenic factor of diet. Well known 'French paradox', regarding the unexplained lower cardiovascular mortality may be explained, beside other factors, by increased prevalence of lactose intolerance and cheese (low lactose) supply more than milk drinking. Insulin resistance syndrome that was almost commonly seen in some African blacks and Pima Indians may be linked with high appearance of lactose intolerance and low milk intake. If milk lactose may be risk factor for the development of atherosclerosis depends mainly on the activity of intestinal lactase, which catalyzes its degradation to glucose and galactose. With regard to chemical properties and affinity to carriers, galactose is reabsorbed from the intestines by using the same carrier (specific for the position of the hydroxyl group on carbon two) as glucose, and then, under physiological conditions it undergoes mainly to specific galactokinase-dependent metabolic pathway. Congenital disorder of galactose metabolism leads to cataracts as a result the osmotic activity of galactose, which after entering the cells, can be converted into galactitol. A number of experimental data about the toxicity of galactose initiated clinical observations related to its cardiovascular risk (Segall, 1994, 2002). Galactose possesses higher affinity for nucleophilic attack and protein nonenzymatic galactosylation than glucose for glycation. Some experimental results documented that cell proteins can be galactosylated about 4-fold faster and more extensively, than glycosylated, compared under equimolar incubation conditions. Insulin makes more effective entry of galactose in a cell than the entry of glucose, contributing to the retention of glucose and development of insulin resistance or other long-term complications (Segall, 2002). Because of that, it is particularly important to distinguish milk types indicated as "lactose-free milk". It should be clarified if it really represents a sugar-free milk, or milk containing digested lactose to their monosaccharides, glucose and galactose. It can be identified clearly on the label packaging, but customers are supposed to know to interpret the packaging. For lactose intolerant (insufficient in lactase) young people, the milk containing digested lactose to glucose and galactose is suitable. If a person suffering from diabetes consumes the milk containing digested lactose, a chance to "better" absorbance of mentioned monosaccharides became higher. Proposed methods of chromatografic adsorption of lactose, proposed by Harju (2007) are now in use by some dairy manufacturers, but new challenges might be to the use of non-toxic natural lectins, according to experimental data of their properties.

3.4 Contaminants of Milk Packing

During the post-processing period, the dairy products may aquire potentially toxic substances as well. Phthalates, as unwanted lipophilic toxic compounds (plasticizers), are usually added to PVC for softening (Sorensen, 2006). They may appear in milk as a result of migration from plastic packaging. Exposure to phthalates may be linked to lungs, liver, kidneys injury and cancer development, increased blood pressure and endocrine system alteration (endocrine-disrupting chemicals-EDCs), accompanied with reduced male fertility. One way to reduce phthalate exposure includes avoiding products made of PVC plastics ("phthalate-free" on packaging), by choosing natural packing materials. In our study by using commercial 1.5% fat UHT cow milk, we isolated about 36 potentially toxic volatile chemically different aliphatic and aromatic aldehydes, ketones, carboxilic acids, sulphur compounds and six phthalate types. Technological procedure by using our device system was able to eliminate (adsorb) more than 90% of identified phthalates, supporting technology innovation in the dairy processing (Kocic et al., 2014a).

4. Conclusion

In conclusion, some milk ingredients or contaminants may directly affect human health, by having the influence on genomic, proteomic and metabolomic events, and these in turn may lead to disease initiation, development, or progression. Good agricultural practices (GAPs) concerning dairy industry should be directed to integrate different preventive strategies to ensure each stage of dairy processing. Since each body may respond differently to the same dairy product, the dairy industry may be supposed to produce different milk types, to satisfy personalized nutrition concept.

Acknowledgements

The study was supported by Ministry of Education and Science Republic Serbia project TR31060.

References

Abedini, M., Falahi, E., & Roosta, S. (2015). Dairy product consumption and the metabolic syndrome. *Diabetes Metab. Syndr., 9*(1), 34-37. http://dx.doi.org/10.1016/j.dsx.2014.04.027

Awasthi, V., Bahman, S., Thakur, L. K., Singh, S. K., Dua, A., & Ganguly, S. (2012). Contaminants in milk and impact of heating: an assessment study. *Indian J. of Public Health, 56*(1), 95-99. http://dx.doi.org/10.4103/0019-557X.96985

Bachmann, M., Theus, R., Lüthy, J., & Schlatter, C. (1985). The occurrence of goitrogenic substances in milk. 1. Release of goitrin in the milk of cows fed on rapeseed extract cakes. *Z Lebensm Unters Forsch., 181*(5), 375-378. http://dx.doi.org/10.1007/BF01027401

Bainbridge, S, A., & Roberts, J. M. (2008). Uric Acid as a Pathogenic Factor in Preeclampsia. *Placenta, 29*(Suppl. 1), 67-72. http://dx.doi.org/10.1016/j.placenta.2007.11.001

Belles, M., Linares, V., Perello, G., & Domingo, J. L. (2013). Human dietary exposure to uranium in Catalonia, Spain. *Biol. Trace Elem. Res., 152*(1), 1-8. http://dx.doi.org/10.1007/s12011-012-9587-2

Bischoff, K., Higgins, W., Thompson, B., & Ebel, J. G. (2014). Lead excretion in milk of accidentally exposed dairy cattle. *Food Addit Contam Part A Chem Anal Control Expo Risk Assess., 31*(5), 839-844. http://dx.doi.org/10.1080/19440049.2014.888787

Cheeke, P. R. (1995). Endogenous toxins and mycotoxins in forage grasses and their effects on livestock. *Journal of Animal Science, 73*, 909-918. Retrieved from http://www.ncbi.nlm.nih.gov/pubmed/7608026

Cheng, T. H., Lin, J. W., Chao, H. H., Chen, Y. L., Chen, C. H., Chan, P., & Liu, J. C. (2008). Uric acid activates extracellular signal-regulated kinases and thereafter endothelin-1 expression in rat cardiac fibroblasts. *Int. J. Cardiol., 139*(1), 42-49. http://dx.doi.org/10.1016/j.ijcard.2008.09.004

Das, A., Bhattacharya, S., Palaniswamy, M., & Angayarkanni, J. (2014). Biodegradation of aflatoxin B1 in contaminated rice straw by Pleurotus ostreatus MTCC 142 and Pleurotus ostreatus GHBBF10 in the presence of metal salts and surfactants. *World J Microbiol Biotechnol., 30*(8), 2315-2324. http://dx.doi.org/10.1007/s11274-014-1657-5

Dincer, E. H., Dincer, A. P., & Levinson, D. J. (2002). Asymptomatic hyperuricemia: To treat or not to treat. *Cleve Clin J Med, 69*(8), 594-608. http://dx.doi.org/10.3949/ccjm.69.8.594

Elliott, R. B., Harris, D. P., Hill, J. P., Bibby, N. J., & Wasmuth, H. E. (1999). Type 1(insulin-dependent) diabetes mellitus and cow milk: Casein variant consumption. *Diabetologia, 42*, 292-296. http://dx.doi.org/10.1007/s001250051153

Feig, D. I., Kang, D. H., & Johnson, R. J. (2008). Uric acid and cardiovascular risk. *N. Engl. J. Med., 359*, 1811-1821. http://dx.doi.org/10.1056/NEJMra0800885

Ferreiro, T., Gayoso, L., & Rodríguez-Otero, J. L. (2015). Milk phospholipids: Organic milk and milk rich in conjugated linoleic acid compared with conventional milk. *J. Dairy Sci., 98*(1), 9-14. http://dx.doi.org/10.3168/jds.2014-8244

Fink-Gremmels, J. (2008). Mycotoxins in cattle feeds and carry-over to dairy milk: a review. *Food Addit Contam Part A Chem Anal Control Expo Risk Assess.*, *25*(2), 172-180. http://dx.doi.org/10.1080/02652030701823142

Foell, D., Wittkowski, H., & Roth, J. (2007). Mechanisms of disease: A 'DAMP' view of inflammatory arthritis. *Nat .Clin. Pract. Rheumatol.*, *3*(7), 382-390. http://dx.doi.org/10.1038/ncprheum0531

Gonzalez, D., Marquina, R., Rondon, N., Antonio, J., Rodriguez-Malaver, A., & Reyes, R. (2008). Effects of aerobic exercise on uric acid, total antioxidant activity, oxidative stress and nitric oxide in human saliva. *Research in Sports Medicine, 16*(2), 128-137. http://dx.doi.org/10.1080/15438620802103700

Harju, M. (2007). *Chromatographic separation of lactose and its applications in the dairy industry IDF Symposium Lactose & its Derivatives*. Moscow, Valio Ltd. Retrieved May 14-16, 2007, from http://lactose.ru/present/7Matti_Harju.pdf

Harrison, R. (2002). Milk Xanthine Oxidase: Hazard or Benefit? *J. Nutritional & Environmental Medicine, 12*(3), 231-238.

Harrison, R. (2006). Milk xanthine oxidase: Properties and physiological roles. *Int. Dairy J. 16*, 546-554. http://dx.doi.org/10.1016/j.idairyj.2005.08.016

Iseki, K., Ikemiya, Y., Inoue, T., Iseki, C., Kinjo, K., & Takishita, S. (2004). Significance of hyperuricemia as a risk factor for developing ESRD in a screened cohort. *Am. J. Kidney Dis.*, *44*, 642-650. http://dx.doi.org/10.1053/j.ajkd.2004.06.006

Khaniki, G. R. J. (2007). Chemical contaminants of human milk and public health concerns: A review. *Internat. J. Dairy Sci.*, *2*(2), 104-115. http://dx.doi.org/10.3923/ijds.2007.104.115

Kiermeier, F. (1973). Aflatoxin residues in fluid milk. *Pure Appl. Chem.*, *35*, 271-274. http://dx.doi.org/10.1351/pac197335030271

Kocic, G., Pavlovic, R., Nikolic, G., Veljkovic, A., Panseri, S., Chiesa, L. M., ... Nikolic, R. (2014). Effect of commercial or depurinized milk on rat liver growth-regulatory kinases, nuclear factor-kappa B, and endonuclease in experimental hyperuricemia: Comparison with allopurinol therapy. *J. Dairy Sci.*, *97*, 4029-4042. http://dx.doi.org/10.3168/jds.2013-7416

Kocic, G., Sokolovic, D., Jevtovic, T., Cvetkovic, T., Veljkovic, A., Kocic, H., ... Zivkovic, P. (2014). Short communication: Effect of commercial or depurinized milk diet on plasma advanced oxidation protein products, cardiovascular markers, and bone marrow CD34(+) stem cell potential in rat experimental hyperuricemia. *J Dairy Sci*, *97*, 6823-6827. http://dx.doi.org/10.3168/jds.2014-8556

Kostyra, E., Sienkievicz-Szlapka, E., Jarmolowska, B., Crawczuk, S., & Kostyra, H. (2004). Opioid peptides derived from milk proteins. *Polish J. Food Nutrition Sci.*, *13*(54), 25-35. Retrieved from http://journal.pan.olsztyn.pl/fd.php?f=484

Laugesen, M., & Elliott, R. (2003). Ischaemic heart disease, Type 1 diabetes, and cow milk A1 beta-casein. *N. Z. Med. J.*, *116*(1168), U295.

Markey, O., Vasilopoulou, D., Givens, D. I., & Lovegrove, J. A. (2014). Dairy and cardiovascular health: Friend or foe? *Nutr. Bull.*, *39*(2), 161-171. http://dx.doi.org/10.1111/nbu.12086

Marphy, P. A., Hendrich, S., Landgren, C., & Bryant, C. M. (2006). Food mycotoxins: an update. *J Food Science, 71*(5), R51-R65. http://dx.doi.org/10.1111/j.1750-3841.2006.00052.x

Martemucci, G., & D'Alessandro, A. G. (2013). Progress in nutritional and health profile of milk and dairy products: a novel drug target. *Endocr. Metab. Immune Disord. Drug Targets, 13*(3), 209-233. http://dx.doi.org/10.2174/18715303113136660045

Mead, M. N. (2007). Nutrigenomics: The Genome–Food Interface. Environ. *Health Perspect, 115*(12), A582-A589. http://dx.doi.org/10.1289/ehp.115-a582

Meisel, H., & FitzGerald, R. J. (2000). Opioid peptides encrypted in intact milk protein sequences. *British Journal of Nutrition, 83*(Suppl. 1), S27-S31. http://dx.doi.org/10.1017/S000711450000221X

Michalski, M. C. (2007). On the supposed influence of milk homogenization on the risk of CVD, diabetes and allergy. *Br. J. Nutr.*, *97*, 598-610. http://dx.doi.org/10.1017/S0007114507657900

Miyazaki, H., Tsuchiyama, T., & Terada, H. (2013). Examination of radioactive contamination in foods. *Shokuhin Eiseigaku Zasshi, 54*(2), 156-64. http://dx.doi.org/10.3358/shokueishi.54.156

Nawaz, S., Scudamore, K. A., & Rainbird, S. C. (1997). Mycotoxins in ingredients of animal feeding stuffs: I. Determination of Alternaria mycotoxins in oilseed rape meal and sunflower seed meal. *Food Additives and Contaminants, 14*, 249-262. http://dx.doi.org/10.1080/02652039709374522

Nyman, A. K., Ekman, T., Emanuelson, U., Gustafsson, A. H., Holtenius, K., Waller, K. P., & Sandgren, C. H. (2007). Risk factors associated with the incidence of veterinary-treated clinical mastitis in Swedish dairy herds with a high milk yield and a low prevalence of subclinical mastitis. *Preventive Veterinary Medicine, 78*, 142-160. http://dx.doi.org/10.1016/j.prevetmed.2006.10.002

Parodi, P. W. (2004). Milk fat in human nutrition. *Austral. J. Dairy Technol., 59*(1), 3-59. Retrieved from ftp://s173-183-201-52.ab.hsia.telus.net/Inetpub

Qin, L. Q., Xu, J. Y., Han, S. F., Zhang, Z. L., Zhao, Y. Y., & Szeto, I. M. (2015). Dairy consumption and risk of cardiovascular disease: an updated meta-analysis of prospective cohort studies. *Asia Pac J Clin Nutr, 24*(1), 90-100. http://dx.doi.org/10.6133/apjcn.2015.24.1.09

Salminen, S., Nybom, S., Meriluoto, J., Collado, M. C., Vesterlund, S., & El-Nezami, H. (2010). Interaction of probiotics and pathogens – Benefits to human health? *Curr Opin Biotechnol, 21*(2), 157-67. http://dx.doi.org/10.1016/j.copbio.2010.03.016

Schretlen, D. J., Inscore, A. B., Jinnah, H. A., Rao, V., Gordon, B., & Pearlson, G. D. (2007). Serum uric acid and cognitive function in community-dwelling older adults. *Neuropsychology, 21*, 136-140. http://dx.doi.org/10.1037/0894-4105.21.1.136

Sculley, D. G., Dawson, P. A., Emmerson, B. T., & Gordon, R. B. (1992). A review of the molecular basis of hypoxanthine guanine phosphoribosyl transferase (HGPRT) deficiency. *Human Genet., 90*, 195-207. http://dx.doi.org/10.1007/BF00220062

Segall, J. J. (1994). Dietary lactose as a possible risk factor for ischaemic heart disease: Review of epidemiology. *Int. J. Cardiol., 46*, 197-207. http://dx.doi.org/10.1016/0167-5273(94)90242-9

Segall, J. J. (2002). Plausibility of Dietary Lactose as a Coronary Risk Factor. *Journal of Nutritional & Environmental Medicine, 12*(3), 217-229. http://dx.doi.org/10.1080/1359084021000006885

Simopoulos, A. P. (2010). Nutrigenetics/Nutrigenomics. *Ann. Rev. Public Health, 31*, 53-68. http://dx.doi.org/10.1146/annurev.publhealth.031809

Sorensen, L. K. (2006). Determination of phthalates in milk and milk products by liquid chromatography/tandem mass spectrometry. *Rapid Commun. Mass Spectrom., 20*, 1135-1143. http://dx.doi.org/10.1016/j.chroma.2013.11.017

Stevens, C. R., Millar, T. M., Kanczler, J. M., Bodamyali, T., & Blake, D. R. (2000). Antibacterial properties of xanthine oxidase in human milk. *Lancet, 356*(9232), 829-830. http://dx.doi.org/10.1016/S0140-6736(00)02660-X

Tatli, E., Aktoz, M., Buyuklu, M., & Altun, A. (2008). The relationship between coronary artery disease and uric acid levels in young patients with acute myocardial infarction. *Cardiol. J., 15*, 21-25. Retrieved from http://www.ncbi.nlm.nih.gov/pubmed/18651381

Villarroel, A., & Halliburton, M. K. (2013). Control of extensive chorioptic mange natural infection in lactating dairy cattle without milk withdrawal. *Vet J., 197*(2), 233-237. http://dx.doi.org/10.1016/j.tvjl.2013.01.003

Warensjo, E., Jansson, J. H., Cederholm, T., Boman, K., Eliasson, M., Hallmans, G., & Sjogren, P. (2010). Biomarkers of milk fat and the risk of myocardial infarction in men and women: A prospective, matched case-control study. *Am. J. Clin. Nutr., 92*, 194-202. http://dx.doi.org/10.3945/ajcn.2009.29054

Witczak, A., Mituniewicz-Małek, A., & Dmytrow, I. (2013). Assessment of daily intake of organochlorine pesticides from milk in different regions of Poland. *J Environ Sci Health B, 48*(2), 83-91. http://dx.doi.org/10.1080/03601234.2013.726589

Zhou, X. W., & Zhao, X. H. (2015). Susceptibility of nine organophosphorus pesticides in skimmed milk towards inoculated lactic acid bacteria and yogurt starters. *J Sci Food Agric, 95*(2), 260-266. http://dx.doi.org/10.1002/jsfa.6710

Permissions

All chapters in this book were first published in JAS, by Canadian Center of Science and Education; hereby published with permission under the Creative Commons Attribution License or equivalent. Every chapter published in this book has been scrutinized by our experts. Their significance has been extensively debated. The topics covered herein carry significant findings which will fuel the growth of the discipline. They may even be implemented as practical applications or may be referred to as a beginning point for another development.

The contributors of this book come from diverse backgrounds, making this book a truly international effort. This book will bring forth new frontiers with its revolutionizing research information and detailed analysis of the nascent developments around the world.

We would like to thank all the contributing authors for lending their expertise to make the book truly unique. They have played a crucial role in the development of this book. Without their invaluable contributions this book wouldn't have been possible. They have made vital efforts to compile up to date information on the varied aspects of this subject to make this book a valuable addition to the collection of many professionals and students.

This book was conceptualized with the vision of imparting up-to-date information and advanced data in this field. To ensure the same, a matchless editorial board was set up. Every individual on the board went through rigorous rounds of assessment to prove their worth. After which they invested a large part of their time researching and compiling the most relevant data for our readers.

The editorial board has been involved in producing this book since its inception. They have spent rigorous hours researching and exploring the diverse topics which have resulted in the successful publishing of this book. They have passed on their knowledge of decades through this book. To expedite this challenging task, the publisher supported the team at every step. A small team of assistant editors was also appointed to further simplify the editing procedure and attain best results for the readers.

Apart from the editorial board, the designing team has also invested a significant amount of their time in understanding the subject and creating the most relevant covers. They scrutinized every image to scout for the most suitable representation of the subject and create an appropriate cover for the book.

The publishing team has been an ardent support to the editorial, designing and production team. Their endless efforts to recruit the best for this project, has resulted in the accomplishment of this book. They are a veteran in the field of academics and their pool of knowledge is as vast as their experience in printing. Their expertise and guidance has proved useful at every step. Their uncompromising quality standards have made this book an exceptional effort. Their encouragement from time to time has been an inspiration for everyone.

The publisher and the editorial board hope that this book will prove to be a valuable piece of knowledge for researchers, students, practitioners and scholars across the globe.

List of Contributors

Evans Atuti Atera
Department of Technical Services, Lake Basin Development Authority, Kenya

John Collins Onyango
Department of Botany, Maseno University, Maseno, Kenya

Pham Thien Thanh
Graduate School of Agricultural Science, Kobe University, Japan

Takashige Ishii
Graduate School of Agricultural Science, Kobe University, Japan

Kazuyuki Itoh
Graduate School of Agricultural Science, Kobe University, Japan

Joe L. Parcell
Department of Agricultural Economics, University of Missouri, Columbia, United States

Joe Horner
Department of Agricultural Economics, University of Missouri, Columbia, United States

Ryan Milhollin
Department of Agricultural Economics, University of Missouri, Columbia, United States

Jason Franken
Agricultural and Industrial Sciences, Sam Houston State University, Huntsville, United States

H. R. L. Benzon
School of Applied Biosciences, Kyungpook National University, Daegu, Korea
National Institute of Molecular Biology and Biotechnology, UPLB, College, Laguna, Philippines

M. R. U. Rubenecia
School of Applied Biosciences, Kyungpook National University, Daegu, Korea

Jr. V. U. Ultra
College of Natural Sciences, Catholic University of Daegu, Gyeongsan, Korea

Sang Chul Lee
School of Applied Biosciences, Kyungpook National University, Daegu, Korea

Guy Stéphane Padzys
Département de Biologie, Faculté des Sciences, Université des Sciences et Technique de Masuku, Franceville, Gabon

Joseph Privat Ondo
Département de Chimie, Faculté des Sciences, Université des Sciences et Technique de Masuku, Franceville, Gabon

Linda Priscilla Omouendze
Département de Biologie, Faculté des Sciences, Université des Sciences et Technique de Masuku, Franceville, Gabon

Antonio Vestrella
IRTA, (Research and Technology, Food and Agriculture Institute) Torre Marimon, Caldes De Montbui, Barcelona, Spain

Robert Savé
IRTA, (Research and Technology, Food and Agriculture Institute) Torre Marimon, Caldes De Montbui, Barcelona, Spain

Carmen Biel
IRTA, (Research and Technology, Food and Agriculture Institute) Torre Marimon, Caldes De Montbui, Barcelona, Spain

Z. Anwar
Department of Environmental Sciences, COMSATS Institute of Information Technology, Abbottabad, Pakistan

M. Irshad
Department of Environmental Sciences, COMSATS Institute of Information Technology, Abbottabad, Pakistan

I. Fareed
Resources Engineering and Management, University of Kurdistan, Hewler, Iraq

A. Saleem
Department of Environmental Sciences, COMSATS Institute of Information Technology, Abbottabad, Pakistan

B. E. Udom
Faculty of Agriculture, Department of Crop and Soil Science, University of Port Harcourt, Port Harcourt, Nigeria

B. O. Nuga
Faculty of Agriculture, Department of Crop and Soil Science, University of Port Harcourt, Port Harcourt, Nigeria
Nukot Consulting Services Inc., Winnipeg, Manitola, Canada

Rajiv Sharma
International Maize & Wheat Improvement Centre (CIMMYT), Kabul, Afghanistan

Kai Sonder
International Maize & Wheat Improvement Centre (CIMMYT), Mexico

G. Sika
International Maize & Wheat Improvement Centre (CIMMYT), Nairobi, Kenya

Sara Saadatmand
Department of Biology, Science and Research Branch, Islamic Azad University, Tehran, Iran

Atefeh Niazi
Department of Biology, Science and Research Branch, Islamic Azad University, Tehran, Iran

Joanna Majkowska-Gadomska
Department of Horticulture, Faculty of Environmental Management and Agriculture, University of Warmia and Mazury in Olsztyn, Olsztyn, Poland

Emilia Mikulewicz
Department of Horticulture, Faculty of Environmental Management and Agriculture, University of Warmia and Mazury in Olsztyn, Olsztyn, Poland

Artur Dobrowolski
Department of Horticulture, Faculty of Environmental Management and Agriculture, University of Warmia and Mazury in Olsztyn, Olsztyn, Poland

Anna Dziedzic
Department of Horticulture, Faculty of Environmental Management and Agriculture, University of Warmia and Mazury in Olsztyn, Olsztyn, Poland

Barnabas M. Msongaleli
Department of Geography and Environmental Studies, School of Social Sciences, University of Dodoma, Dodoma, Tanzania

Filbert Rwehumbiza
Department of Soil Science, Sokoine University of Agriculture, Morogoro, Tanzania

Siza D. Tumbo
Department of Agricultural Engineering and Land Planning, Sokoine University of Agriculture, Morogoro, Tanzania

Nganga Kihupi
Department of Agricultural Engineering and Land Planning, Sokoine University of Agriculture, Morogoro, Tanzania

Ruojin Zhang
School of Insurance, Southwestern University of Finance and Economics, China

Thomas L. Marsh
School of Economic Sciences, Washington State University, Pullman, WA, USA

David H. Gent
Department of Botany and Plant Pathology, Oregon State University, Corvallis, OR, USA

Saad Drissi
Plant Production, Protection and Biotechnology Department, Hassan II Agronomy and Veterinary Institute, Rabat, Morocco

Abdelhadi Aït Houssa
Department of Agronomy and Plant Breeding, National Agricultural School of Meknes, Meknes, Morocco

Ahmed Bamouh
Plant Production, Protection and Biotechnology Department, Hassan II Agronomy and Veterinary Institute, Rabat, Morocco

El Madani Daoudi
Moroccan Laboratory of Agriculture (LABOMAG), Casablanca, Morocco

Mohamed Benbella
Department of Agronomy and Plant Breeding, National Agricultural School of Meknes, Meknes, Morocco

Ebraheem Suliman Altahat
Department of Agricultural Economics and Extension, Faculty of Agriculture, Jerash University, Jerash, Jordan

A. S. Coster
Department of Agricultural Economics, University of Ibadan, Nigeria

A. I. Adeoti
Department of Agricultural Economics, University of Ibadan, Nigeria

Zhixiong Liu
Institute of Chemical Ecology, Shanxi Agricultural University, Taigu, Shanxi, China

Hongxia Liu
Institute of Chemical Ecology, Shanxi Agricultural University, Taigu, Shanxi, China

Jintong Zhang
Institute of Chemical Ecology, Shanxi Agricultural University, Taigu, Shanxi, China

Xupeng Chen
Institute of Chemical Ecology, Shanxi Agricultural University, Taigu, Shanxi, China

Stephanie Leithner
Organismic Biology, University of Salzburg, Salzburg, Austria

Josef Wanzenböck
Research Institute for Limnology, University of Innsbruck, Mondsee, Austria

Abhinaya Subedi
Civil and Environmental Engineering Department, Colorado State University, Fort Collins, Colorado, USA

José L. Chávez
Civil and Environmental Engineering Department, Colorado State University, Fort Collins, Colorado, USA

I. B. Cuvaca
Department of Biosystems Engineering & Soil Science, University of Tennessee Institute of Agriculture, Knoxville, Tennessee, USA

N. S. Eash
Department of Biosystems Engineering & Soil Science, University of Tennessee Institute of Agriculture, Knoxville, Tennessee, USA

S. Zivanovic
Department of Food Science and Technology, University of Tennessee Institute of Agriculture, Knoxville, Tennessee, USA

D. M. Lambert
Department of Agricultural & Resource Economics, University of Tennessee Institute of Agriculture, Knoxville, Tennessee, USA

F. Walker
Department of Biosystems Engineering & Soil Science, University of Tennessee Institute of Agriculture, Knoxville, Tennessee, USA

B. Rustrick
Eastern Produce Kenya Ltd, Nandi Hills, Kenya

Leah Sandler
Lee Greenley Jr. Memorial Research Center, University of Missouri, Novelty, MO, USA

Kelly A. Nelson
Division of Plant Sciences, University of Missouri, Novelty, MO, USA

Christopher J. Dudenhoeffer
Lee Greenley Jr. Memorial Research Center, University of Missouri, Novelty, MO, USA

Phatela M. Raleting
Department of Agricultural Economics & Extension, University of Fort Hare, Alice, South Africa

Ajuruchukwu Obi
Department of Agricultural Economics & Extension, University of Fort Hare, Alice, South Africa

Gordana M. Kocic
University Nis, Medical Faculty, Nis, Serbia

Tatjana Jevtovic-Stoimenov
University Nis, Medical Faculty, Nis, Serbia

Dusan Sokolovic
University Nis, Medical Faculty, Nis, Serbia

Hristina Kocic
University Maribor, Medical Faculty, Maribor, Slovenia